Humanoid Robots: Applied Principles and Research

Humanoid Robots: Applied Principles and Research

Edited by **Paula Sutton**

New Jersey

Published by Clanrye International,
55 Van Reypen Street,
Jersey City, NJ 07306, USA
www.clanryeinternational.com

Humanoid Robots: Applied Principles and Research
Edited by Paula Sutton

International Standard Book Number: 978-1-63240-300-1 (Hardback)

Printed in the United States of America.

Contents

Preface

Advanced engineering and scientific research results and growth in the area of humanoid robotics as well as its applications have all been elucidated in this book. Humanoids are believed to transform the way we communicate with machines and possess the ability to blend smoothly into an environment already designed for humans. The information provided in this book aims at discovering the future abilities of humanoid robots by evaluating a variety of integrated research in distinct engineering and scientific areas like perception, human-robot interaction, locomotion, machine learning, adaptive behavior and neuroscience. It is a practical book as it lays stress on important information for the people engaged in the fields of robotics, computational methods, artificial and cognitive science as well as other scientific fields directly or indirectly connected with the applications and advancement of future humanoid robots. The information in this book is contributed by experts holding significant research and development experience, patents as well as publications in the field of humanoid robotics.

The researches compiled throughout the book are authentic and of high quality, combining several disciplines and from very diverse regions from around the world. Drawing on the contributions of many researchers from diverse countries, the book's objective is to provide the readers with the latest achievements in the area of research. This book will surely be a source of knowledge to all interested and researching the field.

In the end, I would like to express my deep sense of gratitude to all the authors for meeting the set deadlines in completing and submitting their research chapters. I would also like to thank the publisher for the support offered to us throughout the course of the book. Finally, I extend my sincere thanks to my family for being a constant source of inspiration and encouragement.

Editor

Part 1

Periodic Tasks and Locomotion Control

1

Performing Periodic Tasks: On-Line Learning, Adaptation and Synchronization with External Signals

Andrej Gams, Tadej Petrič, Aleš Ude and Leon Žlajpah
Jožef Stefan Institute, Ljubljana
Slovenia

1. Introduction

One of the central issues in robotics and animal motor control is the problem of trajectory generation and modulation. Since in many cases trajectories have to be modified on-line when goals are changed, obstacles are encountered, or when external perturbations occur, the notions of trajectory generation and trajectory modulation are tightly coupled.
This chapter addresses some of the issues related to trajectory generation and modulation, including the supervised learning of periodic trajectories, and with an emphasis on the learning of the frequency and achieving and maintaining synchronization to external signals. Other addressed issues include robust movement execution despite external perturbations, modulation of the trajectory to reuse it under modified conditions and adaptation of the learned trajectory based on measured force information. Different experimental scenarios on various robotic platforms are described.
For the learning of a periodic trajectory without specifying the period and without using traditional off-line signal processing methods, our approach suggests splitting the task into two sub-tasks: (1) frequency extraction, and (2) the supervised learning of the waveform. This is done using two ingredients: nonlinear oscillators, also combined with an adaptive Fourier waveform for the frequency adaptation, and nonparametric regression [1] techniques for shaping the attractor landscapes according to the demonstrated trajectories. The systems are designed such that after having learned the trajectory, simple changes of parameters allow modulations in terms of, for instance, frequency, amplitude and oscillation offset, while keeping the general features of the original trajectory, or maintaining synchronization with an external signal.
The system we propose in this paper is based on the motion imitation approach described in (Ijspeert et al., 2002; Schaal et al., 2007). That approach uses two dynamical systems like the system presented here, but with a simple nonlinear oscillator to generate the phase and the amplitude of the periodic movements. A major drawback of that approach is that it requires the frequency of the demonstration signal to be explicitly specified. This means that the frequency has to be either known or extracted from the recorded signal by signal

[1] The term "nonparametric" is to indicate that the data to be modeled stem from very large families of distributions which cannot be indexed by a finite dimensional parameter vector in a natural way. It does not mean that there are no parameters.

processing methods, e.g. Fourier analysis. The main difference of our new approach is that we use an adaptive frequency oscillator (Buchli & Ijspeert, 2004; Righetti & Ijspeert, 2006), which has the process of frequency extraction and adaptation totally embedded into its dynamics. The frequency does not need to be known or extracted, nor do we need to perform any transformations (Righetti et al., 2006). This simplifies the process of teaching a new task/trajectory to the robot. Additionally, the system can work incrementally in on-line settings. We use two different approaches. One uses several frequency oscillators to approximate the input signal, and thus demands a logical algorithm to extract the basic frequency of the input signal. The other uses only one oscillator and higher harmonics of the extracted frequency. It also includes an adaptive fourier series.
Our approach is loosely inspired from dynamical systems observed in vertebrate central nervous systems, in particular central pattern generators (Ijspeert, 2008a). Additionally, our work fits in the view that biological movements are constructed out of the combination of "motor primitives" (Mataric, 1998; Schaal, 1999), and the system we develop could be used as blocks or motor primitives for generating more complex trajectories.

1.1 Overview of the research field

One of the most notable advantages of the proposed system is the ability to synchronize with an external signal, which can effectively be used in control of rhythmic periodic task where the dynamic behavior and response of the actuated device are critical. Such robotic tasks include swinging of different pendulums (Furuta, 2003; Spong, 1995), playing with different toys, i.e. the yo-yo (Hashimoto & Noritsugu, 1996; Jin et al., 2009; Jin & Zacksenhouse, 2003; Žlajpah, 2006) or a gyroscopic device called the Powerball (Cafuta & Curk, 2008; Gams et al., 2007; Heyda, 2002; Petrič et al., 2010), juggling (Buehler et al., 1994; Ronsse et al., 2007; Schaal & Atkeson, 1993; Williamson, 1999) and locomotion (Ijspeert, 2008b; Ilg et al., 1999; Morimoto et al., 2008). Rhythmic tasks are also handshaking (Jindai & Watanabe, 2007; Kasuga & Hashimoto, 2005; Sato et al., 2007) and even handwriting (Gangadhar et al., 2007; Hollerbach, 1981). Performing these tasks with robots requires appropriate trajectory generation and foremost precise frequency tuning by determining the basic frequency. We denote the lowest frequency relevant for performing a given task, with the term "basic frequency".
Different approaches that adjust the rhythm and behavior of the robot, in order to achieve synchronization, have been proposed in the past. For example, a feedback loop that locks onto the phase of the incoming signal. Closed-loop model-based control (An et al., 1988), as a very common control of robotic systems, was applied for juggling (Buehler et al., 1994; Schaal & Atkeson, 1993), playing the yo-yo (Jin & Zackenhouse, 2002; Žlajpah, 2006) and also for the control of quadruped (Fukuoka et al., 2003) and in biped locomotion (Sentis et al., 2010; Spong & Bullo, 2005). Here the basic strategy is to plan a reference trajectory for the robot, which is based on the dynamic behavior of the actuated device. Standard methods for reference trajectory tracking often assume that a correct and exhaustive dynamic model of the object is available (Jin & Zackenhouse, 2002), and their performance may degrade substantially if the accuracy of the model is poor.
An alternative approach to controlling rhythmic tasks is with the use of nonlinear oscillators. Oscillators and systems of coupled oscillators are known as powerful modeling tools (Pikovsky et al., 2002) and are widely used in physics and biology to model phenomena as diverse as neuronal signalling, circadian rhythms (Strogatz, 1986), inter-limb coordination (Haken et al., 1985), heart beating (Mirollo et al., 1990), etc. Their properties, which include robust limit cycle behavior, online frequency adaptation (Williamson, 1998)

and self-sustained limit cycle generation on the absence of cyclic input (Bailey, 2004), to name just a few, make them suitable for controlling rhythmic tasks.

Different kinds of oscillators exist and have been used for control of robotic tasks. The van der Pol non-linear oscillator (van der Pol, 1934) has successfully been used for skill entrainment on a swinging robot (Veskos & Demiris, 2005) or gait generation using coupled oscillator circuits, e.g. (Jalics et al., 1997; Liu et al., 2009; Tsuda et al., 2007). Gait generation has also been studied using the Rayleigh oscillator (Filho et al., 2005). Among the extensively used oscillators is also the Matsuoka neural oscillator (Matsuoka, 1985), which models two mutually inhibiting neurons. Publications by Williamson (Williamson, 1999; 1998) show the use of the Matsuoka oscillator for different rhythmic tasks, such as resonance tuning, crank turning and playing the slinky toy. Other robotic tasks using the Matsuoka oscillator include control of giant swing problem (Matsuoka et al., 2005), dish spinning (Matsuoka & Ooshima, 2007) and gait generation in combination with central pattern generators (CPGs) and phase-locked loops (Inoue et al., 2004; Kimura et al., 1999; Kun & Miller, 1996).

On-line frequency adaptation, as one of the properties of non-linear oscillators (Williamson, 1998) is a viable alternative to signal processing methods, such as fast Fourier transform (FFT), for determining the basic frequency of the task. On the other hand, when there is no input into the oscillator, it will oscillate at its own frequency (Bailey, 2004). Righetti et al. have introduced adaptive frequency oscillators (Righetti et al., 2006), which preserve the learned frequency even if the input signal has been cut. The authors modify non-linear oscillators or pseudo-oscillators with a learning rule, which allows the modified oscillators to learn the frequency of the input signal. The approach works for different oscillators, from a simple phase oscillator (Gams et al., 2009), the Hopf oscillator, the Fitzhugh-Nagumo oscillator, etc. (Righetti et al., 2006). Combining several adaptive frequency oscillators in a feedback loop allows extraction of several frequency components (Buchli et al., 2008; Gams et al., 2009). Applications vary from bipedal walking (Righetti & Ijspeert, 2006) to frequency tuning of a hopping robot (Buchli et al., 2005). Such feedback structures can be used as a whole imitation system that both extracts the frequency and learns the waveform of the input signal.

Not many approaches exist that combine both frequency extraction and waveform learning in imitation systems (Gams et al., 2009; Ijspeert, 2008b). One of them is a two-layered imitation system, which can be used for extracting the frequency of the input signal in the first layer and learning its waveform in the second layer, which is the basis for this chapter. Separate frequency extraction and waveform learning have advantages, since it is possible to independently modulate temporal and spatial features, e.g. phase modulation, amplitude modulation, etc. Additionally a complex waveform can be anchored to the input signal. Compact waveform encoding, such as splines (Miyamoto et al., 1996; Thompson & Patel, 1987; Ude et al., 2000), dynamic movement primitives (DMP) (Schaal et al., 2007), or Gaussian mixture models (GMM) (Calinon et al., 2007), reduce computational complexity of the process. In the next sections we first give details on the two-layered movement imitation system and then give the properties. Finally, we propose possible applications.

2. Two-layered movement imitation system

In this chapter we give details and properties of both sub-systems that make the two-layered movement imitation system . We also give alternative possibilities for the canonical dynamical system.

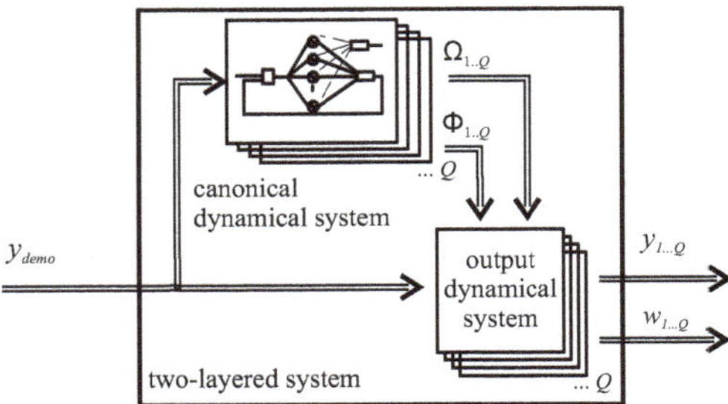

Fig. 1. Proposed structure of the system. The two-layered system is composed of the *Canonical Dynamical System* as the first layer for the frequency adaptation, and the *Output Dynamical System* for the learning as the second layer. The input signal $y_{demo}(t)$ is an arbitrary Q-dimensional periodic signal. The Canonical Dynamical System outputs the fundamental frequency Ω and phase of the oscillator at that frequency, Φ, for each of the Q DOF, and the Output Dynamical System learns the waveform.

Figure 1 shows the structure of the proposed system for the learning of the frequency and the waveform of the input signal. The input into the system $y_{demo}(t)$ is an arbitrary periodic signal of one or more degrees of freedom (DOF).

The task of frequency and waveform learning is split into two separate tasks, each performed by a separate dynamical system. The frequency adaptation is performed by the *Canonical Dynamical System*, which either consists of several adaptive frequency oscillators in a feedback structure, or a single oscillator with an adaptive Fourier series. Its purpose is to extract the basic frequency Ω of the input signal, and to provide the phase Φ of the signal at this frequency.

These quantities are fed into the *Output Dynamical System*, whose goal is to adapt the shape of the limit cycle of the Canonical Dynamical System, and to learn the waveform of the input signal. The resulting output signal of the Output Dynamical System is not explicitly encoded but generated during the time evolution of the Canonical Dynamical System, by using a set of weights learned by Incremental Locally Weighted Regression (ILWR) (Schaal & Atkeson, 1998).

Both frequency adaptation and waveform learning work in parallel, thus accelerating the process. The output of the combined system can be, for example, joint coordinates of the robot, position in task space, joint torques, etc., depending on what the input signal represents.

In the next section we first explain the second layer of the system - the output dynamical system - which learns the waveform of the input periodic signal once the frequency is determined.

2.1 Output dynamical system

The output dynamical system is used to learn the waveform of the input signal. The explanation is for a 1 DOF signal. For multiple DOF, the algorithm works in parallel for all the degrees of freedom.

The following dynamics specify the attractor landscape of a trajectory y towards the anchor point g, with the Canonical Dynamical System providing the phase Φ to the function Ψ_i of the control policy:

$$\dot{z} = \Omega \left(\alpha_z \left(\beta_z \left(g - y \right) - z \right) + \frac{\sum_{i=1}^{N} \Psi_i w_i r}{\sum_{i=1}^{N} \Psi_i} \right) \tag{1}$$

$$\dot{y} = \Omega z \tag{2}$$

$$\Psi_i = \exp \left(h \left(\cos \left(\Phi - c_i \right) - 1 \right) \right) \tag{3}$$

Here Ω (chosen amongst the ω_i) is the frequency given by canonical dynamical system, Eq. (10), α_Z and β_z are positive constants, set to $\alpha_z = 8$ and $\beta_z = 2$ for all the results; the ratio 4:1 ensures critical damping so that the system monotonically varies to the trajectory oscillating around g - an anchor point for the oscillatory trajectory. N is the number of Gaussian-like periodic kernel functions Ψ_i, which are given by Eq. (3). w_i is the learned weight parameter and r is the amplitude control parameter, maintaining the amplitude of the demonstration signal with $r = 1$. The system given by Eq. (1) without the nonlinear term is a second-order linear system with a unique globally stable point attractor (Ijspeert et al., 2002). But, because of the periodic nonlinear term, this system produces stable periodic trajectories whose frequency is Ω and whose waveform is determined by the weight parameters w_i.

In Eq. (3), which determines the Gaussian-like kernel functions Ψ_i, h determines their width, which is set to $h = 2.5N$ for all the results presented in the paper unless stated otherwise, and c_i are equally spaced between 0 and 2π in N steps.

As the input into the learning algorithm we use triplets of position, velocity and acceleration $y_{demo}(t)$, $\dot{y}_{demo}(t)$, and $\ddot{y}_{demo}(t)$ with *demo* marking the input or demonstration trajectory we are trying to learn. With this Eq. (1) can be rewritten as

$$\frac{1}{\Omega}\dot{z} - \alpha_z \left(\beta_z \left(g - y \right) - z \right) = \frac{\sum_{i=1}^{N} \Psi_i w_i r}{\sum_{i=1}^{N} \Psi_i} \tag{4}$$

and formulated as a supervised learning problem with on the right hand side a set of local models $w_i r$ that are weighted by the kernel functions Ψ_i, and on the left hand side the target function f_{targ} given by $f_{targ} = \frac{1}{\Omega^2}\ddot{y}_{demo} - \alpha_z \left(\beta_z \left(g - y_{demo} \right) - \frac{1}{\Omega}\dot{y}_{demo} \right)$, which is obtained by matching y to y_{demo}, z to $\frac{\dot{y}_{demo}}{\Omega}$, and $\dot{z}$ to $\frac{\ddot{y}_{demo}}{\Omega}$.

Locally weighted regression corresponds to finding, for each kernel function Ψ_i, the weight vector w_i, which minimizes the quadratic error criterion [2]

$$J_i = \sum_{t=1}^{P} \Psi_i(t) \left(f_{targ}(t) - w_i r(t) \right)^2 \tag{5}$$

where t is an index corresponding to discrete time steps (of the integration). The regression can be performed as a *batch* regression, or alternatively, we can perform the minimization of the J_i cost function incrementally, while the target data points $f_{targ}(t)$ arrive. As we want continuous learning of the demonstration signal, we use the latter. Incremental regression is done with the use of recursive least squares with a forgetting factor of λ, to determine the parameters (or weights) w_i. Given the target data $f_{targ}(t)$ and $r(t)$, w_i is updated by

$$w_i(t+1) = w_i(t) + \Psi_i P_i(t+1) r(t) e_r(t) \tag{6}$$

[2] LWR is derived from a piecewise linear function approximation approach (Schaal & Atkeson, 1998), which decouples a nonlinear least-squares learning problem into several locally linear learning problems, each characterized by the local cost function J_i. These local problems can be solved with standard weighted least squares approaches.

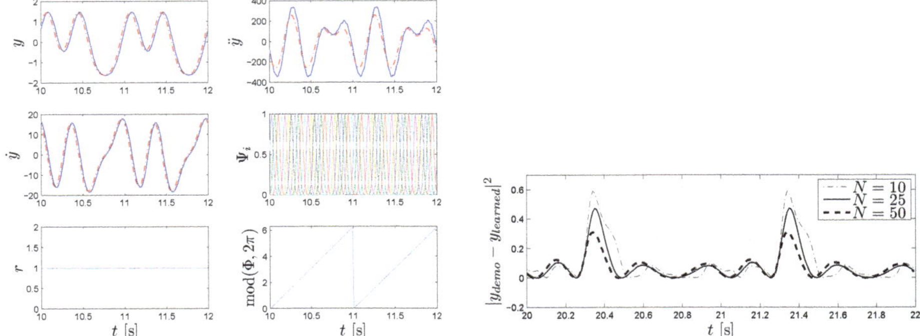

Fig. 2. *Left*: The result of Output Dynamical System with a constant frequency input and with continuous learning of the weights. In all the plots the input signal is the dash-dot line while the learned signal is the solid line. In the middle-right plot we can see the evolution of the kernel functions. The kernel functions are a function of Φ and do not necessarily change uniformly (see also Fig. 7). In the bottom right plot the phase of the oscillator is shown. The amplitude is here $r = 1$, as shown bottom-left. *Right*: The error of learning decreases with the increase of the number of Gaussian-like kernel functions. The error, which is quite small, is mainly due to a very slight (one or two sample times) delay of the learned signal.

$$P_i(t+1) = \frac{1}{\lambda}\left(P_i(t) - \frac{P_i(t)^2 r(t)^2}{\frac{\lambda}{\Psi_i} + P_i(t) r(t)^2}\right) \tag{7}$$

$$e_r(t) = f_{targ}(t) - w_i(t) r(t). \tag{8}$$

P, in general, is the inverse covariance matrix (Ljung & Söderström, 1986). The recursion is started with $w_i = 0$ and $P_i = 1$. Batch and incremental learning regressions provide identical weights w_i for the same training sets when the forgetting factor λ is set to one. Differences appear when the forgetting factor is less than one, in which case the incremental regression gives more weight to recent data (i.e. tends to forget older ones). The error of weight learning e_r (Eq. (8)) is not "related" to e when extracting frequency components (Eq. (11)). This allows for complete separation of frequency adaptation and waveform learning.

Figure 2 left shows the time evolution of the Output Dynamical System anchored to a Canonical Dynamical System with the frequency set at $\Omega = 2\pi$ rad/s, and the weight parameters w_i adjusted to fit the trajectory $y_{demo}(t) = \sin(2\pi t) + \cos(4\pi t) + 0.4\sin(6\pi t)$. As we can see in the top-left plot, the input signal and the reconstructed signal match closely. The matching between the reconstructed signal and the input signal can be improved by increasing the number of Gaussian-like functions.

Parameters of the Output Dynamical System

When tuning the parameters of the Output Dynamical System, we have to determine the number of Gaussian-like Kernel functions N, and specially the forgetting factor λ. The number N of Gaussian-like kernel functions could be set automatically if we used the locally weighted learning (Schaal & Atkeson, 1998), but for simplicity it was here set by hand. Increasing the number increases the accuracy of the reconstructed signal, but at the same time also increases the computational cost. Note that LWR does not suffer from problems of overfitting when the

number of kernel functions is increased.[3] Figure 2 right shows the error of learning e_r when using $N = 10$, $N = 25$, and $N = 50$ on a signal $y_{demo}(t) = 0.65\sin(2\pi t) + 1.5\cos(4\pi t) + 0.3\sin(6\pi t)$. Throughout the paper, unless specified otherwise, $N = 25$.
The forgetting factor $\lambda \in [0,1]$ plays a key role in the behavior of the system. If it is set high, the system never forgets any input values and learns an average of the waveform over multiple periods. If it is set too low, it forgets all, basically training all the weights to the last value. We set it to $\lambda = 0.995$.

2.2 Canonical dynamical system

The task of the Canonical Dynamical System is two-fold. Firstly, it has to extract the fundamental frequency Ω of the input signal, and secondly, it has to exhibit stable limit cycle behavior in order to provide a phase signal Φ, that is used to anchor the waveform of the output signal. Two approaches are possible, either with a pool of oscillators (PO), or with an adaptive Fourier Series (AF).

2.2.1 Using a pool of oscillators

As the basis of our canonical dynamical system we use a set of phase oscillators, see e.g. (Buchli et al., 2006), to which we apply the adaptive frequency learning rule as introduced in (Buchli & Ijspeert, 2004) and (Righetti & Ijspeert, 2006), and combine it with a feedback structure (Righetti et al., 2006) shown in Figure 3. The basic idea of the structure is that each of the oscillators will adapt its frequency to one of the frequency components of the input signal, essentially "populating" the frequency spectrum.
We use several oscillators, but are interested only in the fundamental or lowest non-zero frequency of the input signal, denoted by Ω, and the phase of the oscillator at this frequency, denoted by Φ. Therefore the feedback structure is followed by a small logical block, which chooses the correct, lowest non-zero, frequency. Determining Ω and Φ is important because with them we can formulate a supervised learning problem in the second stage - the Output Dynamical System, and learn the waveform of the full period of the input signal.

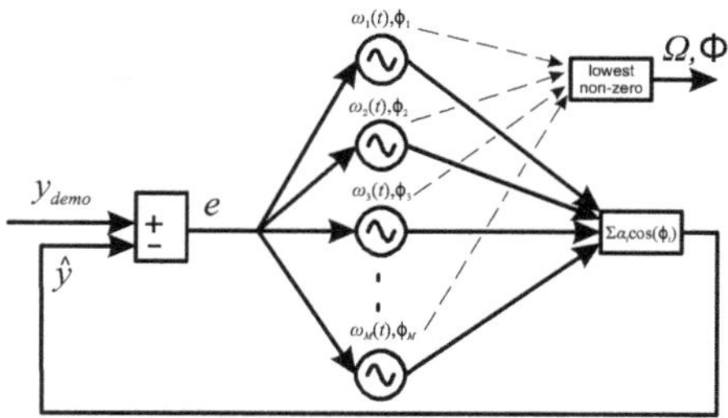

Fig. 3. Feedback structure of a network of adaptive frequency phase oscillators, that form the Canonical Dynamical System. All oscillators receive the same input and have to be at different starting frequencies to converge to different final frequencies. Refer also to text and Eqs. (9-13).

The feedback structure of M adaptive frequency phase oscillators is governed by the following equations:

[3] This property is due to solving the bias-variance dilemma of function approximation locally with a closed form solution to leave-one-out cross-validation (Schaal & Atkeson, 1998).

$$\dot{\phi}_i = \omega_i - Ke\sin(\phi_i) \tag{9}$$

$$\dot{\omega}_i = -Ke\sin(\phi_i) \tag{10}$$

$$e = y_{demo} - \hat{y} \tag{11}$$

$$\hat{y} = \sum_{i=1}^{M} \alpha_i \cos(\phi_i) \tag{12}$$

$$\dot{\alpha}_i = \eta \cos(\phi_i) e \tag{13}$$

where K is the coupling strength, ϕ_i is the phase of oscillator i, e is the input into the oscillators, y_{demo} is the input signal, $\hat{y}$ is the weighted sum of the oscillators' outputs, M is the number of oscillators, α_i is the amplitude associated to the i-th oscillator, and η is a learning constant. In the experiments we use $K = 20$ and $\eta = 1$, unless specified otherwise.

Eq. (9) and (10) present the core of the Canonical Dynamical System – the adaptive frequency phase oscillator. Several (M) such oscillators are used in a feedback loop to extract separate frequency components. Eq. (11) and (12) specify the feedback loop, which needs also amplitude adaptation for each of the frequency components (Eq. (13)).

As we can see in Figure 3, each of the oscillators of the structure receives the same input signal, which is the difference between the signal to be learned and the signal already learned by the feedback loop, as in Eq. (11). Since a negative feedback loop is used, this difference approaches zero as the weighted sum of separate frequency components, Eq. (12), approaches the learned signal, and therefore the frequencies of the oscillators stabilize. Eq. (13) ensures amplitude adaptation and thus the stabilization of the learned frequency. Such a feedback structure performs a kind of dynamic Fourier analysis. It can learn several frequency components of the input signal (Righetti et al., 2006) and enables the frequency of a given oscillator to converge as $t \to \infty$, because once the frequency of a separate oscillator is set, it is deducted from the demonstration signal y_{demo}, and disappears from e (due to the negative feedback loop). Other oscillators can thus adapt to other remaining frequency components. The populating of the frequency spectrum is therefore done without any signal processing, as the whole process of frequency extraction and adaptation is totally embedded into the dynamics of the adaptive frequency oscillators.

Frequency adaptation results for a time-varying signal are illustrated in Figure 4, left. The top plot shows the input signal y_{demo}, the middle plot the extracted frequencies, and the bottom plot the error of frequency adaptation. The figure shows results for both approaches, using a pool of oscillators (PO) and for using one oscillator and an adaptive Fourier series (AF), explained in the next section. The signal itself is of three parts, a non-stationary signal (presented by a chirp signal), followed by a step change in the frequency of the signal, and in the end a stationary signal. We can see that the output frequency stabilizes very quickly at the (changing) target frequency. In general the speed of convergence depends on the coupling strength K (Righetti et al., 2006). Besides the use for non-stationary signals, such as chirp signals, coping with the change in frequency of the input signal proves especially useful when adapting to the frequency of hand-generated signals, which are never stationary. In this particular example, a single adaptive frequency oscillator in a feedback loop was enough, because the input signal was purely sinusoidal.

The number of adaptive frequency oscillators in a feedback loop is therefore a matter of design. There should be enough oscillators to avoid missing the fundamental frequency and to limit the variation of frequencies described below when the input signal has many

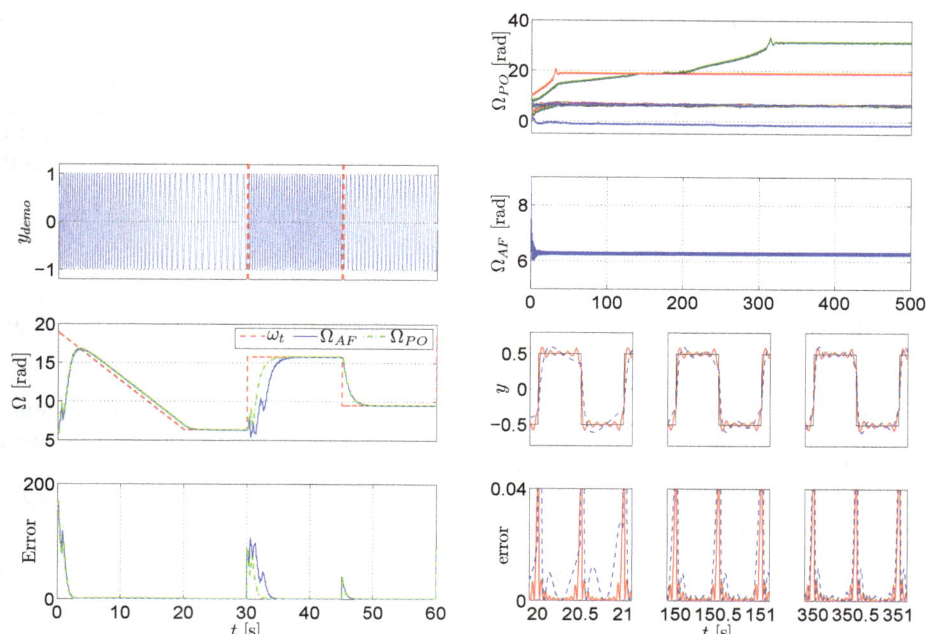

Fig. 4. *Left*: Typical convergence of an adaptive frequency oscillator combined with an adaptive Fourier series (-) compared to a system with a poll of i oscillators (-.-). One oscillator is used in both cases. The input is a periodic signal ($y = sin(\omega_t t)$, with $\omega_t = (6\pi - \pi/5t)$ rad/s for $t < 20$ s, followed by $\omega_t = 2\pi$ rad/s for $t < 30$ s, followed again by $\omega_t = 5\pi$ rad/s for $t < 45$ s and finally $\omega_t = 3\pi$ rad/s). Frequency adaptation is presented in the middle plot, starting at $\Omega_0 = \pi$ rad/s, where ω_t is given by the dashed line and Ω by the solid line. The square error between the target and the extracted frequency is shown in the bottom plot. We can see that the adaptation is successful for non-stationary signals, step changes and stationary signals. *Right*: Comparison between using the PO and the AF approaches for the canonical dynamical system. The first plot shows the evolution of frequency distribution using a pool of 10 oscillators. The second plot shows the extracted frequency using the AF approach. The comparison of the target and the approximated signals is presented in the third plot. The thin solid line presents the input signal y_{demo}, the thick solid line presents the AF approach $\hat{y}$ and the dotted line presents the PO approach $\hat{y}_o$. The square difference between the input and the approximated signals is presented in the bottom plot.

frequencies components. A high number of oscillators can be used. Beside the almost negligible computational costs, using too many oscillators does not affect the solution. A practical problem that arises is that the oscillators' frequencies might come too close together, and then lock onto the same frequency component. To solve this we separate their initial frequencies ω_0 in a manner that suggests that (preferably only) one oscillator will go for the offset, one will go for the highest frequency, and the others will "stay between".

With a high number of oscillators, many of them want to lock to the offset (0 Hz). With the target frequency under 1 rad/s the oscillations of the estimated frequency tend to be higher, which results in longer adaptation times. This makes choosing the fundamental frequency

without introducing complex decision-making logic difficult. Results of frequency adaptation for a complex waveform are presented in Fig. 4, where results for both PO and AF approach are presented.
Besides learning, we can also use the system to repeat already learned signals. It this case, we cut feedback to the adaptive frequency oscillators by setting $e(t) = 0$. This way the oscillators continue to oscillate at the frequency to which they adapted. We are only interested in the fundamental frequency, determined by

$$\dot{\Phi} = \Omega \tag{14}$$

$$\dot{\Omega} = 0 \tag{15}$$

which is derived from Eqs. (9 and 10). This is also the equation of a normal phase oscillator.

2.3 Using an adaptive Fourier series

In this section an alternative, novel architecture for the canonical dynamical system is presented. As the basis of the canonical dynamical system one single adaptive frequency phase oscillator is used. It is combined with a feedback structure based on an adaptive Fourier series (AF). The feedback structure is shown in Fig. 5. The feedback structure of an adaptive frequency phase oscillator is governed by

$$\dot{\phi} = \Omega - Ke\sin\Phi, \tag{16}$$

$$\dot{\Omega} = -Ke\sin\Phi, \tag{17}$$

$$e = y_{demo} - \hat{y}, \tag{18}$$

where K is the coupling strength, Φ is the phase of the oscillator, e is the input into the oscillator and y_{demo} is the input signal. If we compare Eqs. (9, 10) and Eqs. (16, 17), we can see that the basic frequency Ω and the phase Φ are in Eqs. (16, 17) clearly defined and no additional algorithm is required to determine the basic frequency. The feedback loop signal $\hat{y}$ in (18) is given by the Fourier series

$$\hat{y} = \sum_{i=0}^{M} (\alpha_i \cos(i\phi) + \beta_i \sin(i\phi)), \tag{19}$$

and not by the sum of separate frequency components as in Eq. (12). In Eq. (19) M is the number of components of the Fourier series and α_i, β_i are the amplitudes associated with the Fourier series governed by

$$\dot{\alpha}_i = \eta\cos(i\phi)e, \tag{20}$$

$$\dot{\beta}_i = \eta\sin(i\phi)e, \tag{21}$$

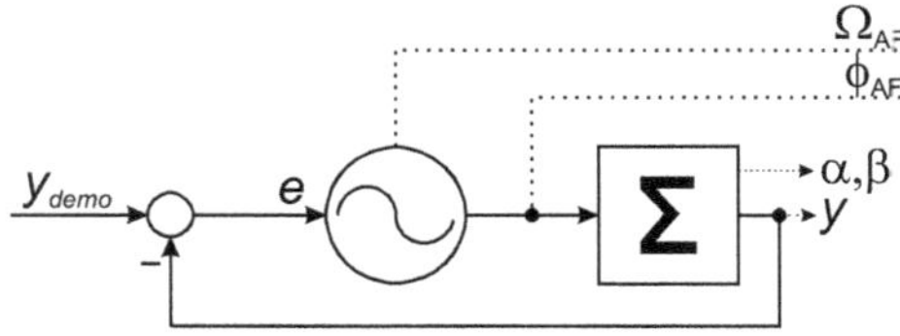

Fig. 5. Feedback structure of an adaptive frequency oscillator combined with a dynamic Fourier series. Note that no logical algorithm is needed.

where η is the learning constant and $i = 0...M$. As shown in Fig. 5, the oscillator input is the difference between the input signal y_{demo} and the Fourier series $\hat{y}$. Since a negative feedback loop is used, the difference approaches zero when the Fourier series representation $\hat{y}$ approaches the input signal y. Such a feedback structure performs a kind of adaptive Fourier analysis. Formally, it performs only a Fourier series approximation, because input signals may drift in frequency and phase. General convergence remains an open issue. The number of harmonic frequency components it can extract depends on how many terms of the Fourier series are used.

As it is able to learn different periodic signals, the new architecture of the canonical dynamical system can also be used as an imitation system by itself. Once e is stable (zero), the periodic signal stays encoded in the Fourier series, with an accuracy that depends on the number of elements used in Fourier series. The learning process is embedded and is done in real-time. There is no need for any external optimization process or other learning algorithm.

It is important to point out that the convergence of the frequency adaptation (i.e. the behavior of Ω) should not be confused with locking behavior (Buchli et al., 2008) (i.e. the classic phase locking behavior, or synchronization, as documented in the literature (Pikovsky et al., 2002)). The frequency adaptation process is an extension of the common oscillator with a fixed intrinsic frequency. First, the adaptation process changes the intrinsic frequency and not only the resulting frequency. Second, the adaptation has an infinite basin of attraction (see (Buchli et al., 2008)), third the frequency stays encoded in the system when the input is removed (e.g. set to zero or $e \approx 0$). Our purpose is to show how to apply the approach for control of rhythmic robotic task. For details on analyzing interaction of multiple oscillators see e.g. (Kralemann et al., 2008).

Augmenting the system with an output dynamical system makes it possible to synchronize the movement of the robot to a measurable periodic quantity of the desired task. Namely, the waveform and the frequency of the measured signal are encoded in the Fourier series and the desired robot trajectory is encoded in the output dynamical system. Since the adaptation of the frequency and learning of the desired trajectory can be done simultaneously, all of the system time-delays, e.g. delays in communication, sensor measurements delays, etc., are automatically included. Furthermore, when a predefined motion pattern for the trajectory is used, the phase between the input signal and output signal can be adjusted with a phase lag parameter ϕ_l (see Fig. 9). This enables us to either predefine the desired motion or to teach the robot how to preform the desired rhythmic task online.

Even though the canonical dynamical system by itself can reproduce the demonstration signal, using the output dynamical system allows for easier modulation in both amplitude and frequency, learning of complex patterns without extracting all frequency components and acts as a sort of a filter. Moreover, when multiple output signals are needed, only one canonical system can be used with several output systems which assure that the waveforms of the different degrees-of-freedom are realized appropriately.

3. On-line learning and modulation

3.1 On-line modulations

The output dynamical system allows easy modulation of amplitude, frequency and center of oscillations. Once the robot is performing the learned trajectory, we can change all of these by changing just one parameter for each. The system is designed to permit on-line modulations of the originally learned trajectories. This is one of the important motivations behind the use of dynamical systems to encode trajectories.

Changing the parameter g corresponds to a modulation of the baseline of the rhythmic movement. This will smoothly shift the oscillation without modifying the signal shape. The results are presented in the second plot in Figure 6 left. Modifying Ω and r corresponds to the changing of the frequency and the amplitude of the oscillations, respectively. Since our differential equations are of second order, these abrupt changes of parameters result in smooth variations of the trajectory y. This is particularly useful when controlling articulated robots, which require trajectories with limited jerks. Changing of the parameter Ω only comes into consideration when one wants to repeat the learned signal at a desired frequency that is different from the one we adapted to with our Canonical Dynamical System. Results of changing the frequency Ω are presented in the third plot of Figure 6 left. Results of modulating the amplitude parameter r are presented in the bottom plot of Figure 6 left.

3.2 Perturbations and modified feedback

3.2.1 Dealing with perturbations

The Output Dynamical System is inherently robust against perturbations. Figure 6 right illustrates the time evolution of the system repeating a learned trajectory at the frequency of 1 Hz, when the state variables y, z and Φ are randomly changed at time $t = 30$ s. From the results we can see that the output of the system reverts smoothly to the learned trajectory. This is an important feature of the approach: the system essentially represents a whole landscape in the space of state variables which not only encode the learned trajectory but also determine how the states return to it after a perturbation.

3.2.2 Slow-down feedback

When controlling the robot, we have to take into account perturbations due to the interaction with the environment. Our system provides *desired* states to the robot, i.e. desired joint angles or torques, and its state variables are therefore not affected by the *actual* states of the robot, unless feedback terms are added to the control scheme. For instance, it might happen that, due to external forces, significant differences arise between the actual position $\tilde{y}$ and the desired position y. Depending on the task, this error can be fed back to the system in order to modify on-line the generated trajectories.

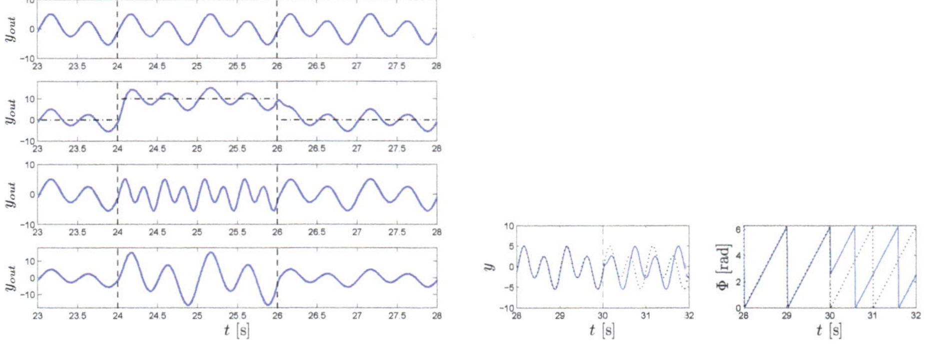

Fig. 6. *Left*: Modulations of the learned signal. The learned signal (top), modulating the baseline for oscillations g (second from top), doubling the frequency Ω (third from top), doubling the amplitude r (bottom). *Right*: Dealing with perturbations – reacting to a random perturbation of the state variables y, z and Φ at $t = 30$ s.

One type of such feedback is the "slow-down-feedback" that can be applied to the Output Dynamical System. This type of feedback affects both the Canonical and the Output Dynamical System. The following explanation is for the replay of a learned trajectory as perturbing the robot while learning the trajectory is not practical.
For the process of repeating the signal, for which we use a phase oscillator, we modify Eqs. (2 and 14) to:

$$\dot{y} = \Omega \left(z + \alpha_{py} \left(\tilde{y} - y \right) \right) \tag{22}$$

$$\dot{\Phi} = \frac{\Omega}{1 + \alpha_{p\Phi} |\tilde{y} - y|} \tag{23}$$

where α_{py} and $\alpha_{p\Phi}$ are positive constants.
With this type of feedback, the time evolution of the states is gradually halted during the perturbation. The desired position y is modified to remain close to the actual position $\tilde{y}$, and as soon as the perturbation stops, rapidly resumes performing the time-delayed planned trajectory. Results are presented in Figure 7 left. As we can see, the desired position y and the actual position $\tilde{y}$ are the same except for the short interval between $t = 22.2$ s and $t = 23.9$ s. The dotted line corresponds to the original unperturbed trajectory. The desired trajectory continues from the point of perturbation and does not jump to the unperturbed desired trajectory.

3.2.3 Virtual repulsive force

Another example of a perturbation can be the presence of boundaries or obstacles, such as joint angle limits. In that case we can modify the Eq. (2) to include a repulsive force $l(y)$ at the limit by:

$$\dot{y} = \Omega \left(z + l(y) \right) \tag{24}$$

For instance, a simple repulsive force to avoid hitting joint limits or going beyond a position in task space can be

$$l(y) = -\gamma \frac{1}{\left(y_L - y \right)^3} \tag{25}$$

where y_L is the value of the limit. Figure 7 right illustrates the effect of such a repulsive force. Such on-line modifications are one of the most interesting properties of using autonomous differential equations for control policies. These are just examples of possible feedback loops, and they should be adjusted depending on the task at hand.

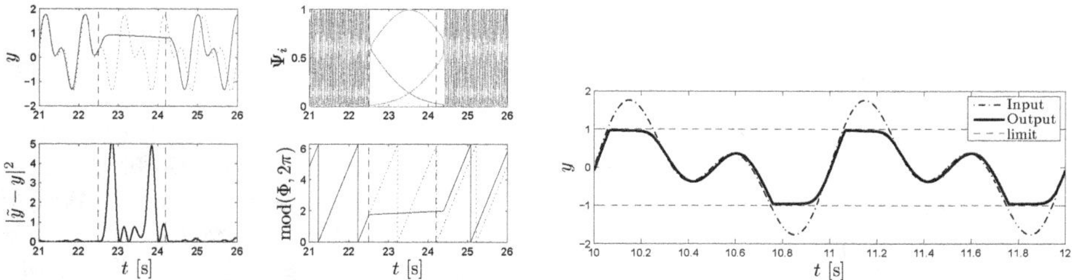

Fig. 7. *Left*: Reacting to a perturbation with a slow-down feedback. The desired position y and the actual position $\tilde{y}$ are the same except for the short interval between $t = 22.2$ s and $t = 23.9$ s. The dotted line corresponds to the original unperturbed trajectory. *Right*: Output of the system with the limits set to $y_l = [-1, 1]$ for the input signal $y_{demo}(t) = \cos(2\pi t) + \sin(4\pi t)$.

3.3 Two-handed drumming

We applied the system for two-handed drumming on a full-sized humanoid robot called CB-i, shown in Fig. 8. The robot learns the waveform and the frequency of the demonstrated movement on-line, and continues drumming with the extracted frequency after the demonstration. The system allows the robot to synchronize to the extracted frequency of music, and thus drum-along in real-time.

CB-i robot is a 51 DOF full-sized humanoid robot, developed by Sarcos. For the task of drumming we used 8 DOF in the upper arms of the robot, 4 per arm. Fig. 8 shows the CB-i robot in the experimental setup.

Fig. 8. Two-handed drumming using the Sarcos CB-i humanoid robot.

The control scheme was implemented in Matlab/Simulik and is presented in Fig. 9. The imitation system provides the desired task space trajectory for the robot's arms. The waveform was defined in advance. Since the sound signal consists usually of several different tones, e.g. drums, guitar, singer, noise etc., it was necessary to pre-process the signal in order to get the periodic signal which represents the drumming. The input signal was modified into short pulses. This pre-processing only modifies the waveform and does not determine the frequency and the phase.

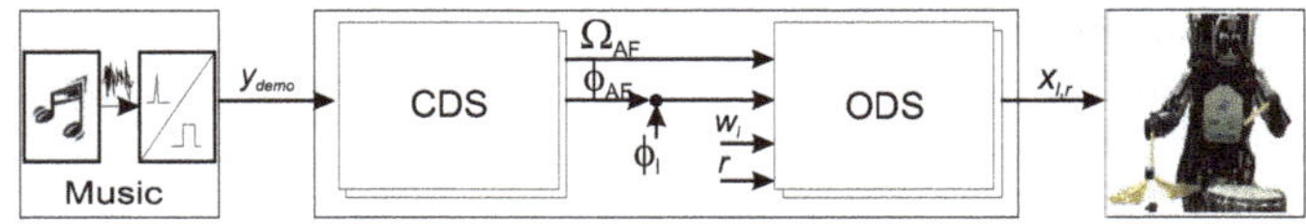

Fig. 9. Proposed two-layered structure of the control system for synchronizing robotic drumming to the music.

Fig. 10 shows the results of frequency adaptation to music. The waveforms for both hands were predefined. The frequency of the imitated motion quickly adapted to the drumming tones. The drumming sounds are presented with a real-time power spectrum. With this particular experiment we show the possibility of using our proposed system for synchronizing robot's motion with an arbitrary periodic signal, e.g. music. Due to the complexity of the audio signal this experiment does require some modification of the measured (audio) signal, but is not pre-processed in the sense of determining the frequency. The drumming experiment shows that the proposed two-layered system is able to synchronize the motion of the robot to the drumming tones of the music.

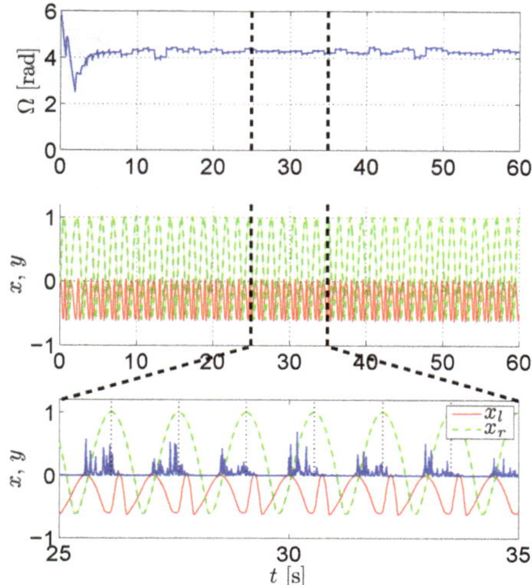

Fig. 10. Extracted frequency Ω of the drumming tones from the music in the top plot. Comparison between the power spectrum of the audio signal (drumming tones) y and robot trajectories for the left (x_l) and the right hand motions (x_r).

3.4 Table wiping

In this section we show how we can use the proposed two-layered system to modify already learned movement trajectories according to the measured force. ARMAR-IIIb humanoid robot, which is kinematically equal to the ARMAR-IIIa (Asfour et al., 2006) was used in the experiment.

From the kinematics point of view, the robot consists of seven subsystems: head, left arm, right arm, left hand, right hand, torso, and a mobile platform. The head has seven DOF and is equipped with two eyes, which have a common tilt and can pan independently. Each arm has 7 DOF and each hand additional 8DOF. The locomotion of the robot is realized using a wheel-based holonomic platform.

In order to obtain reliable motion data of a human wiping demonstration through observation by the robot, we exploited the color features of the sponge to track its motion. Using the stereo camera setup of the robot, the implemented blob tracking algorithm based on color segmentation and a particle filter framework provides a robust location estimation of the sponge in 3D. The resulting trajectories were captured with a frame rate of 30 Hz.

For learning of movements we first define the area of demonstration by measuring the lower-left and the upper-right position within a given time-frame, as is presented in Fig. 11. All tracked sponge-movement is then normalized and given as offset to the central position of this area.

For measuring the contact forces between the object in the hand and the surface of the plane a 6D-force/torque sensor is used, which is mounted at the wrist of the robot.

3.5 Adaptation of the learned trajectory using force feedback

Learning of a movement that brings the robot into contact with the environment must be based on force control, otherwise there can be damage to the robot or the object to which the robot applies its force. In the task of wiping a table, or any other object of arbitrary shape,

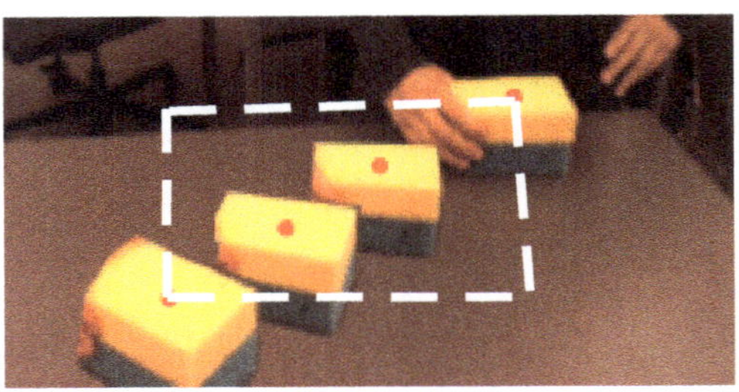

Fig. 11. Area for movement demonstration is determined by measuring the bottom-left most and the top-right most positions within a given time frame. These coordinates make a rectangular area (marked with dashed lines) where the robot tracks the demonstrated movements.

constant contact with the object is required. To teach the robot the necessary movement, we decoupled the learning of the movement from the learning of the shape of the object. We first apply the described two-layered movement imitation system to learn the desired trajectories by means of visual feedback. We then use force-feedback to adapt the motion to the shape of the object that the robot acts upon.

Periodic movements can be of any shape, yet wiping can be effective with simple one dimensional left-right movement, or circular movement. Once we are satisfied with the learned movement, we can reduce the frequency of the movement by modifying the Ω value. The low frequency of movement and consequentially low movement speed reduce the possibility of any damage to the robot. When performing playback we modify the learned movement with an active compliance algorithm. The algorithm is based on the velocity-resolved approach (Villani & J., 2008). The end-effector velocity is calculated by

$$\mathbf{v_r} = \mathbf{S_v v_v} + \mathbf{K_F S_F}(\mathbf{F_m} - \mathbf{F_0}). \tag{26}$$

Here $\mathbf{v_r}$ stands for the resolved velocities vector, $\mathbf{S_v}$ for the velocity selection matrix, $\mathbf{v_v}$ for the desired velocities vector, $\mathbf{K_F}$ for the force gain matrix, $\mathbf{S_F}$ for the force selection matrix, and $\mathbf{F_m}$ for the measured force. $\mathbf{F_0}$ denotes the force offset which determines the behavior of the robot when not in contact with the environment. To get the desired positions we use

$$\mathbf{Y} = \mathbf{Y_r} + \mathbf{S_F} \int \mathbf{v_r} dt. \tag{27}$$

Here $\mathbf{Y_r}$ is the desired initial position and $\mathbf{Y} = (y_j),\ j = 1, ..., 6$ is the actual position/orientation. Using this approach we can modify the trajectory of the learned periodic movement as described below.

Equations (26 – 27) become simpler for the specific case of wiping a flat surface. By using a null matrix for $\mathbf{S_v}$, $\mathbf{K_F} = diag(0, 0, k_F, 0, 0, 0)$, $\mathbf{S_F} = diag(0, 0, 1, 0, 0, 0)$, the desired end-effector height z in each discrete time step Δt becomes

$$\dot{z}(t) = k_F(F_z(t) - F_0), \tag{28}$$

$$z(t) = z_0 + \dot{z}(t)\Delta t. \tag{29}$$

Here z_0 is the starting height, k_F is the force gain (of units kg/s), F_z is the measured force in the z direction and F_0 is the force with which we want the robot to press on the object. Such formulation of the movement ensures constant movement in the $-z$ direction, or constant contact when an object is encountered. Another simplification is to use the length of the force vector $F = \sqrt{F_x^2 + F_y^2 + F_z^2}$ for the feedback instead of F_z in (28). This way the robot can move upwards every time it hits something, for example the side of a sink. No contact should be made from above, as this will make the robot press up harder and harder.

The learning of the force profile is done by modifying the weighs w_i for the selected degree of freedom y_j in every time-step by incremental locally weighted regression (Atkeson et al., 1997), see also Section 2.1.

The $\mathbf{K_F}$ matrix controls the behavior of the movement. The correcting movement has to be fast enough to move away from the object if the robot hand encounters sufficient force, and at the same time not too fast so that it does not produce instabilities due to the discrete-time sampling when in contact with an object. A dead-zone of response has to be included, for example $|F| < 1$ N, to take into account the noise. We empirically set $k_F = 20$, and limited the force feedback to allow maximum linear velocity of 120 mm/s.

Feedback from a force-torque sensor is often noisy due to the sensor itself and mainly due to vibrations of the robot. A noisy signal is not the best solution for the learning algorithm because we also need time-discrete first and second derivatives. The described active compliance algorithm uses the position of the end-effector as input, which is the integrated desired velocity and therefore has no difficulties with the noisy measured signal.

Having adapted the trajectory to the new surface enables very fast movement with a constant force profile at the contact of the robot/sponge and the object, without any time-sampling and instability problems that may arise when using compliance control only. Furthermore, we can still use the compliant control once we have learned the shape of the object. Active compliance, combined with a passive compliance of a sponge, and the modulation and perturbation properties of DMPs, such as slow-down feedback, allow fast and safe execution of periodic movement while maintaining a sliding contact with the environment.

3.5.1 The learning scenario

Our kitchen scenario includes the ARMAR-IIIb humanoid robot wiping a kitchen table. First the robot attempts to learn wiping movement from human demonstration. During the demonstration of the desired wiping movement the robot tracks the movement of the sponge in the demonstrator's hand with his eyes. The robot only reads the coordinates of the movement in a horizontal plane, and learns the frequency and waveform of the movement. The waveform can be arbitrary, but for wiping it can be simple circular or one-dimensional left-right movement. The learned movement is encoded in the task space of the robot, and an inverse kinematics algorithm controls the movement of separate joints of the 7-DOF arm. The robot starts mimicking the movement already during the demonstration, so the demonstrator can stop learning once he/she is satisfied with the learned movement. Once the basic learning of periodic movement is stopped, we use force-feedback to modify the learned trajectory. The term $F - F_0$ in (28) provides velocity in the direction of $-z$ axis, and the hand holding the sponge moves towards the kitchen table or any other surface under the arm. As the hand makes contact with the surface of an object, the vertical velocity adapts. The force profile is learned in a few periods of the movement. The operator can afterwards stop force profile learning and execute the adjusted trajectory at an arbitrary frequency.

Fig. 12 on the left shows the results of learning the force-profile for a flat surface. As the robot grasps the sponge, its orientation and location are unknown to the robot, and the tool center point (TCP) changes. Should the robot simply perform a planar trajectory it would not ensure constant contact with the table. As we can see from the results, the hand initially

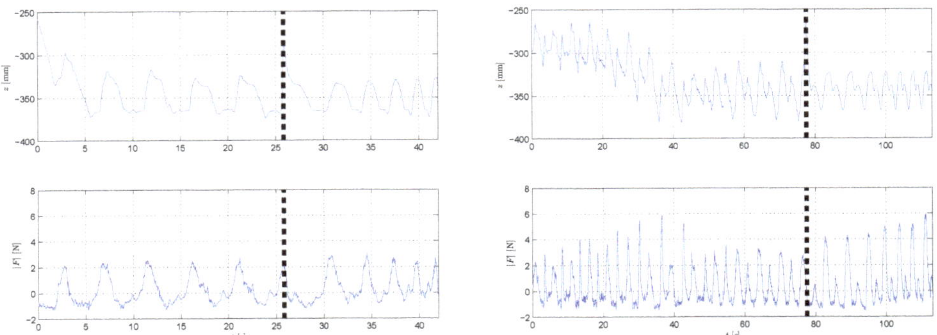

Fig. 12. Results of learning the force profile on a flat surface on the left and on a bowl-shaped surface on the right. For the flat surface we can see that the height of the movement changes for approx. 5 cm during one period to maintain contact with the surface. The values were attained trough robot kinematics. For the bowl shaped surface we can see that the trajectory assumes a bowl-shape with an additional change of direction, which is the result of the compliance of the wiping sponge and the zero-velocity dead-zone. A dashed vertical line marks the end of learning of the force profile. Increase in frequency can be observed in the end of the plot. The increase was added manually.

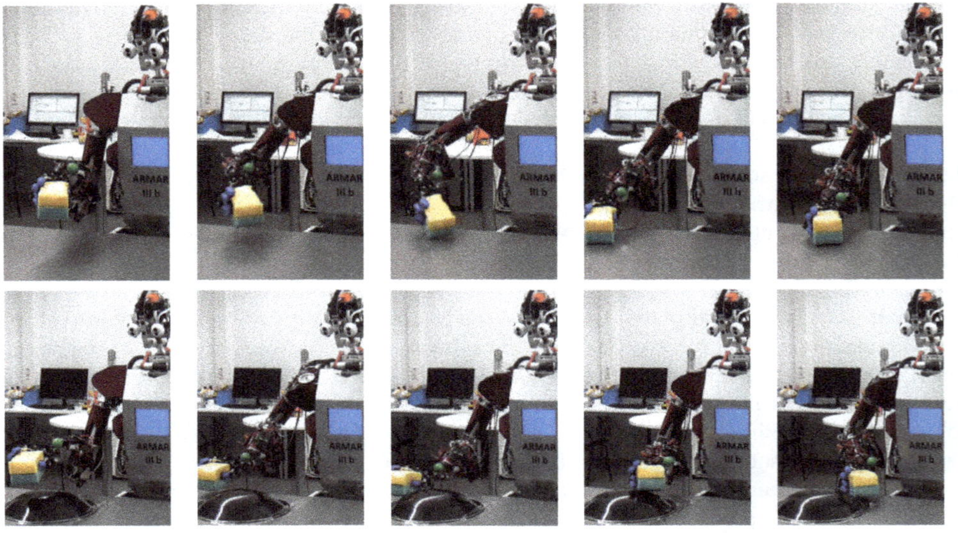

Fig. 13. A sequence of still photos showing the adaptation of wiping movement via force-feedback to a flat surface, as of a kitchen table, in the top row, and adaptation to a bowl-shaped surface in the bottom row.

moves down until it makes contact with the surface. The force profile later changes the desired height by approx. 5 cm within one period. After the learning (stopped manually, marked with a vertical dashed line) the robot maintains such a profile. A manual increase in frequency was introduced to demonstrate the ability to perform the task at an arbitrary frequency. The

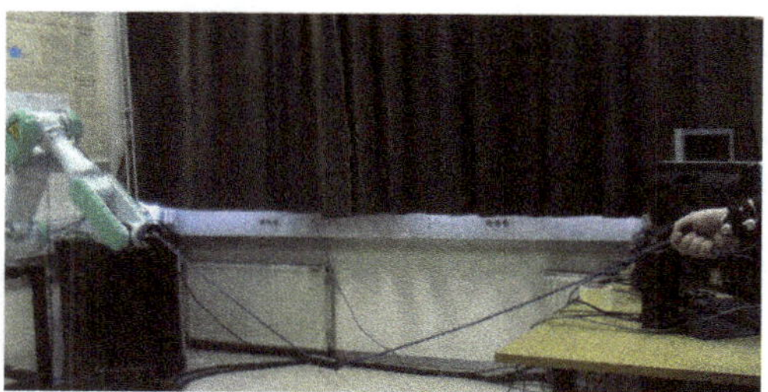

Fig. 14. Experimental setup for cooperative human-robot rope turning.

bottom plot shows the measured length of the force vector $|F|$. As we can see the force vector keeps the same force profile, even though the frequency is increased. No increase in the force profile proves that the robot has learned the required trajectory. Fig. 13 shows a sequence of photos showing the adaptation to the flat and bowl-shaped surfaces.

Fig. 12 on the right shows the results for a bowl-shaped object. As we can see from the results the height of the movement changes for more than 6 cm within a period. The learned shape (after the vertical dashed lined) maintains the shape of a bowl, but has an added local minimum. This is the result of the dead-zone within the active compliance, which comes into effect when going up one side, and down the other side of the bowl. No significant change in the force profile can be observed in the bottom plot after a manual increase in frequency. Some drift, as the consequence of an error of the sensor and of wrist control on the robot, can be observed.

4. Synchronization with external signals

Once the movement is learned we can change its frequency. The new frequency can be determined from an external signal using the canonical dynamical system. This allows easy synchronization to external measured signals, such as drumming, already presented in Section 3.3. In this section we show how we applied the system to a rope turning task, which is task that requires continuous cooperation of a human and a robot. We also show how we can synchronize to an EMG signal, which is inherently very noisy.

4.1 Robotic rope turning

We performed the rope-turning experiment on a Mitsubishi PA-10 robot with a JR-3 force/torque sensor attached to the top of the robot to measure the torques and forces in the string. Additionally, an optical system (Optotrak Certus) was used for validation, i.e. for measuring the motion of a human hand. The two-layered control system was implemented in Matlab/Simulink. The imitation system provides a pre-defined desired circular trajectory for the robot. The motion of a robot is constrained to up-down and left-right motion using inverse kinematics. Figure 14 shows the experimental setup.

Determining the frequency is done using the canonical dynamical system. Fig. 15 left shows the results of frequency extraction (top plot) from the measured torque signal (second plot). The frequency of the imitated motion quickly adapts to the measured periodic signal. When the rotation of the rope is stable, the human stops swinging the rope and maintains the hand in a fixed position. The movement of the human hand is shown in the third plot. In the last plot we show the movement of the robot. By comparing the last two plots in Fig. 15 left, we can see that after 3 s the energy transition to the rope is done only by the motion of the robot.

The frequency of the task depends on the parameters of the rope, i.e. weight, length, flexibility etc., and the energy which is transmitted in to the rope. The rotating frequency of the rope can be influenced by the amplitude of the motion, i.e. how much energy is transmitted to the rope. The amplitude can be easily modified with the amplitude parameter r.

Fig. 15 right shows the behavior of the system, when the distance between the human hand and the top end of the robot (second plot) is changing, while the length of the rope remains the same, consequently the rotation frequency changes. The frequency adaptation is shown in the top plot. As we can see, the robot was able to rotate the rope and maintain synchronized even if disturbances like changing the distance between the human hand and the robot occur. This shows that the system is adaptable and robust.

4.2 EMG based human-robot synchronization

In this section we show the results of synchronizing robot movement to an EMG signal measured from the human biceps muscle. More details can be found at (Petrič et al., 2011). The purpose of this experiment is to show frequency extraction from a signal with a low signal-to-noise ratio. This type of applications can be used for control of periodic movements of limb prosthesis (Castellini & Smagt, 2009) or exoskeletons.

In our experimental setup we attached an array of 3 electrodes (Motion Control Inc.) over the biceps muscle of a subject and asked the subject to flex his arm when he hears a beep. The frequency of beeping was 1 Hz from the start of the experiment, then changed to 0.5 Hz after 30 s, and then back to 1 Hz after additional 30 s. Fig. 16 left shows the results of frequency

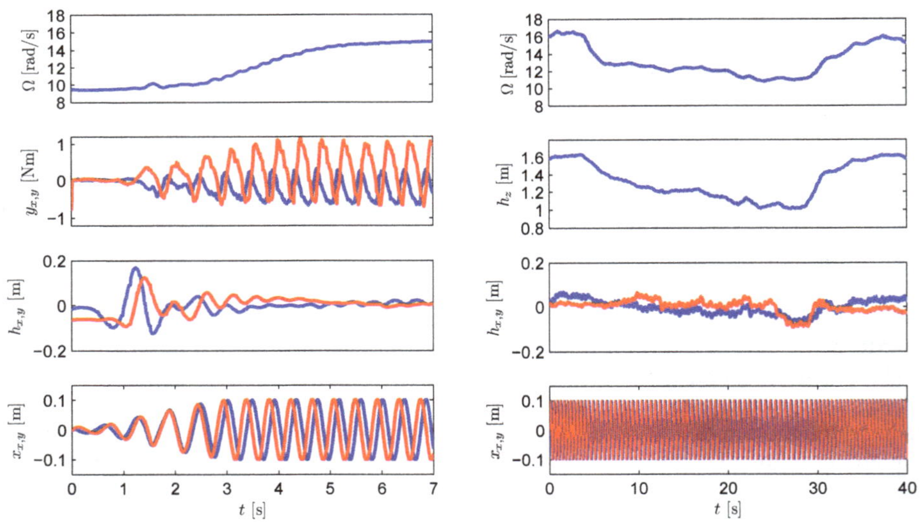

Fig. 15. *Left*: the initial frequency adaptation process (top plot) of the cooperative human-robot rope turning. The second plot shows the measured torque signal. The third plot shows the movement of a human hand, and the bottom plot shows the movement of a robot. *Right*: the behavior of our proposed system when human changes the distance between human hand and top end of the robot. Frequency adaptation is shown in the top plot, and the second plot shows the measured torque signal. The third plot shows the movement of a human hand, and the bottom plot shows the movement of a robot.

extraction (third plot) from the envelope (second plot) of the measured EMG signal (top plot). The bottom plots show the power spectrums of the input signal from 0 s to 30 s, from 30 s to 60 s and from 60 s to 90 s, respectively. The power spectrums were determined off-line.

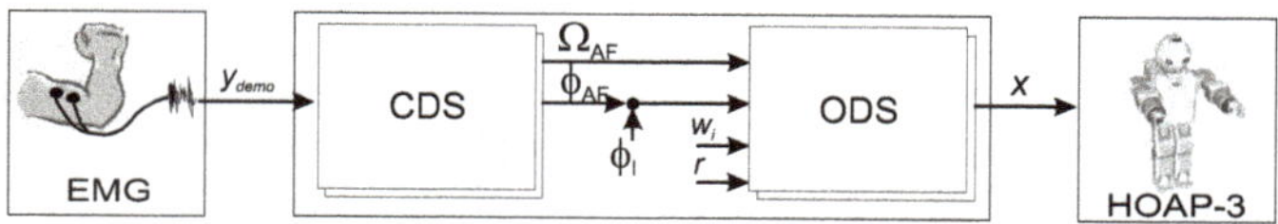

Fig. 16. Proposed two-layered structure of the control system for synchronizing the robotic motion to the EMG signal.

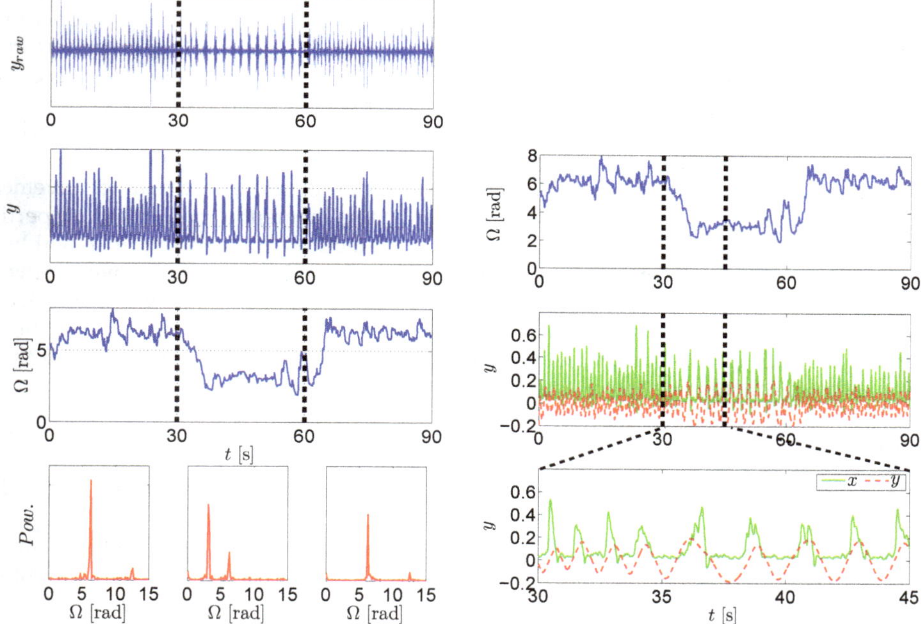

Fig. 17. *Left*:Raw EMG signal in the top plot and envelope of the EMG signal, which is the input into the proposed system, in the second plot. The third plot shows the extracted frequency Ω. The bottom plots show the power spectrum of the signal at different times (determined off-line). *Right*: Extracted frequency in the top plot. Comparison between the envelope of the rectified EMG signal (y), which is used as the input into the frequency-extraction system, and the generated output trajectory for the robot arm (x) is shown in the middle and bottom plots.

Fig. 17 right show the comparison between the envelope of the measured and rectified EMG signal (y), and the generated movement signal (x). As we can see the proposed system matches the desired movement of the robot with the measured movement.

5. Conclusion

We have shown the properties and the possible use of the proposed two-layered movement imitation system. The system can be used for both waveform learning and frequency extraction. Each of these has preferable properties for learning and controlling of robots by imitation. On-line waveform learning can be used for effective and natural learning of robotic movement, with the demonstrator in the loop. During learning the demonstrator can observe the learned behavior of the robot and if necessary adapt his movement to achieve better robot performance. The table wiping task is an example of such on-line learning, where not only is the operator in the loop for the initial trajectory, but online learning adapts the trajectory based on the measured force signal.

Online frequency extraction, essential for waveform learning, can additionally be used for synchronization to external signals. Examples of such tasks are drumming, cooperative rope turning and synchronization to a measured EMG signal. Specially the synchronization to an EMG signal shows at great robustness of the system. Furthermore, the process of synchronizing the movement of the robot and the actuated device can be applied in a similar manner to different tasks and can be used as a common generic algorithm for controlling periodic tasks. It is also easy to implement, with hardly any parameter tuning at all.

The overall structure of the system, based on nonlinear oscillators and dynamic movement primitives is, besides computationally extremely light, also inherently robust. The properties of dynamic movement primitives, which allow on-line modulation, and return smoothly to the desired trajectory after perturbation, allow learning of whole families of movement with one demonstrated trajectory. An example of such is learning of circular movement, which can be increased in amplitude, or we can change the center of the circular movement. Additional modifications, such as slow-down feedback and virtual repulsive forces even expand these properties into a coherent and complete block for control of periodic robotic movements.

6. References

An, C., Atkeson, C. & Hollerbach, J. (1988). *Model-based control of a robot manipulator*, MIT Press.

Asfour, T., Regenstein, K., Azad, P., Schröder, J., Bierbaum, A., Vahrenkamp, N. & Dillmann, R. (2006). ARMAR-III: An Integrated Humanoid Platform for Sensory-Motor Control, *IEEE-RAS Int. Conf. on Humanoid Robots (Humanoids 2006)*, Genoa, Italy.

Atkeson, C., Moore, A. & Schaal, S. (1997). Locally weighted learning, *AI Review* 11: 11–73.

Bailey, A. (2004). *Biomimetic control with a feedback coupled nonlinear oscillator: insect experiments, design tools, and hexapedal robot adaptation results*, PhD thesis, Stanford University.

Buchli, J. & Ijspeert, A. (2004). A simple, adaptive locomotion toy-system, *in* S. Schaal, A. Ijspeert, A. Billard, S. Vijayakumar, J. Hallam & J. Meyer (eds), *From Animals to Animats 8. Proceedings of the Eighth International Conference on the Simulation of Adaptive Behavior (SAB'04)*, MIT Press, pp. 153–162.

Buchli, J., Righetti, L. & Ijspeert, A. (2006). Engineering entrainment and adaptation in limit cycle systems – from biological inspiration to applications in robotics, *Biological Cybernetics* 95(6): 645–664.

Buchli, J., Righetti, L. & Ijspeert, A. J. (2005). A dynamical systems approach to learning: A frequency-adaptive hopper robot, *in* M. S. Capcarrere, A. A. Freitas, P. J. Bentley, C. G. Johnson & J. Timmis (eds), *Advances in Artificial Life*, Vol. 3630 of *Lecture Notes in Computer Science*, Springer Berlin / Heidelberg, pp. 210–220.

Buchli, J., Righetti, L. & Ijspeert, A. J. (2008). Frequency Analysis with coupled nonlinear Oscillators, *Physica D: Nonlinear Phenomena* 237: 1705–1718.

Buehler, M., Koditschek, D. & Kindlmann, P. (1994). Planning and Control of Robotic Juggling and Catching Tasks, *The International Journal of Robotics Research* 13(2): 101–118.

Cafuta, P. & Curk, B. (2008). Control of nonholonomic robotic load, *Advanced Motion Control, 2008. AMC '08. 10th IEEE International Workshop on*, pp. 631 –636.

Calinon, S., Guenter, F. & Billard, A. (2007). On learning, representing, and generalizing a task in a humanoid robot, *Systems, Man, and Cybernetics, Part B: Cybernetics, IEEE Transactions on* 37(2): 286 –298.

Castellini, C. & Smagt, P. v. d. (2009). Surface emg in advanced hand prosthetics, *Biological Cybernetics* 100(1): 35–47.

Filho, A. C. d. P., Dutra, M. S. & Raptopoulos, L. S. C. (2005). Modeling of a bipedal robot using mutually coupled rayleigh oscillators, *Biological Cybernetics* 92: 1–7. 10.1007/s00422-004-0531-1.

Fukuoka, Y., Kimura, H. & Cohen, A. H. (2003). Adaptive Dynamic Walking of a Quadruped Robot on Irregular Terrain Based on Biological Concepts, *The International Journal of Robotics Research* 22(3-4): 187–202.

Furuta, K. (2003). Control of pendulum: from super mechano-system to human adaptive mechatronics, *Decision and Control, 2003. Proceedings. 42nd IEEE Conference on*, Vol. 2, pp. 1498 – 1507 Vol.2.

Gams, A., Ijspeert, A. J., Schaal, S. & Lenarcic, J. (2009). On-line learning and modulation of periodic movements with nonlinear dynamical systems, *Auton. Robots* 27(1): 3–23.

Gams, A., Žlajpah, L. & Lenarčič, J. (2007). Imitating human acceleration of a gyroscopic device, *Robotica* 25(4): 501–509.

Gangadhar, G., Joseph, D. & Chakravarthy, V. (2007). An oscillatory neuromotor model of handwriting generation, *International Journal on Document Analysis and Recognition* 10: 69–84. 10.1007/s10032-007-0046-0.

Haken, H., Kelso, J. A. S. & Bunz, H. (1985). A theoretical model of phase transitions in human hand movements, *Biological Cybernetics* 51: 347–356. 10.1007/BF00336922.

Hashimoto, K. & Noritsugu, T. (1996). Modeling and control of robotic yo-yo with visual feedback, *Robotics and Automation, 1996. Proceedings., 1996 IEEE International Conference on*, Vol. 3, pp. 2650 –2655 vol.3.

Heyda, P. G. (2002). Roller ball dynamics revisited, *American Journal of Physics* 70(10): 1049–1051.

Hollerbach, J. M. (1981). An oscillation theory of handwriting, *Biological Cybernetics* 39: 139–156. 10.1007/BF00336740.

Ijspeert, A. J. (2008a). Central pattern generators for locomotion control in animals and robots: a review, *Neural Networks* 21(4): 642–653.

Ijspeert, A. J. (2008b). Central pattern generators for locomotion control in animals and robots: A review, *Neural Networks* 21(4): 642 – 653. Robotics and Neuroscience.

Ijspeert, A. J., Nakanishi, J. & Schaal, S. (2002). Movement imitation with nonlinear dynamical systems in humanoid robots, *Proc. IEEE Int. Conf. Robotics and Automation*, pp. 1398–1403.

Ilg, W., Albiez, J., Jedele, H., Berns, K. & Dillmann, R. (1999). Adaptive periodic movement control for the four legged walking machine bisam, *Robotics and Automation, 1999. Proceedings. 1999 IEEE International Conference on*, Vol. 3, pp. 2354 –2359 vol.3.

Inoue, K., Ma, S. & Jin, C. (2004). Neural oscillator network-based controller for meandering locomotion of snake-like robots, *Robotics and Automation, 2004. Proceedings. ICRA '04. 2004 IEEE International Conference on*, Vol. 5, pp. 5064 – 5069 Vol.5.

Jalics, L., Hemami, H. & Zheng, Y. (1997). Pattern generation using coupled oscillators for robotic and biorobotic adaptive periodic movement, *Robotics and Automation, 1997. Proceedings., 1997 IEEE International Conference on*, Vol. 1, pp. 179 –184 vol.1.

Jin, H.-L., Ye, Q. & Zacksenhouse, M. (2009). Return maps, parameterization, and cycle-wise planning of yo-yo playing, *Trans. Rob.* 25(2): 438–445.

Jin, H.-L. & Zackenhouse, M. (2002). Yoyo dynamics: Sequence of collisions captured by a restitution effect, *Journal of Dynamic Systems, Measurement, and Control* 124(3): 390–397.

Jin, H.-L. & Zacksenhouse, M. (2003). Oscillatory neural networks for robotic yo-yo control, *Neural Networks, IEEE Transactions on* 14(2): 317 – 325.

Jindai, M. & Watanabe, T. (2007). Development of a handshake robot system based on a handshake approaching motion model, *Advanced intelligent mechatronics, 2007 IEEE/ASME international conference on*, pp. 1 –6.

Kasuga, T. & Hashimoto, M. (2005). Human-robot handshaking using neural oscillators, *Robotics and Automation, 2005. ICRA 2005. Proceedings of the 2005 IEEE International Conference on*, pp. 3802 – 3807.

Kimura, H., Akiyama, S. & Sakurama, K. (1999). Realization of dynamic walking and running of the quadruped using neural oscillator, *Autonomous Robots* 7: 247–258. 10.1023/A:1008924521542.

Kralemann, B., Cimponeriu, L., Rosenblum, M., Pikovsky, A. & Mrowka, R. (2008). Phase dynamics of coupled oscillators reconstructed from data, *Phys. Rev. E* 77(6): 066205.

Kun, A. & Miller, W.T., I. (1996). Adaptive dynamic balance of a biped robot using neural networks, *Robotics and Automation, 1996. Proceedings., 1996 IEEE International Conference on*, Vol. 1, pp. 240 –245 vol.1.

Liu, C., Chen, Q. & Zhang, J. (2009). Coupled van der pol oscillators utilised as central pattern generators for quadruped locomotion, *Control and Decision Conference, 2009. CCDC '09. Chinese*, pp. 3677 –3682.

Ljung, L. & Söderström, T. (1986). *Theory and Practice of Recursive Identification*, MIT Press.

Mataric, M. (1998). Behavior-based robotics as a tool for synthesis of artificial behavior and analysis of natural behavior, *Trends in Cognitive Science* 2(3): 82–87.

Matsuoka, K. (1985). Sustained oscillations generated by mutually inhibiting neurons with adaptation, *Biological Cybernetics* 52: 367–376. 10.1007/BF00449593.

Matsuoka, K., Ohyama, N., Watanabe, A. & Ooshima, M. (2005). Control of a giant swing robot using a neural oscillator, *ICNC (2)*, pp. 274–282.

Matsuoka, K. & Ooshima, M. (2007). A dish-spinning robot using a neural oscillator, *International Congress Series* 1301: 218 – 221.

Mirollo, R. E., Steven & Strogatz, H. (1990). Synchronization of pulse-coupled biological oscillators, *SIAM J. Appl. Math* 50: 1645–1662.

Miyamoto, H., Schaal, S., Gandolfo, F., Gomi, H., Koike, Y., Osu, R., Nakano, E., Wada, Y. & Kawato, M. (1996). A kendama learning robot based on bi-directional theory, *Neural Networks* 9(8): 1281–1302.

Morimoto, J., Endo, G., Nakanishi, J. & Cheng, G. (2008). A biologically inspired biped locomotion strategy for humanoid robots: Modulation of sinusoidal patterns by a coupled oscillator model, *Robotics, IEEE Transactions on* 24(1): 185 –191.

Petrič, T., Curk, B., Cafuta, P. & Žlajpah, L. (2010). Modeling of the robotic powerball: a nonholonomic, underactuated, and variable structure-type system, *Mathematical and Computer Modelling of Dynamical Systems* . 10.1080/13873954.2010.484237.

Petrič, T., Gams, A., Tomšič, M. & Žlajpah, L. (2011). Control of rhythmic robotic movements through synchronization with human muscle activity, *2011 IEEE International Conference on Robotics and Automation, ICRA 2011, May 9-13, 2011, Shanghai, China, Proceedings*, IEEE.

Pikovsky, A., Author, Rosenblum, M., Author, Kurths, J., Author, Hilborn, R. C. & Reviewer (2002). Synchronization: A universal concept in nonlinear science, *American Journal of Physics* 70(6): 655–655.

Righetti, L., Buchli, J. & Ijspeert, A. (2006). Dynamic hebbian learning in adaptive frequency oscillators, *Physica D* 216(2): 269–281.

Righetti, L. & Ijspeert, A. J. (2006). Programmable Central Pattern Generators: an application to biped locomotion control, *Proceedings of the 2006 IEEE International Conference on Robotics and Automation*.

Ronsse, R., Lefevre, P. & Sepulchre, R. (2007). Rhythmic feedback control of a blind planar juggler, *Robotics, IEEE Transactions on* 23(4): 790 –802.

Sato, T., Hashimoto, M. & Tsukahara, M. (2007). Synchronization based control using online design of dynamics and its application to human-robot interaction, *Robotics and Biomimetics, 2007. ROBIO 2007. IEEE International Conference on*, pp. 652 –657.

Schaal, S. (1999). Is imitation learning the route to humanoid robots?, *Trends in cognitive sciences* (3): 233–242.

Schaal, S. & Atkeson, C. (1993). Open loop stable control strategies for robot juggling, *Robotics and Automation, 1993. Proceedings., 1993 IEEE International Conference on*, pp. 913–918.

Schaal, S. & Atkeson, C. (1998). Constructive incremental learning from only local information, *Neural Computation* 10: 2047–2084.

Schaal, S., Mohajerian, P. & Ijspeert, A. (2007). *Dynamics systems vs. optimal control – a unifying view*, Vol. 165 of *Progress in Brain Research*, Elsevier, pp. 425 – 445. 10.1016/S0079-6123(06)65027-9.

Sentis, L., Park, J. & Khatib, O. (2010). Compliant control of multicontact and center-of-mass behaviors in humanoid robots, *Robotics, IEEE Transactions on* 26(3): 483 –501.

Spong, M. (1995). The swing up control problem for the acrobot, *Control Systems Magazine, IEEE* 15(1): 49 –55.

Spong, M. & Bullo, F. (2005). Controlled symmetries and passive walking, *Automatic Control, IEEE Transactions on* 50(7): 1025 – 1031.

Strogatz, S. H. (1986). *The mathematical structure of the human sleep-wake cycle*, Springer-Verlag New York, Inc., New York, NY, USA.

Thompson, S. E. & Patel, R. V. (1987). Formulation of joint trajectories for industrial robots using b-splines, *Industrial Electronics, IEEE Transactions on* IE-34(2): 192 –199.

Tsuda, S., Zauner, K.-P. & Gunji, Y.-P. (2007). Robot control with biological cells, *Biosystems* 87(2-3): 215 – 223. Papers presented at the Sixth International Workshop on Information Processing in Cells and Tissues, York, UK, 2005 - IPCAT 2005, Information Processing in Cells and Tissues.

Ude, A., Atkeson, C. G. & Riley, M. (2000). Planning of joint trajectories for humanoid robots using b-spline wavelets, *ICRA*, pp. 2223–2228.

van der Pol, B. (1934). The nonlinear theory of electric oscillations, *Proceedings of the Institute of Radio Engineers* 22(9): 1051 –1086.

Veskos, P. & Demiris, Y. (2005). Developmental acquisition of entrainment skills in robot swinging using van der pol oscillators, Vol. 123, Lund University Cognitive Studies, pp. 87–93.

Villani, L. & J., D. S. (2008). *Handbook of Robotics*, Springer, chapter Force Control.

Žlajpah, L. (2006). Robotic yo-yo: modelling and control strategies, *Robotica* 24(2): 211–220.

Williamson, M. (1999). Designing rhythmic motions using neural oscillators, *Intelligent Robots and Systems, 1999. IROS '99. Proceedings. 1999 IEEE/RSJ International Conference on*, Vol. 1, pp. 494 –500 vol.1.

Williamson, M. M. (1998). Neural control of rhythmic arm movements, *Neural Networks* 11(7-8): 1379 – 1394.

2

Design of Oscillatory Neural Network for Locomotion Control of Humanoid Robots

Riadh Zaier

Department of Mechanical and Industrial Engineering, Sultan Qaboos University
Sultanate of Oman

1. Introduction

Standing and walking are very important activities for daily living, so that their absence or any abnormality in their performance causes difficulties in doing regular task independently. Analysis of human motion has traditionally been accomplished by subjectively through visual observations. By combining advanced measurement technology and biomechanical modeling, the human gait is today objectively quantified in what is known as Gait analysis. Gait analysis research and development is an ongoing activity. New models and methods continue to evolve. Recently, humanoid robotics becomes widely developing world-wide technology and currently represents one of the main tools not only to investigate and study human gaits but also to acquire knowledge on how to assist paraplegic walking of patient (Acosta-M´arquez and Bradley, 2000). Towards a better control of humanoid locomotion, much work can be found in the literature that has been focused on the dynamics of the robot using the Zero Moment Point (ZMP) approach (Vukobratovic and Borovac, 2004). More recently, biologically inspired control strategies such as Central Pattern Generators (CPG) have been proposed to generate autonomously adaptable rhythmic movement (Grillner, 1975, 1985; Taga, 1995; Taga *et. al*, 1991). Despite the extensive research focus in this area, suitable autonomous control system that can adapt and interact safely with the surrounding environment while delivering high robustness are yet to be discovered.

In this chapter, we deal with the design of oscillatory neural network for bipedal motion pattern generator and locomotion controller. The learning part of the system will be built based on the combination of simplified models of the system with an extensive and efficient use of sensory feedback (sensor fusion) as the main engine to stabilize and adapt the system against parameters changes. All motions including reflexes will be generated by a neural network (NN) that represents the lower layer of the system. Indeed, we believe that the NN would be the most appropriate code when dealing, to a certain limit, with the system behavior, which can be described by a set of ordinary differential equations (ODEs) (Zaier and Nagashima, 2002, 2004). The neural network will be augmented by neural controllers with sensory connections to maintain the stability of the system. Hence, the proposed learning method is expected to be much faster than the conventional ones. To validate the theoretical results, we used the humanoid robot "HOAP-3" of Fujitsu.

The structure of the chapter is as follows: the first section will present an introduction on the conventional CPG based locomotion control as well as the Van der Pol Based Oscillator;

then the Piecewise Linear Function Based Oscillator as our proposed approach will be detailed. The fourth section will present the Experiment Results and Discussion, and finally the conclusion will highlight the possible of future developments of the method.

2. CPG based locomotion controller

Animal produces may periodic movements such as walking and running, which are called gaits. It was found (Grillner, 1985) that those undirected movements are produced by Central Pattern Generators (CPGs). CPG is a neural network that can produce rhythmic patterned outputs without any input, and they underlie the production of most rhythmic motor patterns. Many researchers developed mathematical CPG models to generate motion in biped robots (Takeshi Mori et. al, 2004), snake robots (Crespi and Ijspeert, 2006), and quadruped robot (Cappelletto et. al, 2007). Those biological inspired controllers insure the production of more natural and simple motion. Also, those oscillators can entrain and adapt the dynamic of the system. For example, Matsuoka, Van-der-pol, Hopf and Rayleigh are oscillators used in robotics control.

Basically, a single neuron is the basic unit of a CPG network. Two neurons when coupled can produce a rhythmic output and this is called a Neural Oscillators (NO). The rhythmic output requires two or more neurons that interact with each other to oscillate and therefore passes by its starting condition (Hooper SL., 2000). Unlike linear systems, nonlinear oscillators can produce stable limit cycles without an oscillatory input of the same frequency, and thus they are very suited to model some parts of the nervous system.

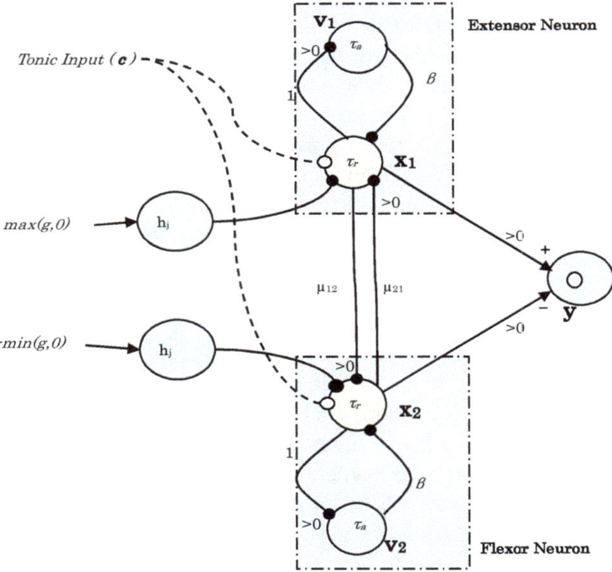

Fig. 1. Matsuoka neural oscillator model. Black dots corresponds to inhibitory connections and the white dots to excitery. Using the notation in (Zaier and Nagashima, 2002), the symbol ">0" reprents a switcher that take the positive part of the input, for example the input from x1 becomes [x1]+

2.1 Matsuoka based oscillator

Matsuoka proposed the neural oscillator shown in Figure 1 that consists of a flexor and extensor neuron (Matsuoka, 1985, 1987). Each neuron is presented by a nonlinear differential equation. Each neuron produces a periodic signal to inhabit the other neuron to control the limbs motion (i.e. extending and flexing the elbow). Compared to other models, Matsuoka model uses significantly less computational resources, has less parameters requiring tuning, and has no need for post-processing of the neural output signals (i.e., filtering of the spikes). The mathematical model of the Matsuoka neural oscillator can be expressed by equations (2-7) and as quoted from Williamson (Williamson, 1999) and illustrated in Fig. 1 using the neuron model of equation 1 and as decribed in (Zaier and Nagashima, 2002):

$$\varepsilon \frac{dx_1}{dt} + x_1 = \sum inputs \tag{1}$$

where ε is the time delay of the neuron.

$$\tau_r \frac{dx_1}{dt} + x_1 = -\beta v_1 - \mu_{21}[x_2]^+ - \sum h_j \left[g_j\right]^+ + c \tag{2}$$

$$\tau_a \frac{dv_1}{dt} + v_1 = [x_1]^+ \tag{3}$$

$$\tau_r \frac{dx_2}{dt} + x_2 = -\beta v_2 - \mu_{12}[x_1]^+ - \sum h_j \left[g_j\right]^- + c \tag{4}$$

$$\tau_a \frac{dv_2}{dt} + v_2 = [x_2]^+ \tag{5}$$

$$y_i = [x_i]^+ = \max(0, x_i) \tag{6}$$

$$y_{out} = [x_1]^+ - [x_2]^+ = y_1 - y_2 \tag{7}$$

$$[x_i]^+ = \max(0, x_i) \tag{8}$$

The extensor neuron 1 in Fig. 1 is governed by equations (2) and (3) and flexor neuron 2 by equations (4) and (5). The variable x_i is the neuron potential or firing rate of the ith neuron, v_i is the variable that represents the degree of adaptation or self-inhibition, c is the external tonic input with a constant rate, β specify the time constant for the adaptation, τ_a and τ_r are the adaptation and rising rates of the neuron potential, μ_{ij} is the weight that represents the strength of inhibition connection between neurons, and g_j is an external input which is usually the sensory feedback weighted by gain h_j. Where the positive part of the input $[g_j]^+$ is applied to one neuron and the negative part $[g_j]^- = -\min(g_j, 0)$ is applied to the other neuron. The positive part of the output of each neuron is denoted by $y_i = [x_1]^+$ and the final output of the oscillator y_{out} is the difference between the two neurons' outputs. However, the parameters should be adjusted correctly to suite the application the oscillator will be used for. This can be done by experiments or by specifying the constraints. According to Williamson (Williamson, 1999), for stable oscillations, τ_r / τ_a should be in the range 0.1-0.5. Keeping the ratio τ_r / τ_a constant makes the natural frequency of the oscillator ω_n (the frequency of the oscillator without an input) proportional to $1/\tau_r$. According to Matsuoka (Matsuoka, 1987), the parameters should be selected based on the following criteria to ensure a stable rhythm $\frac{\mu_{ji}}{(1+\beta)} < 1$ and $\sqrt{\mu_{12} \cdot \mu_{21}} > 1 + \frac{\tau_r}{\tau_a}$. Some researchers used Genetic Algorithm (GA) technique to find and optimize the oscillator's parameters (Inada, 2004).

2.2 Van der Pol based oscillator

The Van der Pol Oscillator, described by second order nonlinear differential equation (13), can be regraded as a mass-spring-damper system. The circuit of this oscillator is shown in Figure 2, where the neuron notation in (Zaier and Nagashima, 2002) is used. It has been adopted by Van de Pol in 1920th to study the oscillations in vacum tube circuits. Recently researchers built CPG model based on Van der Pol oscillator (Senda K, Tanaka T, 2000). Equation (10) is for the forced oscillotor where the right hand side term is the periodic forcing term. The unforced Van der Pol oscillator is investigated using MATLAB/Simulink© block diagram as shown in Figure 3.

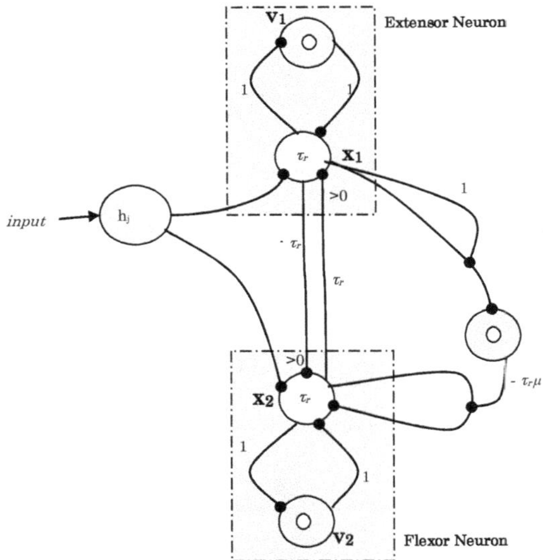

Fig. 2. Van der Pol oscillator model. Black dots corresponds to inhibitory connections.

$$\ddot{x} + \mu\left(x^2 - 1\right)\dot{x} + x = 0 \tag{9}$$

$$\ddot{x} + \mu\left(x^2 - 1\right)\dot{x} + x = a\ \sin(2\pi \upsilon \tau) \tag{10}$$

Where x and $\dot{x}$ are the states of the system and μ is a control parameter that represents the degree of nonlinearity of the system or its damping strength.

The equation 9 has a periodic solution that attracts other solutions except the trivial one at the unique equlibrum point (origin of the phase plote).
In the case of $\mu > 0$, the system will be unstable, oscillate forever and the phase plot will never converge to an equilibrium point, as shown in Figure 4 (a) and (b). While in the case of $\mu < 0$, the system will converge to an equilibrium point as shown in Figure 4 (c) and (d). In this case, the decaying is directly proportional to the negative value of μ, where the oscillation dies faster as the constant goes larger in the negative direction. However, if μ is set to zero, then the system has no damping and will oscillate sinusoidally forever as shown in Figure 4 (e). The phase plot draws an exact circle in this case. In conclusion, the Van der Pol oscillator has the ability to produce a periodic behavior which can represent a periodic locomotion pattern of robots (Veskos P., 2005).

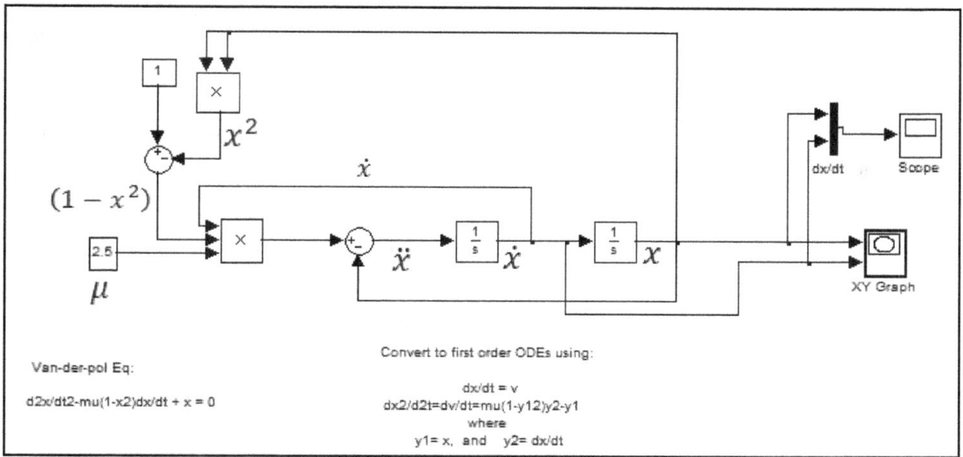

Fig. 3. Simulation of Van der Pol Oscillator Behavior.

3. Piecewise linear function based oscillator

The control strategy here is based on piecewise linear oscillators and inspired by the solution of the Van der Pol system in Figure 4b, and with few parameters that can be easily obtained. Moreover, the motivation is to provide much flexibility to the design of the motion pattern generator so that modulation of the output by other circuits such as circuits generating reflexes can be realized without complexity or re-design of the motion pattern generator (Zaier and Kanda, 2007). The method requires not to satisfy constraints on robot's foot or ZMP stability margin, it simply uses piecewise-linear functions and a first order low-pass filter generated by an original recurrent neural network (RNN), where the "integrate and fire" neuron model (Gerstner, 1995) has been used. Indeed, the piecewise linear control is much easier to analyze than the control based directly on non-linear equations. Moreover, this type of control provides much intuition about the system behavior. Also, our method for generating walking motion, although inspired by the inverted pendulum, it considers not strictly the system dynamics. The stability of the system is assured by a gyro feedback loop, while terrain irregularities are compensated by adjusting the pitching motion using a virtual spring-damper systems. The proposed method is straightforward with respect to three design parameters namely, the slope of the piecewise function, the time delay of the neuron, and the rolling amplitude of the hip and ankle joints. In this section, we investigate the stability of the system simplified as an inverted pendulum, the robustness of the locomotion controller against terrain irregularities.

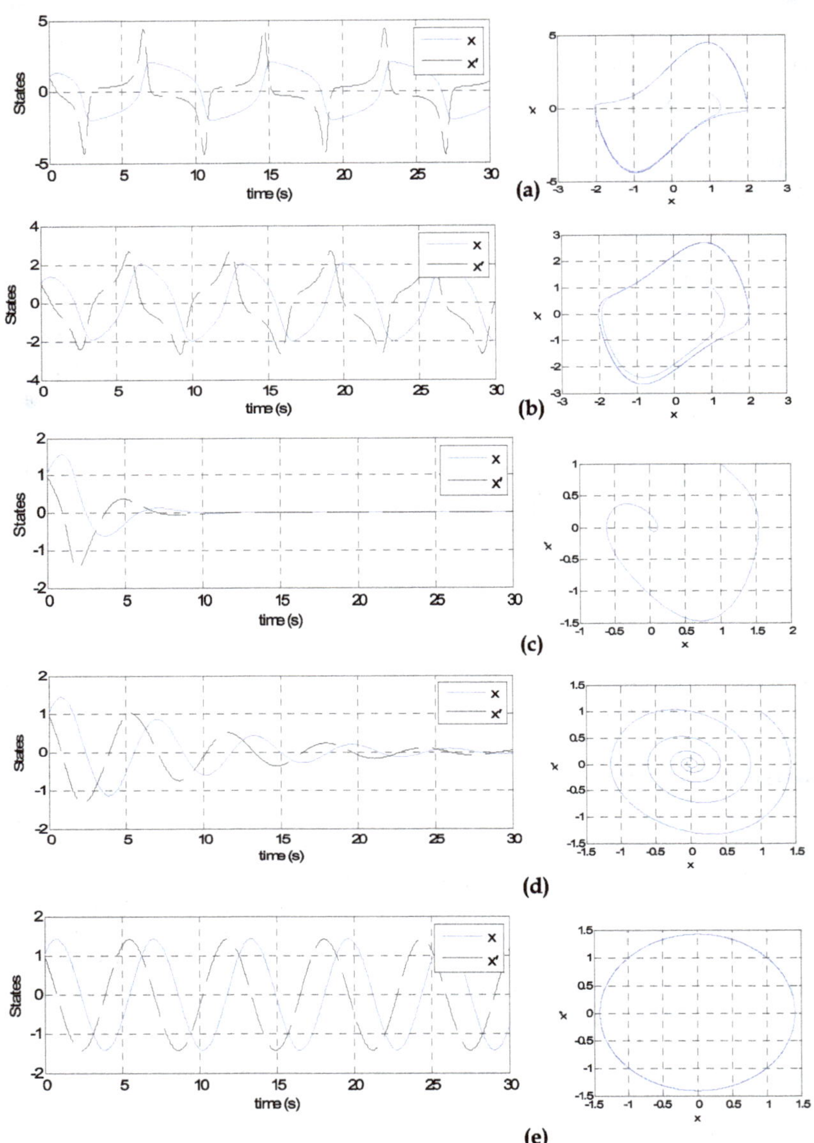

Fig. 4. Van der Pol States over Time (left) and Phase Plot (right) with the (a) μ=2.5, (b) μ=1.0, (c) μ= -1.0, (d) μ= -0.25, and (e) μ= 0.0.

3.1 Motivation

To control the robot locomotion, we consider the general form of neural network as;

$$\begin{aligned}\frac{dx}{dt} &= -Dx + Tg(x) + Sv \\ y &= Cx\end{aligned} \tag{11}$$

where $x \in R^n$ is the neurons' outputs vector, $y \in R^m$ depicts the motors' commands vector, D is an nxn diagonal matrix with strictly positive real numbers, T is an nxn neurons' interconnection matrix, $g(x)$: $R^n \rightarrow R^n$ is the nonlinear vector activation function, C is an mxn output matrix, $v \in R^m$ is a sensory input vector, and S is an nxm matrix. Note that the right side of equation (11) can be considered as decomposition into linear part "Sv" and a nonlinear one "$Tg(x)$". To investigate the stability of equation (11), we consider the behavior of a linearized model of system at the equilibrium points with $Tg(x)=0$. Indeed, a motion can be described topologically as a homotopy mapping at each equilibrium point as well as switching, or jump mapping between these points (Forti, 1996). In this research framework we consider locomotion control of a humanoid robot simplified as an inverted pendulum when standing on one leg. Notice that, while the inverted pendulum model considerably simplifies the control of the humanoid robot, the inertia effect of distributed masses such as the arms presents a limitation to that approach.

3.2 Problem formulation

The proposed control strategy is illustrated in figures 5 and 6, where the neural network modeled by equation (11) has to control the humanoid robot locomotion as a continuous

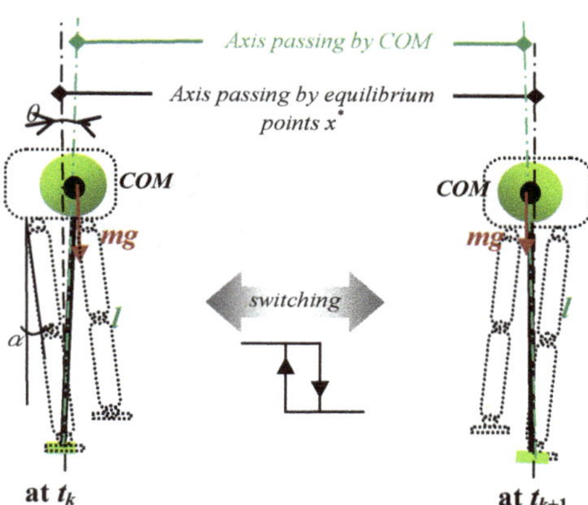

Fig. 5. System state switching between equilibrium points

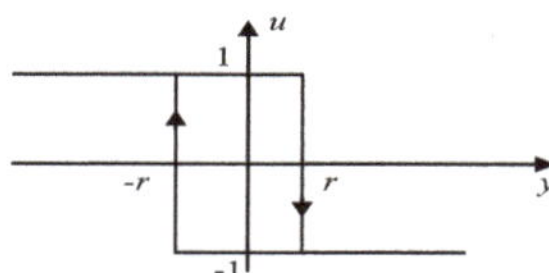

Fig. 6. Relay with hysteresis property.

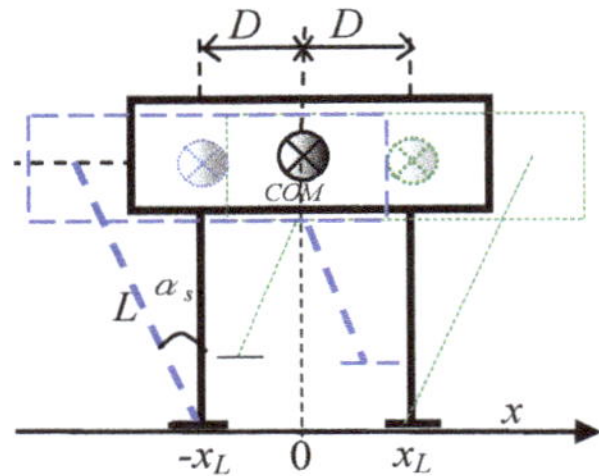

Fig. 7. Lateral component of the *COM*

commutation between the two equilibrium points defined at single support phases. Then, using the output of the switched system, we decide the rolling motion profile. The stability of the equilibrium points during locomotion is maintained using gyro feedback, while the pitching motion is adjusted by virtual dumper-spring system. To begin with, we investigate the stability of the robot simplified as an inverted pendulum when standing at one leg that can be expressed by;

$$\frac{d^2\theta}{dt^2} + 2\varsigma\frac{d\theta}{dt} - \mu\theta = u(\theta) \tag{12}$$

where $\theta \in \boldsymbol{R}$ is the counterclockwise angle of the inverted pendulum from the vertical, $\zeta \in \boldsymbol{R}$ is damping ratio, $\mu = g/l \in \boldsymbol{R}$, and $u \in \boldsymbol{R}$ is the control input to the system. Let $\theta = a - a_s$, where a is the angular position of the hip joint and a_s is its value when the projected center of mass (COM) is inside the support polygon. Then, equation (12) can be rewritten as follows;

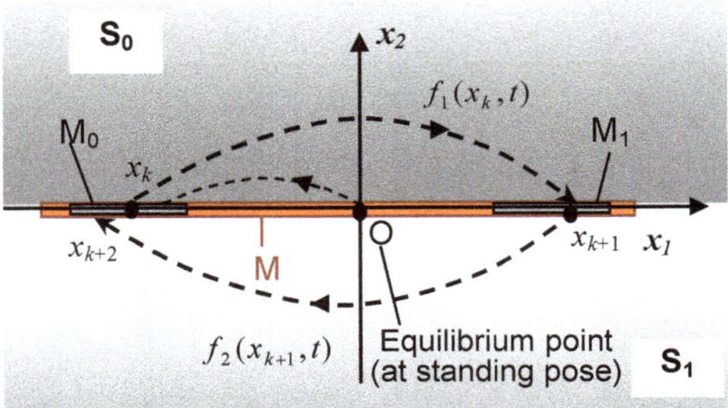

Fig. 8. Phase space and Poincare map.

$$\frac{d^2\alpha}{dt^2} + 2\varsigma\frac{d\alpha}{dt} - \mu\alpha = u(\alpha) \tag{13}$$

Moreover, according to figure 7, the relationship between a and the COM is $x_L(t) = L\sin\alpha_s(t)$. Since a_s is very small (practically it is between -0.14 and 0.14rad), the relationship can be approximated as;

$$x_L(t) = L\alpha_s(t) \tag{14}$$

Hence we can write $a_s = D/L$ in [rad], where D is the lateral component of the distance between the COM and the hip joint. On the other hand, to control the inverted pendulum as illustrated in figures 1 and 2, we propose the control input as follows;

$$u(\alpha) = u_g(\alpha) + u_{sw}(\alpha) \tag{15}$$

where $u_g(a)$ is the gyro feedback signal stabilizing the inverted pendulum around $a=a_s$, and $u_{sw}(a)$ is a relay type feedback signal with a hysteresis property defined by;

$$u_{sw}(y(t)) = \begin{cases} +1 & if\ y(t^-) > -r \\ -1 & if\ y(t^-) < r \\ \{-1,1\} & if \quad t = k\tau\ and\ if\ y(t) = \pm r \end{cases} \tag{16}$$

where t^- is the time just before commutation. Notice that the switching law of equation (16) depends not only on the output $y(t)$ but also on the switching time constant τ. In our design we choose a fixed switching time, which reduces the impact map to a linear one as detailed in (Goncalves et al., 2000).
The control system of equation (15) can be written in the state space form as;

$$\begin{cases} \dot{x} = Ax + Bu \\ y = Cx \end{cases} \tag{17}$$

where $x = (\alpha \quad \dot{\alpha})^T$, A and B are matrices with compatible dimensions, and $C=[r\ 0]$.
Let's further illustrate the stability of the control system by considering the phase space diagram in figure 8, where the x_1-axis represents the lateral component of the COM position, and which divides the phase space into two regions S_0 and S_1. The axis x_2 represents the velocity. The state space trajectory in each region is defined by a different linear differential equation according to the state of relay feedback of equation (16). It will be possible to define Poincare sections where the switching takes place. The origin "O" corresponds to the state of the robot at the double support phase. The motion starts by switching the state of the robot to section M_0, i.e., standing on the left leg (figure 5). The periodic motion is defined as a continuous switching of state between M_0 and M_1, in other words, the state that starts from section M_0 at x_k reaches section M_1 at x_{k+1} following a trajectory described by the linear differential equation defined for region S_1. It can be shown that the system of equation (17) with the feedback control of equation (16) can produce limit cycles (Goncalves et al., 2000).
The stability analysis of the system can be conducted by considering the manifolds of its equilibrium points x^* as follows

$$x^* = -A^{-1}Bu_{sw}\text{, with } u_{sw} = \pm 1 \tag{18}$$

With the feedback signal of equation (15), the trajectories of the system will follow the invariant manifolds of only two equilibrium points of equation (17). To investigate the stability of equation (13) for a close to a_s, reconsider the system in the absence of control input. Then, at each equilibrium point the system will have the same eigenvalues as follows;

$$\lambda_{1,2} = -\varsigma \pm \sqrt{\varsigma^2 + \mu} \tag{19}$$

Since ζ<<μ, the equilibrium points of equation (18) are saddle ones. To generate periodic movement as explained in figure 5 using equation (16), we first stabilize the equilibrium points by a feedback control $u_g(a) = k_1 a + 2k_2 d(a)/dt$ using the angular velocity of the inverted pendulum measured by the gyro sensor. Therefore, combining equations (13) and (16) we can get;

$$\frac{d^2\alpha}{dt^2} + 2(\varsigma + k_2)\frac{d\alpha}{dt} + (k_1 - \mu)\alpha = u_{sw}(\alpha) \tag{20}$$

which can be rewrite as

$$\frac{d^2\alpha(t)}{dt^2} + 2\varsigma'\frac{d\alpha(t)}{dt} + \mu'\alpha(t) = u_{sw}(\alpha) \tag{21}$$

where $\varsigma' = \varsigma + k_2$, and $\mu' = k_1 - \mu$. The equilibrium points becomes stable if $k_2 > -\varsigma$ and $k_1 > \mu$. Besides, if the feedback controller's parameters satisfy the following inequalities:

$$k_2 > -\varsigma \text{ and } k_1 > \mu + (\varsigma + k_2)^2 \tag{22}$$

then the eigenvalues will be complex conjugate with a negative real part. Now, consider the trajectories in both regions S_0 and S_1 defined by $u_{sw}(\alpha) = \pm 1$, which are governed by the invariant manifolds of stable virtual equilibrium points. The state space trajectory will follow the invariant manifold of a stable equilibrium point but a switching will occurs before reaching this point, then the trajectory will be governed by the invariant manifold of the other stable virtual equilibrium point and so on. This consecutive switching will result in a limit cycle in the phase space. Without loss of generality we assume that the time origin is when the state space is at M_0. The trajectory of the state space of system is given by

$$x(t) = \begin{cases} e^{-\varsigma' t}(c_1 \cos\gamma t + c_2 \sin\gamma t) + u_{sw}(x_1) \\ e^{-\varsigma' t}[(\gamma c_2 - \varsigma' c_1)\cos\gamma t + (-\gamma c_1 - \varsigma' c_2)\sin\gamma t] \end{cases} \tag{23}$$

where $\gamma = \sqrt{\mu' - \varsigma'^2}$ and the constants c_1 and c_2 depend on the initial conditions $x(0)=(-1,V_0)$. The switching sections are defined by,

$$M_0 = \{x \mid x = (-1,0\} \quad M_1 = \{x \mid x = (1,0)\} \tag{24}$$

and equation (23) can be re-written as follows,

$$x(t) = \begin{cases} \frac{V_0}{\gamma} e^{-\varsigma' t} \sin\gamma t + u_{sw}(x_1) \\ e^{-\varsigma' t}(V_0 \cos\gamma t - \varsigma' \frac{V_0}{\gamma} \sin\gamma t) \end{cases} \tag{25}$$

where the position $x_1(t)$ for $\gamma t \in [0, \pi/2]$ is shown in figure 9. Let T be the period of time the state space required to return to the same section M_0, the behavior of a point of the state trajectory at M_0 with initial velocity $x_2^0 = V_0$ hits the same section according to the Poincare map given by,

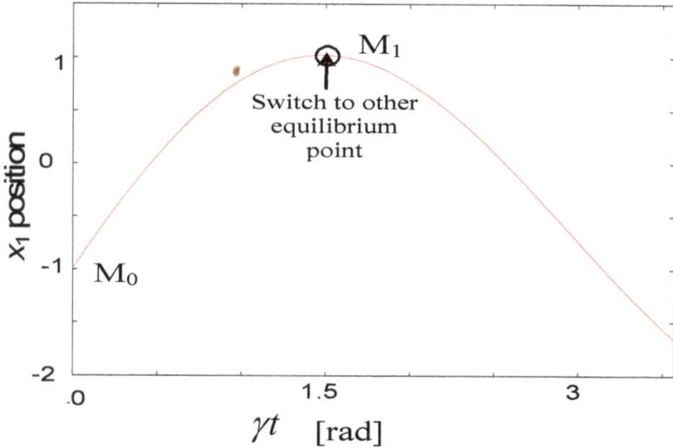

Fig. 9. Plot of x_1 position for $V_0 = 2.326, \varsigma' = 0.1$ and $\gamma = 0.995$

$$\begin{aligned} &\Pi_0 : M_0 \rightarrow M_0 \\ &(1, x_2^i) \rightarrow (1, P(x_2^i)),\ i = 1, 2, 3, \ldots \end{aligned} \tag{26}$$

where

$$P(x_2^i) = x_2^0 e^{-i\varsigma' T} \tag{27}$$

Also, when $\mu < k_1 < \mu + (\varsigma + k_2)^2$, the eigenvalues of equation (13) become pure negative real numbers (λ_1 and λ_2), and instead of (27) we will have the following

$$P(x_2^i) = \frac{x_2^0}{\lambda_1 - \lambda_2}(\lambda_1 e^{\lambda_1 T} - \lambda_2 e^{\lambda_2 T})^i \text{ and } \lambda_1 \neq \lambda_2 \tag{28}$$

It is clear from equations (27) and (28) that for any $u_g(a)$ with given parameters k_1 and k_2 stabilizing the system at the switching point, the Poincare map of equation (26) is a contracting one. Moreover, it can be noticed that when there is no switching, the state trajectory obviously will converge to one of the equilibrium points of equation (18). Moreover, if the switching occurs outside the attraction region of equation (18), the robot will have an unstable walking behavior and will fall down after few steps. Therefore, the choice of the switching time is crucial for the stability of the system. In this paper we approximate the angular position solution of equation (25) for $\gamma t \in [0, \pi/2]$ (during the commutation from M_0 to M_1) as a piecewise linear function modulated by a decaying exponential instead of sine function modulated by a decaying exponential as will be explained later on. The smallest time τ the state trajectory may take when the robot is commutating from one leg to another is such that $\gamma\tau = \pi/2$ and we will have;

$$\tau = 0.5\pi / \gamma \tag{29}$$

where $\gamma = \sqrt{\mu' - \varsigma'^2}$, ζ' and μ' are parameters of the inverted pendulum with the feedback controller as defined in equation (21). Due to the symmetry between the left and right sides of the robot, the walking cycle time is assumed to be 2τ. Moreover, for given commutation duration τ calculated by equation (29) with the feedback controller parameters in equation (22), the initial linear velocity can be expressed from equation (25) by;

$$V_0 = 2\gamma e^{\psi\varsigma'/\gamma} / \sin(\gamma\tau) \tag{30}$$

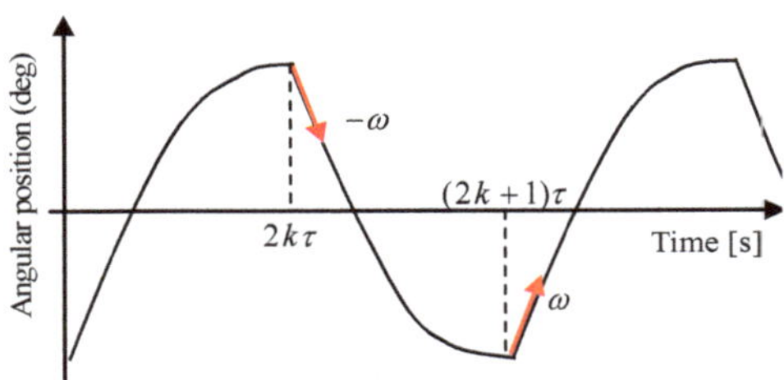

Fig. 10. Hip position command

3.3 Approximate solution of rolling

According to equation (25), the position command to the hip is shown in figure 10, and the control law at switching times can be expressed as;

$$\begin{cases} \theta(t^-) = (-1)^{i+1}\theta_{\max}, & d\theta/dt\big|_{t=t^-} = 0 \\ \theta(t^+) = (-1)^{i+1}\theta_{\max}, & d\theta/dt\big|_{t=t^+} = (-1)^i\omega \end{cases} \quad i = 1,2,3\ldots \tag{31}$$

where i is the number of half walking cycle τ, the t^- and t^+ are the times just before and after commutation, respectively. For $t \neq i\tau$ the state trajectory will follow the desired limit cycle as described in figure 9, where the stability of the equilibrium points is guaranteed by the gyro feedback added to the system. Since the position command to the hip is the same as that to the ankle joint with opposed sign, the robot upper body orientation in the absence of disturbance remains unchanged, in other words the angular velocity around the rolling axis during locomotion is zero. Notice that the control law described by equation (31) is explicit of time. Instead, the state commutation can be constrained by angular velocity $d\theta / dt = 0$, and the control law becomes independent of the cycle time as a pure feedback law. However, the condition of having a fixed commutation time (Goncalves et al., 2000) will be no longer satisfied. For the sake of reducing the computing cost and ease the implementation of the controller, instead of considering the exact solution of the state trajectory of equation (25), we propose an approximate solution $a(t)$ as shown in figures 11 and 12, using a unit piecewise linear function as follows;

$$u(t_i,\omega_i) = \omega_i(t-t_i)\left[u_s(t-t_i) - u_s(t-t_i-1/\omega_i)\right] + u_s(t-t_i-1/\omega_i) \tag{32}$$

where $u_s(.)$ is the unit step function, $\omega_i > 0$ is the slope of the function $u(.)$ between t_i and (t_i+1/ω_i). The approximate solution $a(t)$ can be formulated, therefore, as a function of time delay ε, joint angular velocity ω, walking period T, and the rolling amplitude a_s.

$$\alpha = f(\varepsilon,\omega,T,\alpha_s) \tag{33}$$

Using equation (32), the trajectory in figure 9 for $\gamma t\in[0,\pi/2]$ can be approximated as $\varepsilon\,\dot{\alpha}(t)+\alpha(t)=2(u(0,\omega_r)-1$, with $a(0)$= -1 and $\varepsilon=\zeta'$. Therefore, the rolling motion can be expressed as,

$$\varepsilon\frac{d\alpha(t)}{dt}+\alpha(t)=\alpha_s[u(t_{r0},2\omega_r)-2u(t_{r0}+(n+f_1)T,\omega_r)+ 2u(t_{r0}+(n+f_2)T,\omega_r))-u(t_f,2\omega_r)], \tag{34}$$

where t_{r0}, t_f are the times at the start and the end of the rolling motion, respectivel. n is number of walking steps. f_1 and f_2 are the relative times with respect to the gait. Let $p=t_{r1}-t_{r0}-1/\omega_r$ be the time duration when the robot stays at the maximum rolling, where t_{r1} is the first switching time at the single support phase, then we can write $f_1=(1/\omega_r+p)/T$ and $f_2=2f_1+1/(\omega_rT)$. The commands $\theta^r_{am}(t)$ and $\theta^r_{hm}(t)$ to be sent respectively to the hip and ankle joints are given by,

$$\theta^r_{am}(t)=-\theta^r_{hm}(t)=\alpha(t)+\theta^r_{fb}(t) \tag{35}$$

where $\theta^r_{fb}(t)$ is the PD controller's output stabilizing the system of equation (17).

Numerical Example

Consider the inverted pendulum with the control law of equation (20) as follows;

$$\frac{d^2\alpha}{dt^2}+0.5\frac{d\alpha}{dt}-\alpha=-k_1\alpha-2k_2\frac{d\alpha}{dt}+u_{sw}(\alpha) \tag{36}$$

where we choose an arbitrary high gain controller satisfying equation (22) with k_1 = 32 and k_2 = 0.9, and does not trigger mechanical resonance. We consider the rolling profile equation (34) with ε = 0.15, and a_s =6 deg. The phase portrait of the output signal is shown in figure 13, where case (a) is for non perturbed system, while case (b) is in the presence of perturbation. This result shows that the state trajectory is bounded, and therefore it is stable.

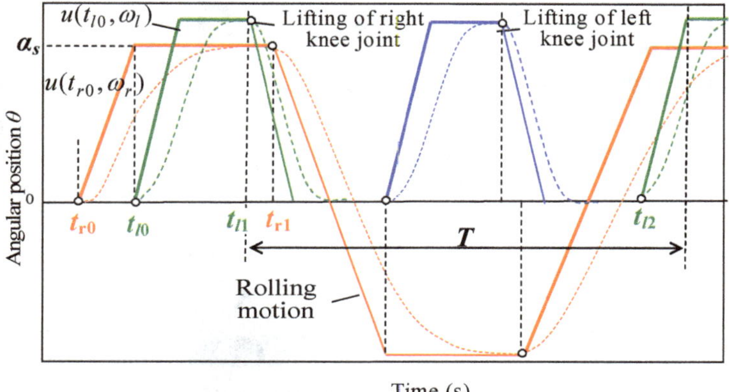

Fig. 11. Rolling motion pattern and design parameters

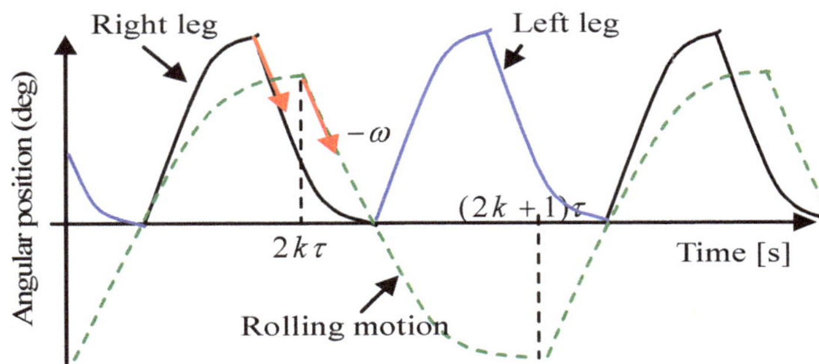

Fig. 12. Profile of the lifting motion (θ_l)

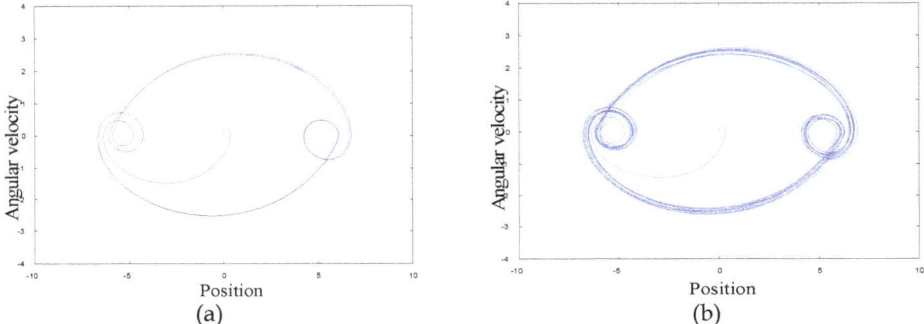

Fig. 13. Phase portrait of the velocity versus position (Simulation); case (a) without perturbation, case (b) with perturbation.

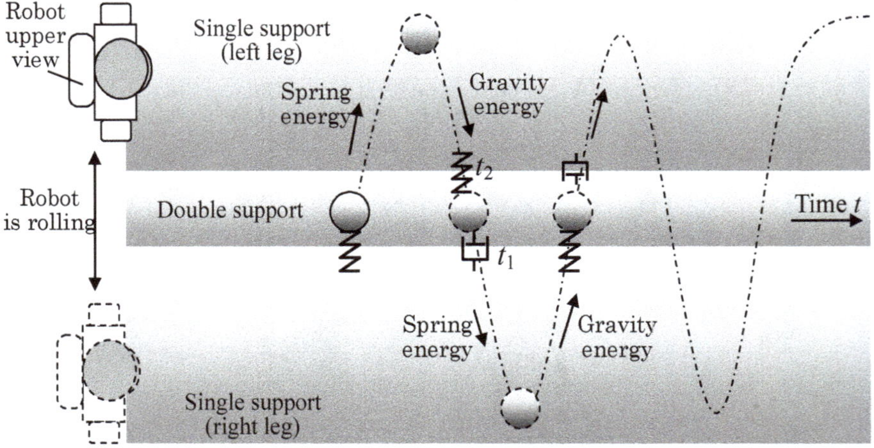

At t_1 the damper is used to land softly, while at t_2 the spring energy is used to lift

Fig. 14. Rolling motion with regards to time, inspired by passive walking.

3.4 Swing and stepping motion

The dynamic of the swing leg can be considered the same as that of a pendulum, and hence it is inherently stable without any compensator. Due to the switching control that we already adopted in generating the rolling motion, the swing motion can be generated in a similar fashion and as illustrated in figure 11, which can be expressed as;

$$\varepsilon\frac{d\theta_l(t)}{dt}+\theta_l(t)=A_l[u(t_{l0},\omega_l)-u(t_{l1}+nT,\omega_l)+u(t_{l2}+nT,\omega_l)-u(t_{lf},\omega_l)], \tag{37}$$

where θ_l is the lifting motion, A_l is the amplitude of lifting, and t_{l0} and t_{l2} are the switching time for the first and second lifting, t_{l1} and t_{lf} are the switching times for the first and the last landing, and ω_l is the joints' angular velocity generating the lifting motion.

To control the motion along the pitching direction, besides the gyro feedback control we consider the robustness against rough ground (ground with some surface irregularity) by controlling the time-varying parameters of the virtual spring damper system $b_s(t)$ and $k_s(t)$ in equation (38) during the gait as described in figure 14. That is, to minimize the force of collision at the landing time t_1, the damping coefficient $b_s(t_1)$ is set at its maximum value while $k_s(t_1)$ is at its minimum value. At the lifting time t_2, we inject spring energy into the system by setting $b_s(t_2)$ at its minimum while $k_s(t_2)$ is at its maximum. We consider linear time-parameter changes between lower/upper limits.

$$m\frac{d^2y_c(t)}{dt^2}+b_s(t)\frac{dy_c(t)}{dt}+k_s(t)y_c(t)=g(t,F_y) \tag{38}$$

where $y_c(t)$is the displacement of the mass m along the vertical axis, and $g(t, F_y)$ is a piecewise function in t and locally Lipschitz in F_y that depicts the external force acting on the supporting leg. We choose g as a saturation function.

If we let $y_c = x_1$, and $dy_c/dt = x_2$, equation (38) can be re-written in the state space form as;

$$dx/dt=\mathbf{A}x+\mathbf{B}u \tag{39}$$

where $A=\begin{bmatrix}0 & 1\\ -k_s/m & -b_s/m\end{bmatrix}$ and $B=\begin{pmatrix}0\\ 1/m\end{pmatrix}$.

Writing the index form of equation (39) yields,

$$\frac{dx_i}{dt}=\sum_{j=1}^{N}a_{ij}x_j+b_iu \tag{40}$$

where N is the number of state space, and using the neuron model given by (1), (40) can be arranged as follows;

$$\frac{\delta_i}{a_{ii}}\frac{dx_i}{dt}+x_i=(1+\delta_i)x_i+\sum_{j=1,j\neq i}^{N}\frac{\delta_i a_{ij}}{a_{ii}}x_j+\frac{\delta_i b_i}{a_{ii}}u \tag{41}$$

where $\delta_i = sign(a_{ii})$. As a result, the right side of (41) represents the inputs to the neuron x_i. The neural controller reflecting (41) is shown in figure 15 for N=2, $k_i = 1+\delta_i$, $\varepsilon_i = \delta_i / a_{ii}$,

$d_i = \delta_i bi / a_{ii}$, and $f_i = \delta_i a_{ij} / a_{ii}$; $i \neq j$. The initial parameters ε_1 and ε_2 of the neural controller in figure 15 are set such that there will be no vibrations at the robot landing leg. An inappropriate choice of these parameters may cause large vibrations of the robot mechanism, which in turn degrades the walking performance.

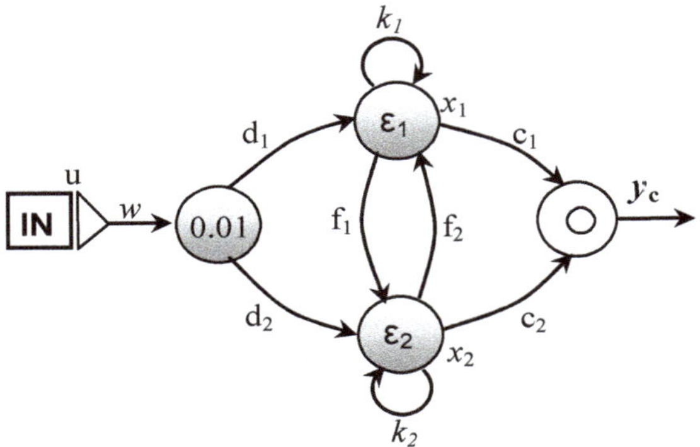

Fig. 15. Neural controller, using the notation in (Zaier and Nagashima, 2002).

The angular positions of pitching motion are as follows;

$$\begin{cases} \theta^p_{am}(t) = \theta_l(t) + \theta^l_{fb}(t) + \theta_s(t) + \theta^y_{zmp}(t) + \theta_c(t) \\ \theta_{km}(t) = -2\theta_l(t) - 2\theta_c(t) \\ \theta^p_{hm}(t) = \theta_l(t) - \theta^l_{fb}(t) - \theta_s(t) + \theta_c(t) \end{cases} \tag{42}$$

where θ^p_{am}, θ_{km}, and θ^p_{hm} are the pitching motor commands to the ankle, the knee and the hip, respectively. θ_s is the angular position that generates the stride, $\theta_c(t) = \arcsin(y_c(t) / L_t)$ is the angular position induced by the virtual damper spring system in equation (22). The L_t is the length of the thigh. The $\theta^l_{fb}(t)$ is the feedback signal satisfying the stability of inverted pendulum in the sagittal plane.

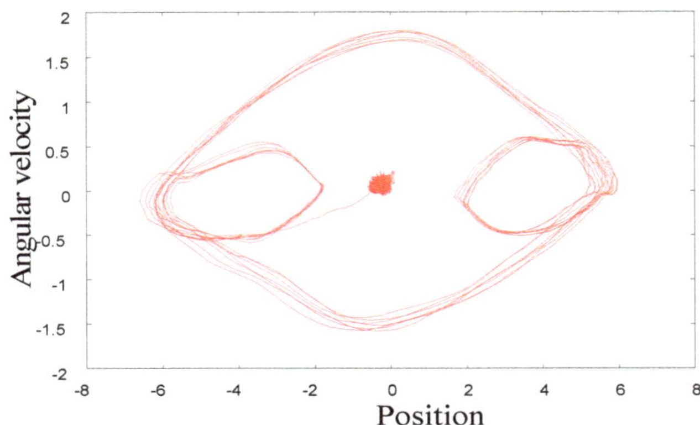

Fig. 16. Phase portrait of the ZMP in the lateral plane (Experiment)

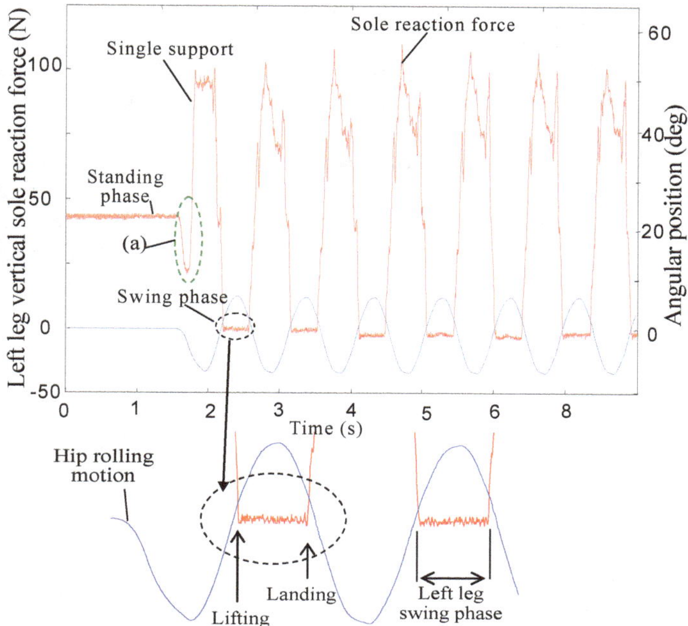

Fig. 17. Hip rolling joint output and sole reaction force acting on the left leg; the robot uses spring energy for lifting and gravity for landing

4. Experiment results and discussion

In the experiment Fujitsu's humanoid HOAP-3 (Murase et al., 2001) is used. First, a gyro feedback loop is implemented to stabilize the robot at the single support phase. Second, the rolling motion parameters $(\varepsilon, \omega, T, \alpha_s)$ are calculated using equation (14), (29), and (30), where $V_0 = \omega$. Moreover, the dynamics of the robot during stepping motion is investigated using the phase portrait of the ZMP position in the lateral plane that is shown in figure 16, which is similar to the simulation result in figure 13. Notice that the more we increase the derivative gain the better convergence we obtain. However, we are limited by the high frequencies resonance that could be triggered at high feedback gain. Figure 17 demonstrates the effect of the virtual damper-spring system of equation (38), which is implemented to work as illustrated in figure 14. Hence, it demonstrates how the robot uses gravity for landing and the spring energy for lifting. Figure 17a shows that the lifting phase starts when the angular position of the rolling joints is almost zero. In other words, the actuators of the hip and ankle rolling joints do not contribute in moving the ZMP to the supporting leg. These joints, therefore, can be considered as locked joints. Figure 18 shows the rolling of the hip and the pitching motions of the knee, hip, and ankle. It demonstrates also how smooth the approximate solution is using the proposed pattern generator with sensory feedback.
This figure shows also how the robot posture is controlled by changing the ankle position at the single support phase according to equation (42). To demonstrate the robustness against

plant perturbations and disturbances, we conducted two experiments; one consists of making the robot walks on hard floor, which can be regarded as walking in the presence of small disturbance. The phase portrait in figure 18 demonstrates that the system exhibits stable limit cycle with three periods. The second experiment consists of locomotion in the presence of larger disturbance and plant perturbation by letting the robot walks on a carpet with surface irregularities. In this case, despite the high damping coefficient of the carpet, the proposed locomotion controller could robustly maintain the stability of the system as shown in figure 19. This result also demonstrates the efficiency of the proposed approach in designing a robust locomotion controller, which is simply based on few parameters, which are the robot mass, COG height, and the distance between hip pitching joints.

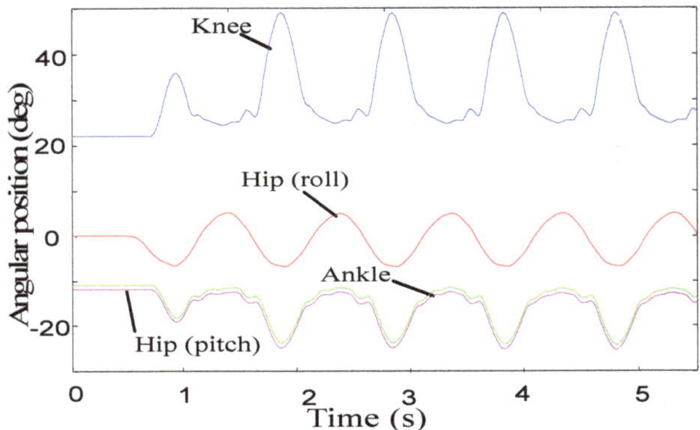

Fig. 18. Outputs of the joints of right leg

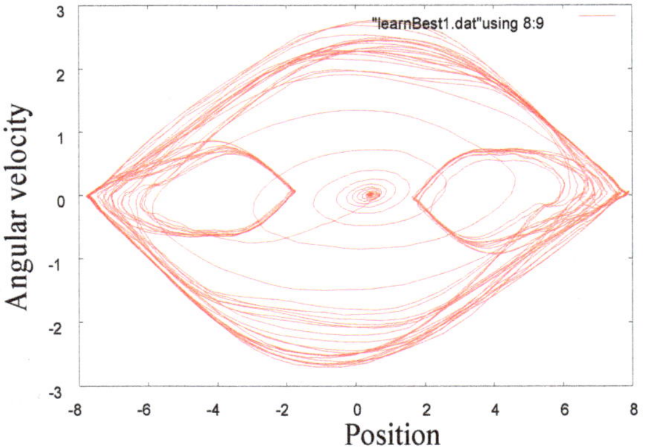

Fig. 19. Phase portrait of the ZMP in the lateral plane (Experiment)

5. Conclusion

In this chapter, we first introduced the Matsuoka and Van der Pol Oscillators. Then we presented our proposed oscillator, which was inspired by the solution of the Van der Pol that leads to stable limit cycle. The oscillator was designed using linear piecewise linear functions and the design parameters were obtained based on the equation of motion of inverted pendulum. Moreover, sensory feedback namely gyro and force sensors feedbacks were considered by adding neural controller to the oscillator's neural network. Therefore, the neural network was augmented by neural controllers with sensory connections to maintain the stability of the system. In addition, the rolling profile parameters were analytically obtained and the approximate solution was implemented giving much modularity to the motion pattern generator to include cuircuts of reflexes and task planning. The locomotion controller became so adaptive that the robot is enabled to walk on floor, carpet, and slope. In order to demonstrate the effectiveness of the proposed system, we conducted experiment using Fujitsu's humanoid robot HOAP-3. It was shown that the proposed pattern generator is robust in the presence of disturbance.

6. Acknowledgements

The experimental work was conducted at Fujitsu Laboratories Limited, Japan.

7. References

Acosta-M´arquez, C. and Bradley, D. (2000). Provision of gait assistance for individuals with lower limb impairments, *In: Proc. of the 7th Mechatronics Forum International Conference,* Pergamon Press.

Bailey S. A. (2004) "Biomimetic Control with a Feedback Coupled Nonlinear Oscillator: Insect Experiments, Design Tools, and Hexapedal Robot Adaptation Results", PhD thesis, Stanford University

Cappelletto J., Estévez P., Fernandez-Lopez G., Grieco J. C. (2007) A CPG with force feedback for a statically stable quadruped gait. *In Proceedings of the 10th International Conference on Climbing and Walking Robots (CLAWAR 2007).*

Crespi and Ijspeert A.J. (2006). AmphiBot II: An amphibious snake robot that crawls and swims using a central pattern generator. *In Proceedings of the 9th International Conference on Climbing and Walking Robots (CLAWAR 2006).*

Forti M. (1996). A note on neural networks with multiple equilibrium points. *IEEE Trans. Circuits Syst. I,* Vol. 43, pp. 487-491.

Gerstner W. (1995) `Time structure of the activity in neural network models`, *Phys. Rev,* E. 51, pp.738-758.

Goncalves J. M., Megretski A., and Dahleh M. A. (2000) `Global stability of relay feedback systems`, *Proc. of the American Control Conf.,* pp. 220-224

Grillner S. (1985). Neurobiological bases of rhythmic motor acts in vertebrates. *Science,* no. 228, pp143-149.

Grillner S. and Zangger P. (1975). How details is the central pattern generation for locomotion. *Brain Research.* pp 367-371.

Hooper Scott L. (1999–2010). Central Pattern Generators. *Encyclopedia of Life Sciences. John Wiley & Sons.* doi:10.1038/npg.els.0000032. ISBN 9780470015902.

Huang W., Chew M., Zheng Y., and Hong G. S. (2009). Bio-Inspired Locomotion Control with Coordination between Neural Oscillators, *International Journal of Humanoid Robotics.* Volume 6, No. 4, pp 585-608

Inada H. and Ishii K. (2004). A Bipedal walk using a Central Pattern Generator. *In Brain-Inspired IT I*, pages 185 - 188. Elsevier.

Kajita S. and Matsumoto O. (2001). Real-time 3D walking pattern generation for a biped robot with telescopic legs. *Proceeding of the 2001 IEEE International Conference on Robotics & Automation*, pp.2299-2306.

Matsuoka, K. (1985). Sustained oscillations generated by mutually inhibiting neurons with adaptation. Journal of Biological Cybernetics. Volume 52, pp. 367–376.

Matsuoka, K. (1987). Mechanisms of frequency and pattern control in the neural rhythm generators. Journal of Biological Cybernetics, Volume 56, pp. 345-353.

Murase Y., Yasukawa Y., Sakai K., Ueki M. (2001). Design of Compact Humanoid Robot as a Platform. *19th Annual Conference of the Robotics Society of Japan*. pp.789-790.

Senda K, Tanaka T. (2000). On nonlinear dynamic that generates rhythmic motion with specific accuracy. *Adaptive Motion of Animals and Machines*. Montreal Canada

Taga G. (1995). A model of the neuro-musculo-skeletal system for human locomotion, I. Emergence of basic gait. *Boil. Cybern*, no.73, pp.97-111

Taga G., Yamaguchi Y., and Shimizu H. (1991) `Self organized control of bipedal locomotion by neural oscillators in unpredictable environment`, *Biological Cybernetics*, vol. 65, pp.147-159.

Takeshi Mori, Yutaka Nakamura, Masa-Aki Sato, Shin Ishii (2004). Reinforcement learning for a CPG-driven biped robot. *Proceedings of the 19th national conference on Artifical intelligence Publisher*. AAAI Press / The MIT Press

Veskos P. and Demiris Y. (2005). Developmental acquisition of entrainment skills in robot swinging using van der Pol oscillators. *Proceedings of the EPIROB-2005*, pp. 87-93.

Vukobratovic M. and Borovac B. (2004). Zero moment point-thirty-five years of its life. *International Journal of Humanoid Robotics*. pp 157–173.

Williamson MM. (1999). Designing rhythmic motions using neural oscillators. *The 2009 IEEE/RSJ International Conference on Intelligent Robots and Systems*. pp 494-500.

Zaier R. and Kanda S. (2007). Piecewise-Linear Pattern Genarator and Reflex System for Humanoid Robots. *IEEE Int. Conf. on Robotics and Automation*. pp 2188-2195.

Zaier R. and Nagashima F. (2002). Recurrent neural network language for robot learning. *The 20th Annual Conf. of the Robotics Society of Japan*.

Zaier R. and Nagashima F. (2004). Motion Generation of Humanoid Robot based on Polynomials Generated by Recurrent Neural Network. *Proceedings of the First Asia International Symposium on Mechatronics*, pp 659-664.

3

Autonomous Motion Adaptation Against Structure Changes Without Model Identification

Yuki Funabora[1], Yoshikazu Yano[2], Shinji Doki[1] and Shigeru Okuma[1]
[1]*Nagoya University*
[2]*Aichi Institute of Technology*
Japan

1. Introduction

It is expected that humanoid robots provide various services to help human daily life such as household works, home security, medical care, welfare and so on(Dominey et al., 2007; Okada et al., 2003; 2005). In order to provide various services, humanoids have multi degree-of-freedom(DOF), sophisticated and complicated structure. These humanoid robots will work under human living environments which are not definable beforehand. So humanoids have to provide their given services under not only the designed environments but also unknown environments. Under unknown environments, robots cannot perform as planned, and they may fall or collide with obstacles. These impacts will wreak several unexpected structure changes such as gear cracks, joint locking, frame distortions and so on. Because of the designed motions are optimized to the robot structure, if the robot structure has changed, the services from robots cannot be provided. Because general users have no expertise knowledge of robots, thus, quick repairs under human living environments cannot be expected. Even in that case, it is expected that the robots should provide services to help human daily life as possible. In the case the humanoid robots cannot get rapid repair service, they have to provide the desired services with their broken body. In addition, using tools to provide some services can be considered as one of the structure changes. Therefore, it is necessary for future humanoids to obtain new motions which can provide the required services with changed structure.

We propose an autonomous motion adaptation method which can be applied to sophisticated and complicated robots represented by humanoids. As a first step, we deal with the simple services based on trajectory control; services can be provided by following the correct path designed by experts. When robot structure has changed, achieving the designed trajectories on changed structure is needed. As the conventional methods, there are two typical approaches. One is the method based on model identification (El-Salam et al., 2005; Groom et al., 1999). Robots locate the occurred changes, identify the changed structure, recalculate inverse kinematics, and then obtain the proper motions. If the changed structure is identified, inverse kinematics leads the proper motions for new properties of changed structure. However, it is so difficult to identify the complicated structure changes in sophisticated robots. In additions, the available solving methods of inverse kinematics for multi DOF robots is non-existent according to the reference (The Robotics Society of Japan, 2005). So model identification method cannot be applied for humanoids. Another approach is the exploration method (Peters & Schaal, 2007); finding the new motions achieving the desired trajectories after structure has changed. In order to obtain the proper motions achieving the desired trajectories, joint angles are varied by trial and error. Injured robots will obtain the proper motions without complicated model identification, but this approach needs huge exploration costs. New motion adaptation method with low exploration costs and without model identification is needed. In this paper, we show one approach to adapt designed motions to changed structure without model identification.

2. Proposed methods

We propose a motion adaptation method to generate new proper motions on changed structure without model identification. Even if robot has unobservable changes in mechanical structure, robot generates new motions which achieve the trajectories matching to the desired ones as much as possible.

Fig.1 shows the outline of proposed method. Left side shows the robot joint coordinate system and right side shows the world coordinate system. The robot has the designed motions achieving the desired trajectory S. To follow the desired trajectory accuracy, T number of target positions $s(t)(1 \leq t \leq T)$ are put on the trajectory. All the designed motions are expressed with time-series joint angle vectors $\vec{\theta}(t)$ which consit of N number of joint angles like this $\vec{\theta}(t) = [\theta_1(t), \theta_2(t), ..., \theta_N(t)]$. Here, we assume the speed and timing should be controled by other control system, we deal with only the position recovery.

When a joint angle vector $\vec{\theta}(t)$ is given to robot, robot will take a posture, and his end-effector indicates the target position $s(t)$(Fig.1(a)). The robot can measure the position of the end-effector through the sensors such as cameras or local GPSs. When structure changes have occurred, robot will take another posture even if the same joint angle vector $\vec{\theta}(t)$ is given to the robot(Fig.1(b)). The robot can detect structure changes indirectly through measuring the changed position of end-effector $s'(t)$. In this case, it is needed to modify all the joint angles to achieve the desired trajectories, but exploration for all the target positions s(t) needs huge exploration costs. So in proposed method, the number of exploration points is reduced using typical points. The robot extracts some combinations of joint angle vectors as typical points $\vec{\theta}_X$ which representative of all the designed joint angle vectors $\vec{\theta}$(Fig.1(c)). By applying conventional exploration method only at typical points, the robot obtains the corresponding joint angle vectors $\vec{\theta}''_X$ achieving the target positions s_X on changed structure(Fig.1(d)). To generate the other corresponding joint angle vectors, the robot estimates all the proper joint angle vectors $\vec{\theta}''(t)$ based on exploration results $\vec{\theta}''_X$ and $\vec{\theta}_X$(Fig.1(e)).

The reachable area where the robot can indicate as target positions depends on the robot structure such as frame length, joint range of motion and so on. So structure changes cause the reachable area to be deteriorated. Also it has possibilities that the robot cannot indicate some target positions in designed joint DOF. Generally, humanoids have multi DOF and they have redundant joints which are not used for expressing the designed motions. If structure changes have occured and the target positions cannot be expressed, expansion of joint DOF using redundant joints is needed to extend the robot reachable area for encompassing all the target positions.

2.1 Recovery of the desired trajectories with low explration

In order to recover the desired trajectories, proposed adaptation method consists of three phases. Vector Quantization(VQ)(Furui et al., 1998) is applied in order to extract typical

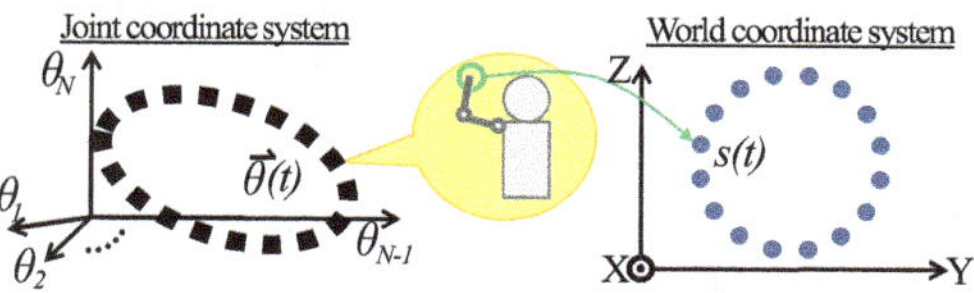

(a) The desired trajectory and the designed joint angle vectors before structure changes

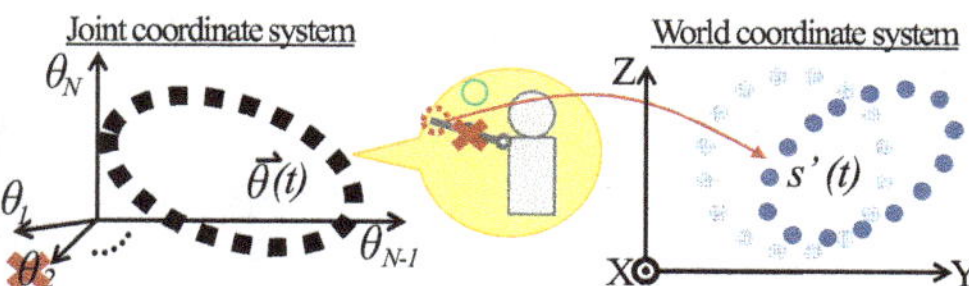

(b) The changed trajectory after structure changes

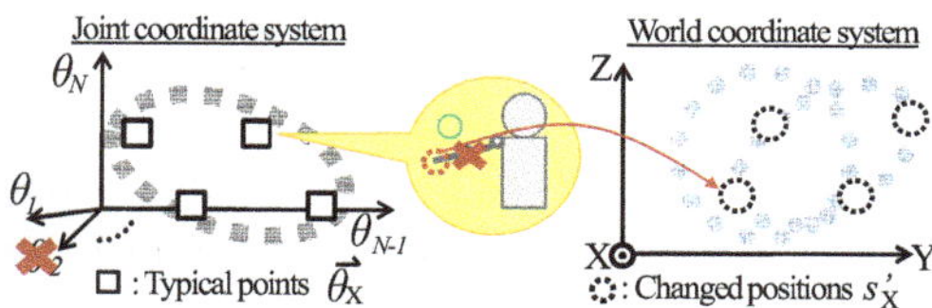

(c) Extraction a small number of typical points from the designed joint angle vectors

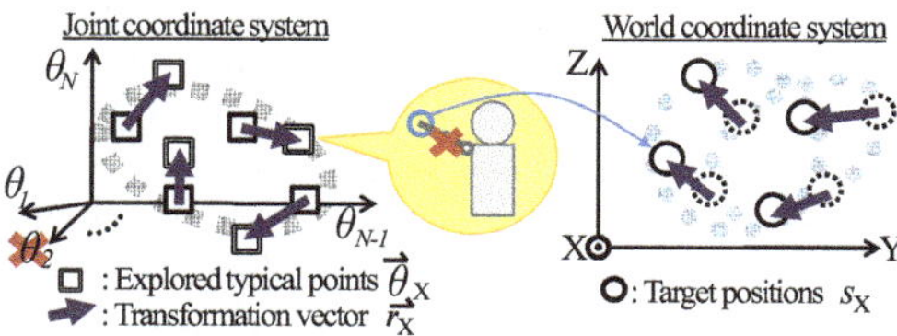

(d) Exploration the corresponding joint angle vectors only at the typical points

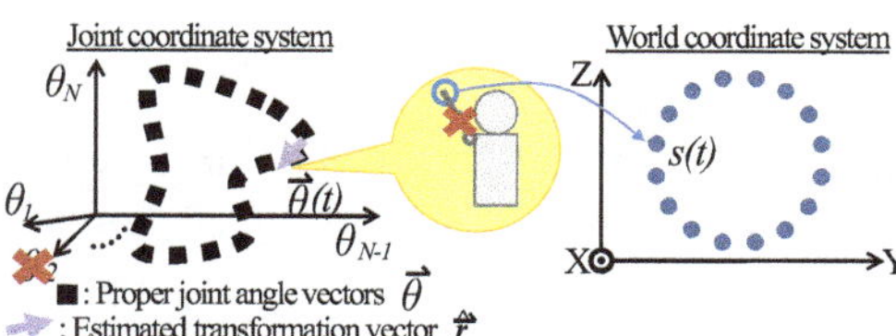

(e) Estimation all the proper joint angle vectors achieving the desired trajectory

Fig. 1. The outline of propsed method

points $\vec{\theta}_X$(Fig.1(c)), Simulated Annearling(SA)(Kirkpatrick, 1984) is applied to explore the corresponding joint angle vectors $\vec{\theta}''_X$ at typical points(Fig.1(d)), and Kriging(Mase & Takeda, 2001) is applied to estimate all the proper joint angle vectors $\vec{\theta}''$(Fig.1(e)).

2.1.1 Extraction of typical points

For cutting costs of exploration, the number of exploration points should be reduced. $K(K \ll T)$ number of typical points $\vec{\theta}_X(k)(1 \leq k \leq K)$ are extracted from the desgined joint angle vectors $\vec{\theta}$. In humanoids, because the new motions would be added from users or adjusted for adaptation to environments and so on, typical points cannot be determined beforehand. It is necessary to extract typical points which are represented distribution of all the designed joint angle vectors $\vec{\theta}$ on joint coordinate system.

In this method, several typical points where the target positions are explored are extracted by VQ. VQ is used in many applications such as image compression, voice compression, voice recognition and so on. Some codewords represent a given set of input vectors on coordinate system. K number of typical points $\vec{\theta}_X(k)$ as codewords resulting from VQ represent all the designed joint angle vectors $\vec{\theta}$. The number K is determined by quantization distortion which is given based on the number of joints, range of joint motion, accuracy of desired trajectories and so on. Robots can get K number of typical points $\vec{\theta}_X$ autonomously based on given quantization distortion.

As mentioned, by applying VQ to all the designed joint angle vectors $\vec{\theta}$ distributing on joint coordinate system, some representative joint angle vectors $\vec{\theta}_X$ are extracted autonomously.

2.1.2 Exploration on typical points

By applying conventional exploration method at a typical points $\vec{\theta}_X(k)$, the corresponding joint angle vectors $\theta''_X(k)$ is explored. There is position error between the target position $s_X(k)$ and the current indicating position $s'_X(k)$. By minimaizing these position errors at all the typical points, the proper joint angle vectors $\vec{\theta}''_X$ are aquired. In proposed method, the proper joint angle vectors are explored by SA. SA is one of the famous global optimization method and often used when the search space is discrete. It is considered that the structure changes cause the joint angle coordinate system to go into some discrete changes.

By applying SA at a typical point $\vec{\theta}_X(t)$, a corresponding joint angle vector $\vec{\theta}''_X(t)$ is aquired. From the difference between the original vector $\vec{\theta}_X(k)$ and the corresponding one $\vec{\theta}''_X(k)$, a transformation vector $\vec{r}_X(\vec{\theta}(k))$ is calculated as (1).

$$\vec{r}(\vec{\theta}_X(k)) = \vec{\theta}''_X(k) - \vec{\theta}_X(k). \tag{1}$$

Each transformation vector is acquired at each typical point. By exploring K target positions s_X, datasets of typical points $\vec{\theta}_X$ and transformation vectors $\vec{r}_X$ can be aquired.

2.1.3 Estimation of the proper joint angle vectors

All the proper joint angle vectors $\vec{\theta}''$ indicating the target positions on changed structure is generated by Kriging; one of the spatial estimation method in geostatistics. Kriging is a technique to predict the value at any point of space from some sample sets of the space points and the values. Mukai(Mukai & Kuriyama, 2005) achieved creating a lot of similar CG motions from limited sample motions by Kriging. By applying interpolation method Kriging, exploration results at typical points can reflect all the designed joint angle vectors.

An estimation transformation vector $\hat{\vec{r}}(t)$ which transforms the designed joint angle vector $\vec{\theta}(t)$ into the proper one $\vec{\theta}''(t)$ is estimated by Kriging based on sets of $\vec{\theta}_X$ and $\vec{r}_X$.
Based on K sets of typical points $\vec{\theta}_X(k)$ and transformation vectors $\vec{r}_X(\vec{\theta}_X(k))$, an estimation transformation vector $\hat{\vec{r}}(\vec{\theta}(t))$ at arbitrary joint angle vector $\vec{\theta}(t)$ is calculated as (2).

$$\hat{\vec{r}}(\vec{\theta}(t)) = \sum_{k=1}^{K} w_k \vec{r}_X(\vec{\theta}_X(k)), \quad \sum_{k=1}^{K} w_k = 1. \tag{2}$$

Here, weight w_k is derived from variogram γ which is calculated from K number of sample sets($\vec{\theta}_X$ and $\vec{r}_X$). A variogram is an indicator of spatial autocorrelation. The variogram for any two samples($\vec{\theta}_X(i)$ and $\vec{\theta}_X(j)(i,j \in \{1, ..., K\})$) is calculated as (3).

$$\gamma_{ij} = \frac{1}{2} Var[\vec{r}_X(\vec{\theta}_X(i)) - \vec{r}_X(\vec{\theta}_X(j))]. \tag{3}$$

Here, we define $\vec{h} = \vec{\theta}_X(i) - \vec{\theta}_X(j)$, the experimental variogram cloud γ^* is shown as a function of $\vec{h}$ like (4).

$$\gamma^*(\vec{h}) = \frac{1}{2} Var[\vec{r}_X(\vec{\theta}_X(j) + \vec{h}) - \vec{r}_X(\vec{\theta}_X(j))], \quad (\forall j \in \{1, ..., K\}). \tag{4}$$

The variogram γ is determined by modeling of the experimental variogram cloud γ^*. We adopt a spherical model which is commonly used in geostatistics. A spherical model is shown as (5).

$$\gamma(\vec{h}; \vec{a}) = \begin{cases} a_0 + a_1(\frac{3}{2}\frac{||\vec{h}||}{a_2} - \frac{1}{2}\{\frac{||\vec{h}||}{a_2}\}^3), & 0 < ||\vec{h}|| \le a_2 \\ a_0 + a_1, & ||\vec{h}|| > a_2 \\ 0, & ||\vec{h}|| = 0 \end{cases} \tag{5}$$

Here, $\vec{a}(a_0, a_1 \ge 0, a_2 > 0)$ shows model parameters and is determined by applying least squares to fit to the experimental variogram cloud. We assume that the joint coordinate system is satisfied locally stationary, an estimation varience of Kriging σ^2 is calculated as (6).

$$\begin{aligned} \sigma^2 &= E\left[(\hat{\vec{r}}(t) - \vec{r}(t))^2\right] \\ &= -\sum_{i=1}^{K}\sum_{j=1}^{K} w_i w_j \gamma(\vec{\theta}_X(i) - \vec{\theta}_X(j)) + 2\sum_{k=1}^{K} w_k \gamma(\vec{\theta}_X(k) - \vec{\theta}(t)). \end{aligned} \tag{6}$$

Hera, E denotes the average. Under the constraints($\sum_{k=1}^{K} w_k = 1$), weights w_k are calculated by applying Lagrange's method of undetermined multipliers to minimize the σ^2. The estimation transformation vectors $\hat{\vec{r}}$ is calculated from aquired w_k and (2).
The proper joint angle vectors $\vec{\theta}''(t)$ is calculaed as (7) using the designed joint angle vector $\vec{\theta}(t)$ and the estimated transformation vector $\hat{\vec{r}}(\vec{\theta}(t))$.

$$\vec{\theta}''(t) = \vec{\theta}(t) + \hat{\vec{r}}(\vec{\theta}(t)). \tag{7}$$

By applying (7) to all the designed joint angles $\vec{\theta}$, the proper joint angle vectors $\vec{\theta}''$ achiving the desired trajectory can be genereted autonomously.

2.2 Expansion of joint DOF

For achieving the desired trajectories on changed structure, the joint DOF should be expanded using redundant joints if necessary. When the generated proper joint angle vectors could not achieve the target positions in designed joint DOF, the new joint angle vectors should be explored in expanded joint DOF adding redundant joints.
The concept of deterioration of the reachable area and expansion of joint DOF is shown as Fig.2. A motion "draw a circle" is designed in 3 DOF consisting of shoulder yaw joint,

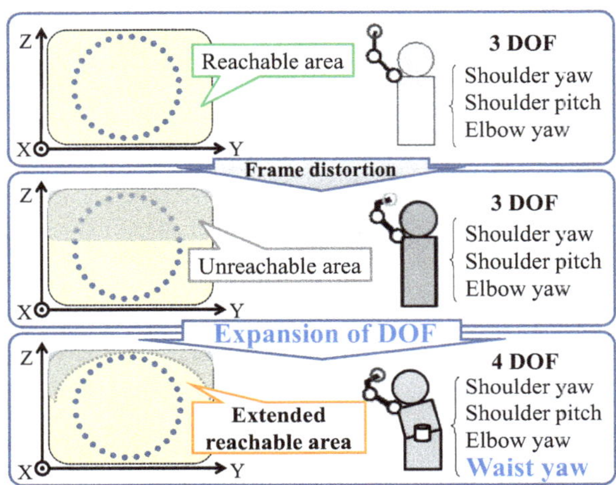

Fig. 2. Deterioration of the reachable area and expansion of joint DOF

shoulder pitch joint and elbow yaw joint. The desired trajectory is achieved in designed 3 DOF before structure changes. When the robot structure has changed by frame distortion, the desired trajectory cannot be achieved in designed DOF, because the reachable area has deteriorated. A redundant waist yaw joint should be added in order to extend the reachable area.
The fewer the joint DOF in which the robot can indicate all the target positions is, the lower exploration costs are. It is difficult for robots to find autonomously the combination of joints in which the robot can indicate all the target positions. In this method, experts of the robots select beforehand the right combination of joints to reach the target positions based on their experimental rule. The prioritization of joints to each end-effector is given in advance. The number of joint DOF is expanded in stages based on the prioritization. An example of expansion of joint DOF to right hand end-effector on a humanoid is shown as Fig.3. If a trajectory cannot be achived, first, robot generates proper joint angle vectors in the designed DOF using the method as mentioned above. When the robot explores the target positions at extracted typical points, he can obtain the differences between the target positions and the explored positions. If these differences are not sufficiently-small, then, the robots explore the target positions again in 4 DOF adding waist yaw joint based on given prioritization of joints. Even if one target position cannot be explored in 4 DOF, the corresponding joint angle vectors are explored in 5 joints adding waist pitch joint. By adding the redundant joints in stages, the corresponding joint angle vectors are acquired with low exploration costs. Proposed method consisting of VQ and Kriging is independent of joint DOF. Only by exploring in expanded

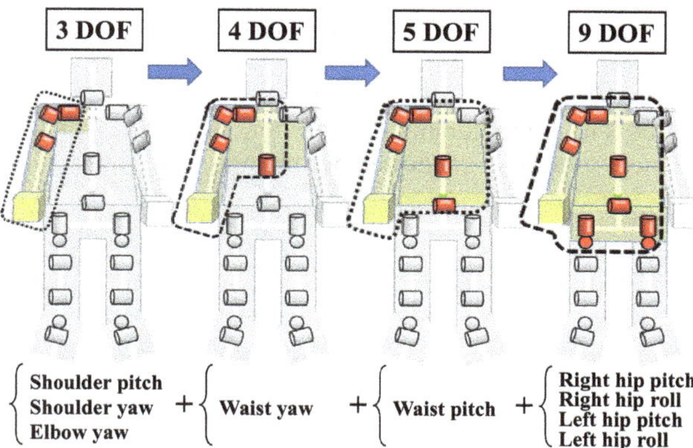

Fig. 3. An example of staged expansion of joint DOF on a humanoid

joint DOF at typical points, the proper joint angle vectors are estimated in expanded joint DOF.

There is a typical point $\vec{\theta}_X(k)$ in designed N_d joint DOF. The expanded typical point $\vec{\theta}^E_X$ adding N_r redundant joints is shown as (8).

$$\vec{\theta}^E_X = (\theta_1, \theta_2, ..., \theta_{N_d}, \theta_{N_d+1}, ..., \theta_{N_d+N_r}) \tag{8}$$

The transformation vectors $\vec{r}^E_X$ are acquired in expanded joint DOF by SA. The estimation transformation vectors $\hat{\vec{r}}^E$ are acquired in expanded joint DOF by Kriging. So the proper joint angle vectors $\vec{\theta}''^E$ are generated in expanded joint DOF.

3. Experiments

The proposed motion adaptation method is evaluated on simulation. The structure of the target humanoid robot is shown in Fig.4. To exclude the discussion of stability, only the upper

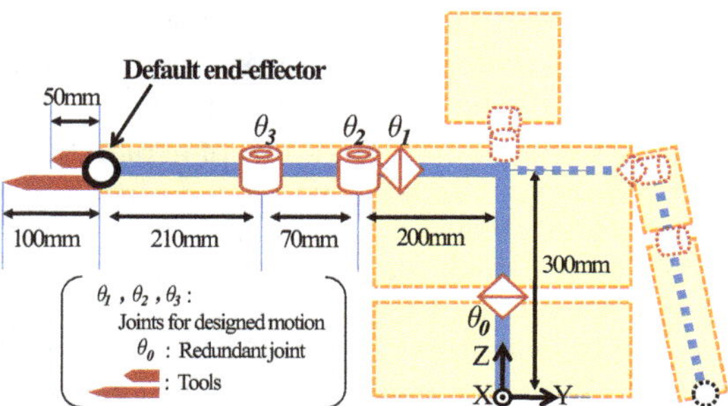

Fig. 4. The structure of the target humanoid

body was modeled. A motion "draw a circle" achieving a desired circular path on Y-Z plane with right hand was designed with 100 time-series joint angle vectors in 3 DOF consisting of shoulder yaw joint θ_1 , shoulder pitch joint θ_2 and elbow yaw joint θ_3. All the range of joint motion was $-130[deg] \leq \theta_1, \theta_2, \theta_3 \leq +130[deg]$. The desired trajectory S was a circular path whose radius was 150mm and center position was (200[mm],-100[mm],300[mm]) on world coorinate system.

So that the quantization distortion on world coordinate system was less than 60mm, 8 typical points were extracted by VQ. If the generated motion in designed 3 DOF cannot achieve the desired trajectory, the new motion is generated according to proposed method in expanded 4 DOF adding waist yaw joint θ_0.

3.1 Adaptation to tools

Assuming the users install the tools like a pen, we extended length of the lower arm by 50mm or 100mm as structure changes(Fig.4).

The results of adaptation to using 50mm tool is shown in Fig.5. In Fig.5, green line shows

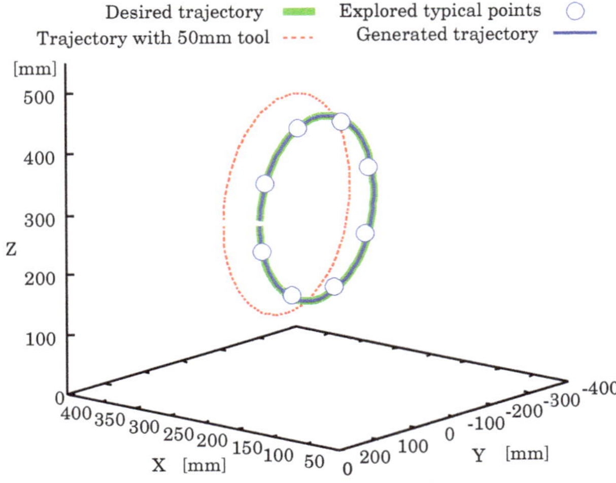

Fig. 5. Observed trajectories with 50mm tool

the desired trajectory, red dots show the changed trajectory with 50mm tools. Undesired trajectory was expressed with changed structure. 8 blue circles show the explored typical points $\vec{\theta}''_X$ which modified by conventional exploration method SA. 100 proper joint angle vectors $\vec{\theta}''$ ware estimated by Kriging based on results of exploration at 8 blue circles. The generated trajectory(blue line) is very consistent with the desired trajectory(green line). The desired trajectory is recovered with about 8% exploration cost on changed structure.

The results of adaptation to using 100mm tool is shown in Fig.6. In Fig.6, red dots show the changed trajectory with 100mm tool. 8 light blue circles show explored typical points in 3 DOF and light blue line shows the generated trajectory in 3 DOF. Attached tool was so long that the robot could not express all the target position in designed 3 DOF. In other words, using the 100mm tool cause deterioration of the reachable area. Even in that case, the generated trajectory(light blue line) represents nearer by the desired one(green line) than the changed one(red dots). The generated trajectory is matched the optimum trajectory acquired from exploration at all the 100 joint angle vectors. It is considered that even if the robot cannot express the desired trajectory on changed structure, by applying porposed method, the robot can generate new motions which achieve the trajectories matching to the desired ones as much as possible.

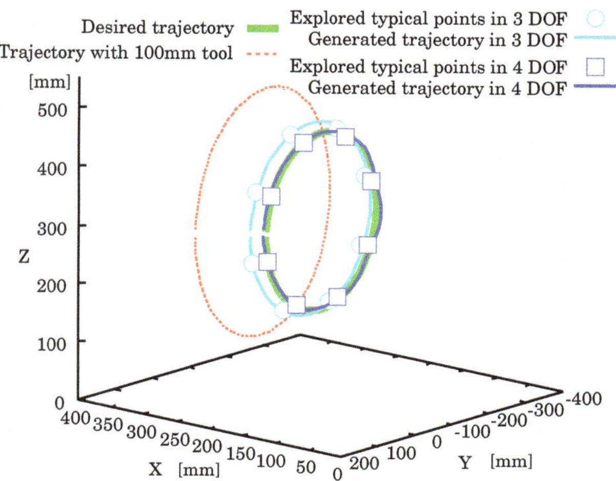

Fig. 6. Observed trajectories with 100mm tool

Proposed method can apply to correspond not only between the same DOF but also higher DOF, as mention above. So, the robot generates a new motion by exploring in 4 DOF adding redundant waist yaw joint θ_0. In Fig.6, blue squares show the explored typical points and blue line shows the generated trajectory in 4 DOF. The generated trajectory in expanded 4 DOF is nearer by the designed one than that of in 3 DOF, and blue line is very consistent with green line. The desired trajectory is recovered on changed structure by expanding joint DOF.

3.2 Adaptation to joint locking

Then, we assumed another change. Another structure change was assumed as elbow yaw joint θ_3 locking to 0 degree. Each line shows each trajectory along with Fig.6. Along with

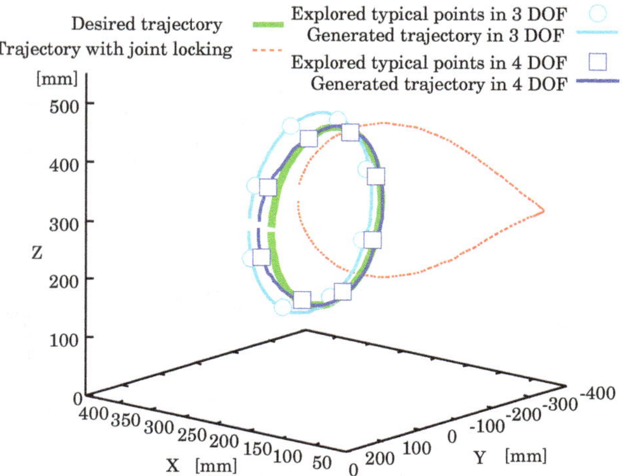

Fig. 7. Observed trajectories with joint locking($\theta_3 = 0[deg]$)

the adaptation to using 100mm tool, in designed 3 DOF, 8 light blue circles are not on the designed trajectory and the generated trajectory(light blue line) is not matched the desired one(green line). It is to be sure that the generated trajectory represents nearer by the designed one than the changed one, satisfactory trajectory could not be obtained. Joint locking causes degrade of joint DOF, and the reachable area of end-effector has deteriorated. Even in this case, the generated trajectory is matched the optimum one acquired from exploration at all the 100 joint angle vectors, too.

So the robot generates a new motion by exploring in 4 DOF adding redundant waist yaw joint θ_0. The new trajectory(blue line) in expanded 4 DOF is very nearer by the designed one than that of in 3 DOF. Because the generated trajectory is well matched the optimum one which is explored at all the 100 joint angle vectors, it is considered that some of the target positions could not be existed on reachable area even if the joint DOF was expanded to 4 DOF. It is expected that adding the other redundant joints will make the generated trajectory nearer and nearer. Average of position error is about 216mm on failure. The difference between the generated trajectory in 3 DOF and the designed one is about 30mm. The difference 7.4mm is observed by proposed method, which is accomplished about 97% similar trajectory with the desired one.

The observed acrual movements of the robot with the designed motion and the generated motions in each joint DOF are shown in Fig.8. The horizontal axis indicates the time of motion t. In 3 DOF, because the robot cannot bend the elbow, the robot cannot express the target

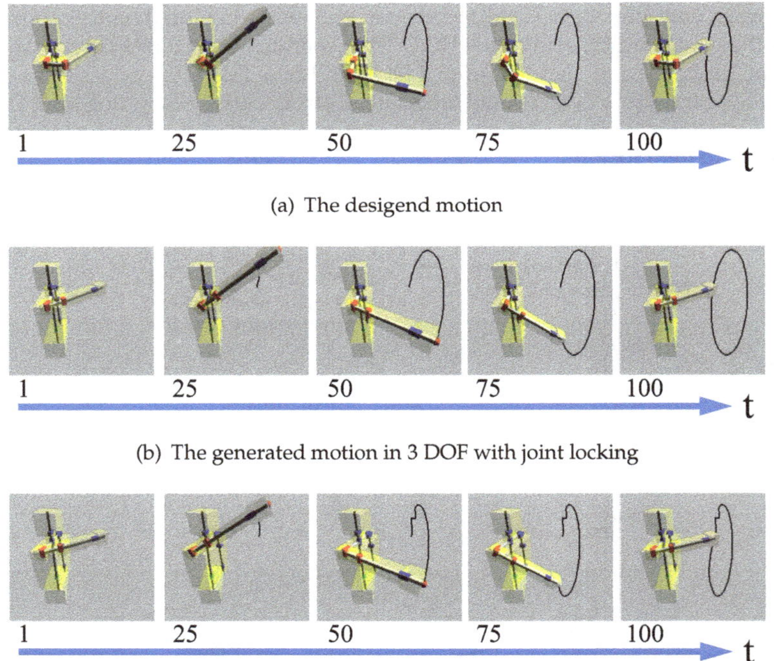

Fig. 8. The obsreved robot motions and trajectories

positions near the body. As the result, the generated trajectory was away from the body. On the other hand, in 4 DOF, in order to express the target positions on near the body, the robot was taking advantage of the redundant waist joint. We looked that robot used redundant waist yaw joint effectively to recover the effect of locked joint.

4. Conclusion

As a case study of the multi DOF humanoid robots, we introduced the study to execute autonomous motion adaptation against robot mechanical structure changes. To apply for humanoids which have multi DOF and sophisticated complicated structure, we proposed an autonomous motion adaptation method without model identification. We showed that the adaptation method could execute efficiency by exploring the corresponding joint angle vectors at a small number of typical points and estimating all the proper joint angles based on the result in exploration. Using proposed method, the generated trajectories are well matched the desired one on unobservable changed structure. Even if the generated motions wouldn't have achieved the designed trajectories, it is expected that the new motion achieving the desired trajectories would be generated by exploring in expanded DOF adding redundant joints.
Now, we showed experiments only for the trajectory "draw a circle". In general, humanoid robots should achieve various trajectories to provide various services. Even if the robot has several designed motions to achieve various trajectories, it is expected using VQ that the effective typical points for modifying are extracted and the designed motions can adapt to changed structure efficiently. It is needed to evaluate the effectiveness of proposed method applying for various motions. Now, we deal with only the motions based on trajectory control. This method is not limited control system. The robot chooses the evaluation function based on control system and explores the corresponding joint angle vectors using the evaluation function at typical points, the robot will generate the adapted motions for another control system. It calls for further researches and experiments. We showed the possibility to adapt motions in expanded DOF, but in this paper, adding redundant joints ware determined by experts. It is necessary to consider the joint selection method to expand joint DOF. This method realized to reduce the number of exploration points, but exploration costs increase exponentially with increasing robot joint DOF. There are some humanoid having over 30 DOF, it is difficult to explore in real time. New adaptation method which is not independent on increasing of joint DOF is needed. It calls for further researches and experiments, too.
In our method, stability problems are not resolved. On typical points, by recovering not only position sensor values but also the sensor values observing the stability such as acceleration, gyro, force and so on, the designed trajectories can be recovered in stable condition. In addition, this problems will be resolved in combination with the other methods such as Maufroy(Maufroy et al., 2010), Otoda(Otoda et al., 2009). Further studies are required.

5. References

Dominey, P., Mallet, A. & Yoshida, E. (2007). Real-time cooperative behavior acquisition by a humanoid apprentice, *Proceedings of IEEE/RAS 2007 International Conference on Humanoid Robotics*.

El-Salam, I. A., El-Haweet, W. & Pertew, A. (2005). Fault-tolerant kinematic controller design for underactuated robot manipulators, *ACSE 05 Conference*, pp. pp.117–123.

Furui, S., Tazaki, S., Kotera, H. & Watanabe, Y. (1998). *Vector Quantization adn Signal Compression*, CORONA PUBLISHING Co., Ltd. (in Japanese).

Groom, K. N., Maciejewski, A. A. & Balakrishnan, V. (1999). Real-time failure-tolerant control of kinematically redundant manipulators, *IEEE Transactions on Robotics and Automation* Vol.15(No. 6).

Kirkpatrick, S. (1984). Optimization by simulated annyealing: Quantitative studies, *Journal of Statistical Physics* Vol.34.

Mase, S. & Takeda, J. (2001). *Spatial Data Modeling*, KYORITSU PUBLISHING Co., Ltd.,. (in Japanese).

Maufroy, C., Kimura, H. & Takase, K. (2010). Integration of posture and rhythmic motion controls in quadrupedal dynamic walking using phase modulations based on leg loading/unloading, *Autonomous Robots* Vol.28: pp.331–353.

Mukai, T. & Kuriyama, S. (2005). Geostatistical motion interpolation, *ACM Transactions on Graphics* Vol.24: pp.1062–1070.

Okada, K., Kino, Y., Inaba, M. & Inoue, H. (2003). Visually-based humanoid remote control system under operator's assistance and it's application to object manipulation, *Proceedings of Third IEEE International Conference on Humanoid Robots*.

Okada, K., Ogura, T., Haneda, A., Fujimoto, J., Gravot, F. & Inaba, M. (2005). Humanoid motion generation system on hrp2-jsk for daily life environment, *in 2005 International Conference on Mechatronics and Automation(ICMA05)*, pp. pp.1772–1777.

Otoda, Y., Kimura, H. & Takase, K. (2009). Construction of gait adaptation model in human splitbelt treadmill walking, *Applied Bionics and Biomechanics* Vol.6: pp.269–284.

Peters, J. & Schaal, S. (2007). Reinforcement learning by reward-weighted regression for operational space control, *Proceedings of the 24th International Conference on Machine Learning*, pp. pp.745–750.

The Robotics Society of Japan (ed.) (2005). *Robotics Handbook - 2nd ed.*, CORONA PUBLISHING Co., Ltd.,. (in Japanese).

Part 2

Grasping and Multi-Fingered Robot Hand

4

Grasp Planning for a Humanoid Hand

Tokuo Tsuji, Kensuke Harada, Kenji Kaneko, Fumio Kanehiro
and Kenichi Maruyama
Intelligent Systems Research Institute, National Institute of Advanced Industrial Science and Technology (AIST)
Japan

1. Introduction

We focus on grasp planning for a humanoid multi-fingered hand attached at the tip of a humanoid robot's arm. The hand has potential possibility to grasp various objects under several situations. Since the multi-fingered hand can use several grasp types such as fingertip grasp, and envelope grasp with taking the advantage of degrees of freedom. We develop grasp planner which selects a feasible grasp type based on the task, and determines contact positions for the fingers and the grasped object surface so that the fingers do not drop the object while staying with limited actuator capacity.

To grasp an object, the robot first measures object position/orientation using vision sensor. Then, the planner plans the body motion to complete the grasping task based on vision sensor information. Even when the object's location is not known beforehand, the robot should complete the grasping task as fast as possible. However, grasp planning with a humanoid robot is complex and often requires long calculation time. Hence, for the grasp planning, a heuristic but fast algorithm is preferred rather than precise but slow algorithms (Shimoga (1989)). Our planner calculates grasp motions within reasonable time by using predefined grasp types which are assigned with contacting finger links, desired sizes of the grasped object. Our planner selects a grasp type according to position/orientation of the grasped object similar to a human. As shown in Fig. 1 , a human grasps the side of the can with all fingers, grasps the top with fewer fingers.

Failing to find feasible grasping posture using arm/hand kinematics alone, our planner attempts to do so using the full body kinematics. Using the degrees of freedom of full body, the planner has adaptable for reaching the object with the several motions such as twisting waist, bending waist, and squatting down.

We demonstrate effectiveness of grasp planning through simulation and experimental results by using humanoid robot HRP-3P (Akachi et al. (2005)) shown in Fig. 2, which has a four-fingered hand HDH (Kaneko et al. (2007)) on its right-arm and a stereo camera system in its head.

We has proposed the grasp planning which was executed within the reasonable time taking into account several constraints imposed on the system such as the feasible grasping style, the friction cone constraint at each contact point, the maximum contact force applied by the fingers, the inverse kinematics of the arm and the fingers (Harada et al. (2008); Tsuji et al. (2008; 2009)). The originality of our method is the following points;

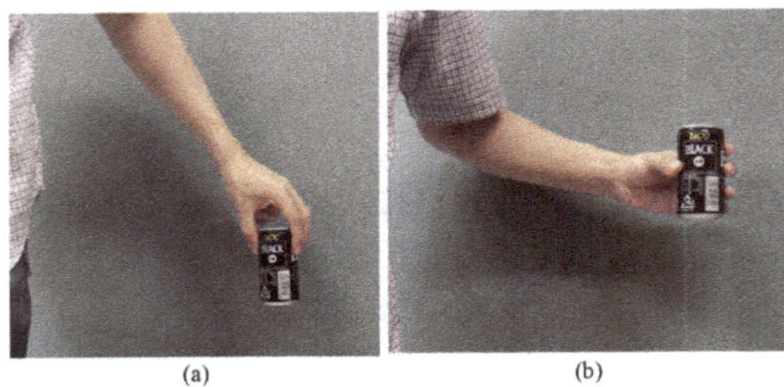

Fig. 1. Grasp of a can by a human

Fig. 2. Humanoid HRP3P with multi-fingered hand

(1) The humanoid hand can use human grasping styles and changes the grasping style depending on the position/orientation of the grasped object like a human. As shown in Fig.1, while a human grasps the side of the can with all the fingers, he/she grasps the top of it with just some of the fingers. Especially for the case of the robotic hands, it is often difficult to put all the fingers on the top of such a can.

(2) If our planner failed in finding the feasible grasping posture only by using the arm/hand kinematics, the planner tries to use the whole body kinematics. For the purpose of grasping an object firmly or reaching it, the robot try to change the shoulder position.

(3) We developed an easy and fast method of checking approximated force closure when a multi-fingered hand grasps an object. Different from previous methods, we consider approximating the friction cone by using an ellipsoid.

(4) Our grasp planner and grasp motion control are modularized as a plugin of Choreonoid, which is an integrated software that allows us to choreograph robot motions and simulate the robot motion dynamics.

(5) We confirmed the effectiveness of our proposed approach by experiment using the robot HRP-3P (Fig. 2).

2. Related works

As for the works on grasp planning, there are a number of works on contact-level grasp synthesis such as Coelho & Grupen (2004); Niparnan & Sudsang (2004); Ponce et al. (1993). As

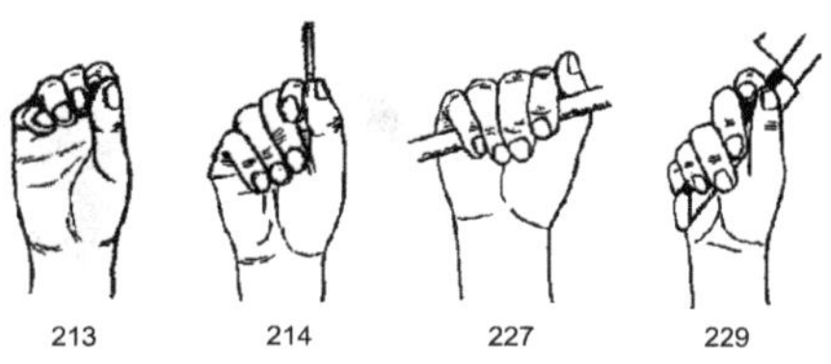

Fig. 3. Human grasping styles

for the grasp planning considering the hand/arm model, Cutkosky (1989) first proposed an expert system to select one of the grasp styles based on grasp quality measures. Pollard (2004) proposed a precise grasp planning algorithm by utilizing heavy computation. Morales et al. (2006) proposed a planning method applicable to the Barret hand. Morales et al. (2006) also proposed a method of planning a grasp by preparing the candidate grasping style corresponding to the grasped object. Prats et al. (2007) proposed a method for planning the hooking task of a door.

Recently, some researchers researched the fast grasp planning problem. **?** proposed planning methods to determine contact positions of a four-fingered hand assuming a fingertip grasp. Miller et al. (2003) used a grasping simulator and proposed a method for determining candidates of a contact point set for grasp planning. Their method calculates $1 \sim 44$ candidates of grasp configuration between 11.4[s] and 478[s] depending on the complexity of the object's shape. Their approach was extended to the learning approach (Pelossof et al. (2004)) and imperfect visual information (Ekvall & Kragic (2007)).

As for the grasp planning based on a random sampling approach, Niparnan & Sudsang (2004) and Borst et al. (2003) proposed methods to realize the force closure using the random sampling. Yashima et al. (2003) proposed a manipulation planning based on the random sampling.

3. Grasp planning

3.1 Reference posture and rectangular convex

When a human grasps an object, a human selects one of the grasp styles according to the shape and the estimated mass of the object. Fig.3 shows some of the grasp styles shown in the book (Kapandji (1985)).

We assume that the shape of the object is given by a polygon model such as VRML. When the shape of the object is given, the grasp planner has to automatically select feasible grasp styles. For realizing this function, we assume the *reference posture* as shown in Fig.4. For each reference posture, we assigned the finger links contacting with the object. In this research, we constructed the reference grasping motion for three-fingered fingertip grasp, four-fingered fingertip grasp, and four-fingered enveloping grasp. We also constructed the reference posture for small, mid and large sized objects for each reference grasping motion. We manually constructed these reference postures.

For each reference posture, we assumed the *grasping rectangular convex* (GRC). The GRC is used for selecting the feasible grasping style for given grasped object. As shown in Fig. 5, we assumed three GRC. The GRCmax is the maximum size of the rectangular-shaped grasped object without interfering the finger links. The GRCmin and the GRCdes are the assumed minimum and desired sized grasped object.

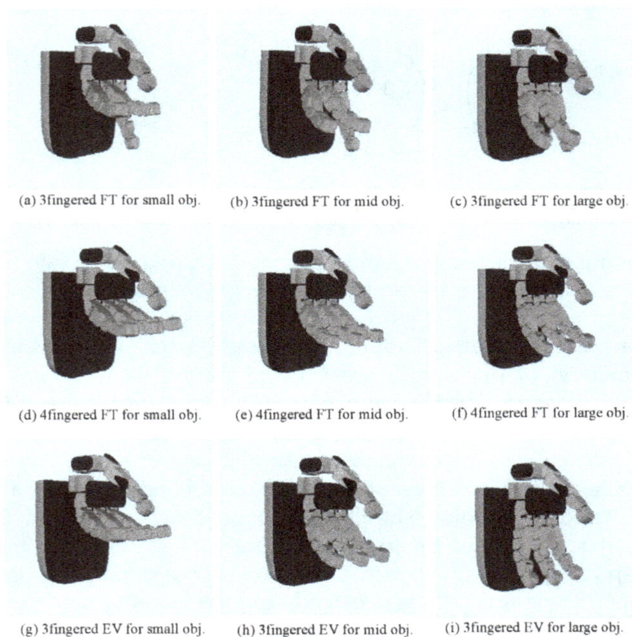

(a) 3fingered FT for small obj. (b) 3fingered FT for mid obj. (c) 3fingered FT for large obj.

(d) 4fingered FT for small obj. (e) 4fingered FT for mid obj. (f) 4fingered FT for large obj.

(g) 3fingered EV for small obj. (h) 3fingered EV for mid obj. (i) 3fingered EV for large obj.

Fig. 4. Reference motion for several fingertip(FT) and envelope(EV) grasps

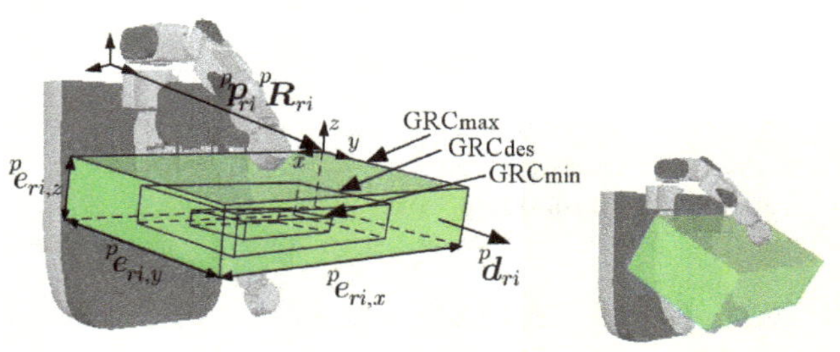

(a) GRC for 3fingered FT for mid obj. (b) GRC for 3fingered EV for large obj.

Fig. 5. Reference motion

By modifying the reference posture, the actual grasping posture is planned by using the method explained in the next section. For the i-th grasping style with the GRCmax which position/orientation is ${}^{p}\boldsymbol{p}_{\max,i}/{}^{p}\boldsymbol{R}_{\max,i}$ $(i = 1, \cdots, n)$, we assume the vector of the edge length ${}^{p}\boldsymbol{e}_{\max,i}$ of the GRC. This vector is used for selecting the grasping style. Also, we assume the approach vector ${}^{p}\boldsymbol{d}_{\max,i}$. This vector defines the approach direction to the object and is one of the outer unit normal vector of the GRC's surface. Furthermore, we assume the maximum and the minimum mass, $m_{\max,i}$ and $m_{\min,i}$ of the object grasped by using the i-th grasping style.

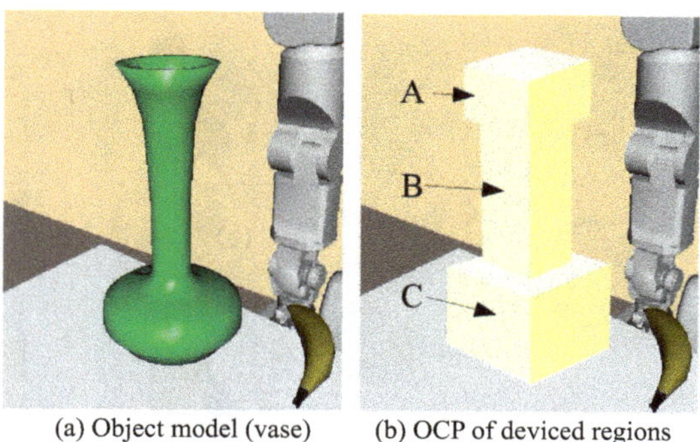

(a) Object model (vase) (b) OCP of deviced regions

Fig. 6. Object and object convex polygon (OCP)

Next, we focus on the object to be grasped. Given the shape of the object, our planner calculates the *object convex polygon* (OCP). The OCP is the minimum-sized convex polygons including the grasped object. In this paper, we consider the rectangular convex as the OCP. For complex shaped objects, we consider splitting the object into some regions and calculate the OCP. Fig.6 shows the OCP of the vase placed on the table. Our planner splits the object into three regions.

Once the polygon model of the grasped object is given, our planner has the set of points included in the surface of the object. By using the eigen vectors of the co-variance matrix of the point set, we can calculate the OCP. For the i-th OCP which position/orientation is $\boldsymbol{p}_{oi}/\boldsymbol{R}_{oi}$ $(i = 1, \cdots, m)$, we assume the vector of edge length $\boldsymbol{e}_{oi}$.

Miller et al. (2003); Pelossof et al. (2004) also used the convex model for the grasped object. In this research, in addition to the OCP, we use the GRC and determine the grasping style and the nominal grasping posture.

3.2 Nominal position/orientation of palm

Let us consider selecting one of the grasping styles and determining the nominal position/orientation of the palm. For such purpose, we introduce some heuristic rules in this subsection. We first focus on the geometrical relationship between the GRC and the OCP. Let $\mathrm{sort}(\boldsymbol{a})$ be the function replacing the elements of the vector $\boldsymbol{a}$ in a decreasing order. We impose the following conditions:

$$\boldsymbol{b}_{\mathrm{max},ij} = \mathrm{sort}(\boldsymbol{e}_{\mathrm{max},i}) - \mathrm{sort}(\boldsymbol{e}_{oj}) > \boldsymbol{0} \quad (1)$$

$$\boldsymbol{b}_{\mathrm{min},ij} = \mathrm{sort}(\boldsymbol{e}_{oj}) - \mathrm{sort}(\boldsymbol{e}_{\mathrm{min},i}) > \boldsymbol{0} \quad (2)$$

$$i = 1, \cdots, n, \quad j = 1, \cdots, m$$

If $\boldsymbol{b}_{\mathrm{max},ij} > \boldsymbol{0}$ and $\boldsymbol{b}_{\mathrm{min},ij} > \boldsymbol{0}$ are satisfied, the OCP can be included inside the GRCmax. In this case, the hand may be able to grasp the object by using the i-th grasping style. Also, we use the following function to select the grasping style:

$$l_{ij} = ||\mathrm{sort}(\boldsymbol{e}_{\mathrm{des},i}) - \mathrm{sort}(\boldsymbol{e}_{oj})|| \quad (3)$$

$$i = 1, \cdots, n, \quad j = 1, \cdots, m$$

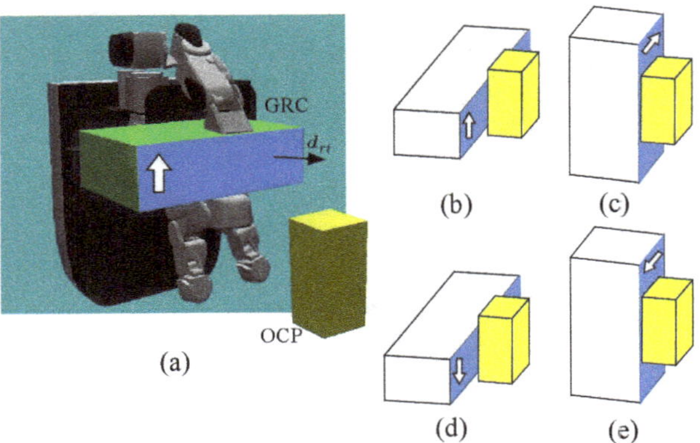

Fig. 7. Four possibilities for palm position/orientation

Let m be the mass of the object. We also impose the following conditions:

$$\delta m_{\max} = m_{\max,i} - m > 0 \tag{4}$$

$$\delta m_{\min} = m - m_{\min,i} > 0 \tag{5}$$

If there are multiple candidates of grasping styles, our planner selects one grasping style according to the nominal palm position/orientation. Due to this function, the robot can select the different grasping style for the same object if the position of the object is changed.

Once the grasping style is determined, our planner next determines the nominal position/orientation of the palm. Fig.7 shows the overview of the method. Let us focus on the surface of the GRCmax having the approach vector $\boldsymbol{d}_{ri}$ as its normal. Assuming that this surface is parallel to one of the surface of OCP and that the center of GRC coincides with the center of the OCP, there are four possibilities for the orientation of the GRC as shown in (b),(c),(d) and (e). For this example, the posture of the GRC shown in (b) and (d) is not feasible since the GRCmax does not include the OCP even if the GRC moves to the direction of the approach vector. On the other hand, for the position/orientation of the palm shown in (a) and (c), we try to solve the inverse kinematics of the arm. If the inverse kinematics problem has a solution, our planner considers it as a candidate of the nominal position/orientation of the palm. We iterate this calculation for all surfaces of the OCP without contacting another OCP.

In case, we have multiple candidates of the nominal position/orientation of the palm, we have to select one. In this research, we select the nominal position/orientation of the palm which has the minimum norm of joint angles of the wrist.

3.3 Force closure condition

3.3.1 Conventional methods

When a multi-fingered hand grasps an object, an appropriate force/moment has to be generated onto the grasped object in order to balance any direction of the external force/moment. This condition is satisfied if the set of resultant force/moment applied by each finger includes the origin. This condition is known as the force closure. The set of resultant force/moment applied to the object can be obtained by calculating the sum of contact force applied by each finger where the contact force is limited inside fiction cone.

Let us consider the contact force $\boldsymbol{f}_i$ $(i = 1, \cdots, m)$ applied at the i-th contact point position $\boldsymbol{p}_i$. The wrench $\boldsymbol{w}_i$ generated by $\boldsymbol{f}_i$ can be calculated as

$$\boldsymbol{w}_i = \begin{bmatrix} \boldsymbol{f}_i \\ \gamma \boldsymbol{p}_i \times \boldsymbol{f}_i \end{bmatrix}. \tag{6}$$

where γ is torque magnitude which is scaled with respect to the force magnitude. The space spanned by all contact forces is called the grasp wrench space (GWS). Ferrari & Canny (1992) proposed two methods of generating GWS. One method generates GWS (W_{L_∞}) by calculating the sum of all combination of contact forces. It is the Minkowski sum of the wrench applied by each finger. The other method generates convex hull of contact wrenches as GWS (W_{L_1}). The set of GWS, W_{L_∞} and W_{L_1} are expressed as,

$$W_{L_\infty} = \{\oplus_{i=1}^{m} \boldsymbol{w}_i | \boldsymbol{w}_i \in W_i\} \tag{7}$$

$$W_{L_1} = ConvexHull(\{\cup_{i=1}^{m} \boldsymbol{w}_i | \boldsymbol{w}_i \in W_i\}) \tag{8}$$

where $\oplus$ is Minkowski sum, W_i is a set of the effect wrench applied at i-th each contact point. When GWS contains the origin of wrench space, we say that the grasp is force closure. If a set of vectors, $\boldsymbol{w}_i \in W_i$ positively spans $\Re^6$, then both W_{L_∞} and W_{L_1} will contain the origin. For testing whether or not, GWS contains the origin, W_{L_1} is useful because of its fast calculation. However, for the purpose of dynamic motion planning of grasped object and more accurate grasp stability evaluation, W_{L_∞} is needed. Since W_{L_∞} space shows the range which can counter external wrench or inertial wrench in any direction.

For constructing W_{L_∞}, the friction cone has been approximated by using a polyhedral convex cone. In this case, if we consider accurately approximating the friction cone by using a larger number of the faces, it results in the dramatic increase of the calculation cost. Let m and n be the number of finger and the number of face of the polyhedral cone, respectively. Its calculation cost is represented as $O(n^m)$. Reduction of the calculation cost of W_{L_∞} is important issue for grasp planning.

As for the research on force closure, Reuleaux (1976) discussed force closure used in classical mechanics. Ohwovoriole (1980) and Salisbury & Roth (1982) introduced it into the research field of robotics. Mishra et al. (1987), Nguyen (1988), and other researchers (Park & Starr (1992)-Ponce & Faverjon (1995)) investigated the construction of force closure grasp by a robotic hand. Kerr & Roth (1986), Nakamura et al. (1989), and Ferrari & Canny (1992) discussed the optimal grasp under force closure.

Linear matrix inequality (LMI) method (Han et al. (2000); Liu & Li (2004)) is a fast method of force closure testing. LMI method can find a solution which satisfying the friction constraint quickly. Some methods are derived from LMI method, such as ray tracing method (Liu (1993); Zhen & Qian (2005)), and heuristics approach (Niparnan & Sudsang (2007)). They are fast method for testing force closure and find optimal solution.

Borst et al. (1999) proposed a method of incremental expansion of subspace of in the weakest direction of W_{L_∞}. It is faster than conventional methods for grasp stablity evaluation of W_{L_∞}.

3.3.2 Ellipsoidal approximation of friction cone

We have proposed a fast method of grasp stability calculation. Our method simply checks inequalities for force closure judgment and does not need to construct convex hull or to solve linear programming problem. If we approximate the friction cone by using a single ellipsoid, the force closure can be confirmed by checking only one inequality.

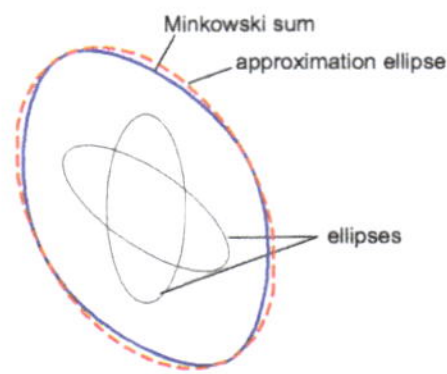

Fig. 8. Minkowski sum of two ellipses and its approximation

By using the nominal position/orientation of the palm, now we determine the final grasping posture. The grasping posture is determined so as to satisfy several constraints imposed on the grasp system.

The grasped object has to resist the external wrench without breaking the contact. For this purpose, we formulate the wrench set generated at a point in the object. In our method, we consider approximating the friction cone constraint by using multiple ellipsoids.

Let us consider the contact force $\boldsymbol{f}_i$ $(i = 1, \cdots, n)$ applied at the i-th contact point position $\boldsymbol{p}_i$. One of the ellipsoids approximating the friction cone can be expressed as

$$(\boldsymbol{f}_i - f_{\max}\boldsymbol{n}_i)^t \boldsymbol{U}_i \boldsymbol{S} \boldsymbol{U}_i^t (\boldsymbol{f}_i - f_{\max}\boldsymbol{n}_i) \leq 1 \tag{9}$$

where $S = \mathrm{diag}[\mu f_{\max}/\sqrt{2} \;\; \mu f_{\max}/\sqrt{2} \;\; \alpha f_{\max}]$, μ is a friction coefficient, and $\boldsymbol{U}_i$ is a 3×3 matrix composed of the unit normal and tangent vectors. By using this equation for n contact points, the set of the wrench $\boldsymbol{w}$ generated by the object can be given by

$$\begin{aligned} f(\boldsymbol{p}_1, \cdots, \boldsymbol{p}_n) &= \\ (\boldsymbol{w} - f_{\max}\boldsymbol{G}\boldsymbol{N})^t (\boldsymbol{G}\boldsymbol{U}^{-1}\boldsymbol{G}^t)^{-1} &(\boldsymbol{w} - f_{\max}\boldsymbol{G}\boldsymbol{N}) \leq n \end{aligned} \tag{10}$$

where

$$\begin{aligned} \boldsymbol{G} &= \begin{bmatrix} \boldsymbol{I} & \cdots & \boldsymbol{I} \\ \gamma\boldsymbol{p}_1 \times & \cdots & \gamma\boldsymbol{p}_n \times \end{bmatrix}, \\ \boldsymbol{N} &= [\boldsymbol{n}_1^t \; \cdots \; \boldsymbol{n}_n^t]^t, \\ \boldsymbol{U} &= \text{block diag}[\boldsymbol{U}_1 \boldsymbol{S} \boldsymbol{U}_1^t \; \cdots \; \boldsymbol{U}_n \boldsymbol{S} \boldsymbol{U}_n^t]. \end{aligned}$$

This method is useful since we can check the force closure condition very quickly just by calculating the left-hand side of eq.(10) (Fig. 8).

3.4 Searching final grasping posture

Failing to find feasible grasping posture using arm/hand kinematics alone, our planner attempts to search the one using full body kinematics. If the robot has joints which can move without losing humanoid body balance, then we add them to an arm joint list. By using the new joint list, inverse kinematics of the redundant joints is solved.

In the case of the robot can squat down, we add the virtual joint which moves shoulder height to the list. If a inverse kinematics of the joint list is solved, the waist height is acquired using the shoulder position.

To obtain the final grasping posture, we use the random sampling technique. Let n be the number of fingers. We use $2 + n$ variables to search for the final grasping posture; $\Delta\boldsymbol{p}_p \in R^3$, $\Delta\boldsymbol{p}_s \in R^3$ and $\Delta\boldsymbol{a}_i \in R^3$ $(i = 1, \cdots, n)$.

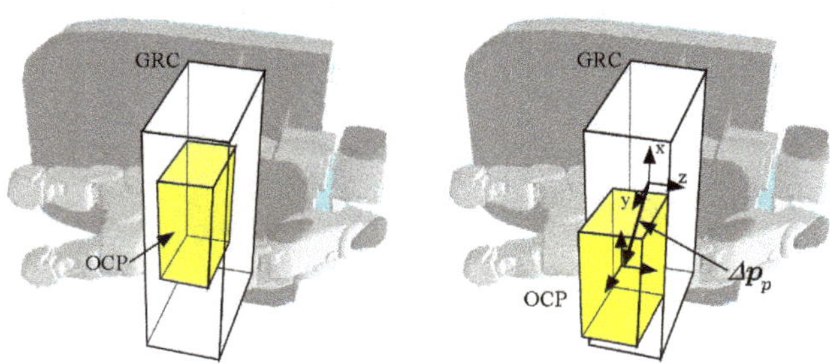

Fig. 9. Sample the position of OCP w.r.t. GRC

As shown in Fig.9, the three dimensional vector $\Delta \boldsymbol{p}_p$ expresses the position of the OCP w.r.t. its nominal position and is used to determine the position of the palm. On the other hand, $\Delta \boldsymbol{p}_s$ and $\Delta \boldsymbol{p}_i$ $(i = 1, \cdots, n)$ express the position of the shoulder and the i-th fingertip, respectively, w.r.t its nominal position. Here, as for $\Delta \boldsymbol{p}_s$ of this paper, we only change the vertical component of this vector if it is difficult to find the grasping posture when $\Delta \boldsymbol{p}_s = \mathbf{0}$.

3.5 Planning algorithm

We assume that we have the polygon model of both the finger and the grasped object. To check the collision between the finger and the object, we used the software PQP (Proximity Query Package). By using PQP, even if two polygon models do not contact each other, we can calculate the distance between two models, the points on both models where the distance is the minimum, and the unit normal vector on the points of the model's surface.

We first explain the method to find the posture of a finger contacting the object. First, for each grasping style, we defined the links of the fingers contacting the object. Then, for each defined link, we assign a joint of a finger compensating its position. We change the angle of the assigned joint by a small amount and checked the collision between the link and the object. We iterate this calculation until the distance between the link and the object is smaller than a predefined value.

By using the pseudo code shown in *Algorithm* 1, we summarize the algorithm explained in this section. In this algorithm, after confirming that $n - f(\boldsymbol{p}_1, \cdots, \boldsymbol{p}_n) \geq \delta f$ is satisfied, we terminate the algorithm.

3.6 Grasping module

We implement our grasp planning module on a software platform Choreonoid (Fig. 10), which is an integrated software that allows us to choreograph robot motions via the interface and simulate the robot motion dynamics. We develop the grasping module has functions of a grasp planner, a motion trajectory planner, and a planning result viewer. We use also RT-middleware which are standardized by OpenRTM (2010). The grasping module can connect other RT-components such as image processing, sound recognition, and sensor network.

We describe user interfaces of the grasping module. There are 5 buttons for executing grasp planning.

SetRobot Assign a working robot to the selected item.

SetObject Assign an grasped object to the selected item.

Algorithm 1 Determination of Grasping Posture for n finger grasp

loop
  loop
    Sample $\Delta \boldsymbol{p}_p$ and $\Delta \boldsymbol{p}_S$.
    if Arm IK is solvable **then break**
  end loop
  for $i = 1$ to n
    for $j = 1$ to m
      Sample $\Delta \boldsymbol{p}_i$.
      if Finger i IK is solvable **then**
        if Found finger i posture contacting object **then break**
      end if
    end for
    if Not found finger i posture **then break**
  end for
  if $n - f(\boldsymbol{p}_1, \cdots, \boldsymbol{p}_n) \geq \delta f$ **then break**
end loop

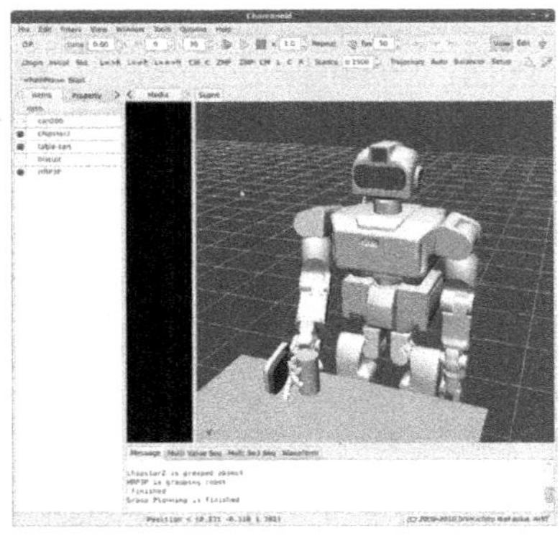

Fig. 10. Grasping module on choreonoid

SetEnv Assign obstructions to the selected items.

Start Start grasp planning.

Stop Abort grasp planning.

The result of grasp planning is showed on a graphics window.

4. Results

4.1 Simulation

To confirm the effectiveness of the proposed method, we show some numerical examples. As a model of the hand/arm system, we use the 7dof arm of the HRP-3P and a developed 4 fingered hand. This 4 fingered hand has the thumb with 5 joints and the index, middle and third fingers with 4 joints. The distal joint of each finger is not directly actuated and moves along with the next joint.

Fig. 11. Simulation environment

We prepared 9 reference motion as shown in Fig.4. The overview of numerical example is shown in Fig.11. We used a can with 0.15[kg] placed on the table as a grasped object. For this weight of the object, the grasp planner selects 3-fingered fingertip grasp or 4-fingered fingertip grasp. For each grasping style, we make only the distal link of finger contacting the object.

As shown in Fig.12, when grasping the can at a breast position, the humanoid robot grasps the can by using the four-fingered fingertip grasp. On the other hand, as shown in Fig.13, when grasping the can at waist height, the finger grasps the top of the can by using the three-fingered fingertip grasp. When grasping the top of the can, it is often difficult to grasp it by using the four-fingered fingertip grasp due to the size of the fingers.

On the other hand, we set the weight of the object as 0.35[kg] for the case of Fig.14. In this case, the humanoid robot grasps the object by using the enveloping grasp with squatting down since it is difficult to grasp the object by simply standing on the ground.

Then we performed experiment. As an grasped object, we used a 200[ml] can. We set $m_{\max,i}$ and $m_{\min,i}$ so that the hand grasps this can by using the enveloping grasp. In case of the enveloping grasp, we determined the finger posture (the 10th line of Algorithm 1) as follows; We first set that the distal (5th) link of the thumb and the 2nd and the 4th link of other fingers contact the object. Then, for the finger links contacting the object, we assigned a joint to each link supposed to make contact and adjusted the angles of this joint. If all the contacts are realized at the same time, 10th line of Algorithm 1 returns **true**. This grasp planning takes less than 1[s] by using Pentium M 2.0[GHz] PC and the calculation time of force closure is less than 1[ms].

4.2 Experiment

Fig. 15(a) shows the image taken by the camera attached at the head of HRP-3P humanoid robot. Three cameras are attached at the head. We used image processing software developed in VVV, AIST (Maruyama et al. (2009)). Given the geometrical model of the can, the position/orientation of it was calculated by using the segmentation based stereo method. Fig. 15(b) shows the result of extracting the segment. Although it takes less than one second

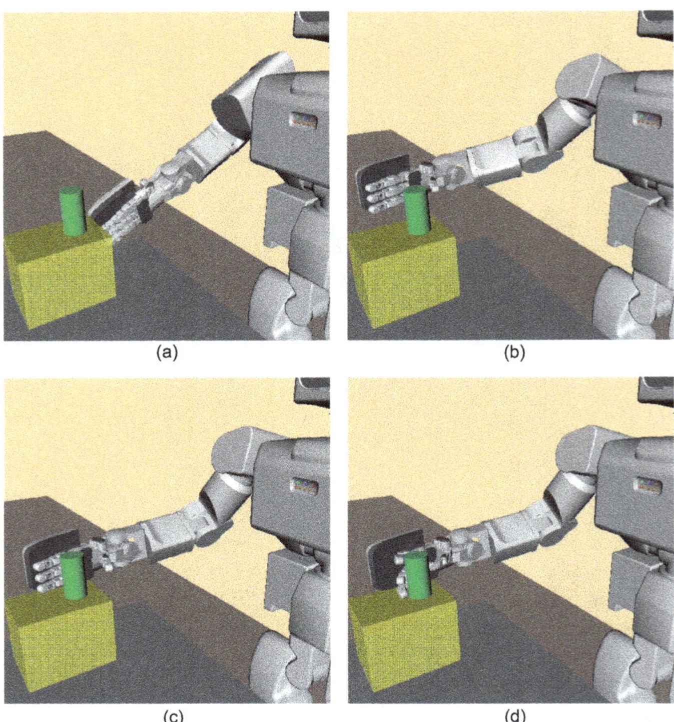

Fig. 12. Four-fingered fingertip grasp when grasping a can at breast height

to obtain the position/orientation of the can, we took the image for a few times to cope with the fault of the image processing.

The image processing was performed on the same PC with the motion planner. This PC was connected to the cameras by using IEEE 1394 cable. The image taken by the camera was processed by using this PC and the position/orientation data was transferred to the motion planner. After the motion planner plans the grasping motion, the joint angle data were transferred to the CPU boards controlling the motion of the robot. Here, we have two CPU boards installed in the chest of the robot where one is used to control the multi-fingered hand and the other is used to control the rest part of the robot. The wireless LAN is equipped with the robot. A directory of the motion planner PC was mounting on the CPU boards and the robot is controlled by using the joint angle data stored to this directory. As shown in Fig.16, HRP-3P successfully grasps objects.

The robot takes out pet bottle from fridge as shown in Fig.17. At first, the robot opens the fridge door using settled motion. Vision sensor measures the object position and the planner cannot find a feasible posture only by using the arm/hand kinematics. The planner searches a feasible posture by changing the shoulder height and finds it. The robot squats down and reaches the object and grasps it successfully.

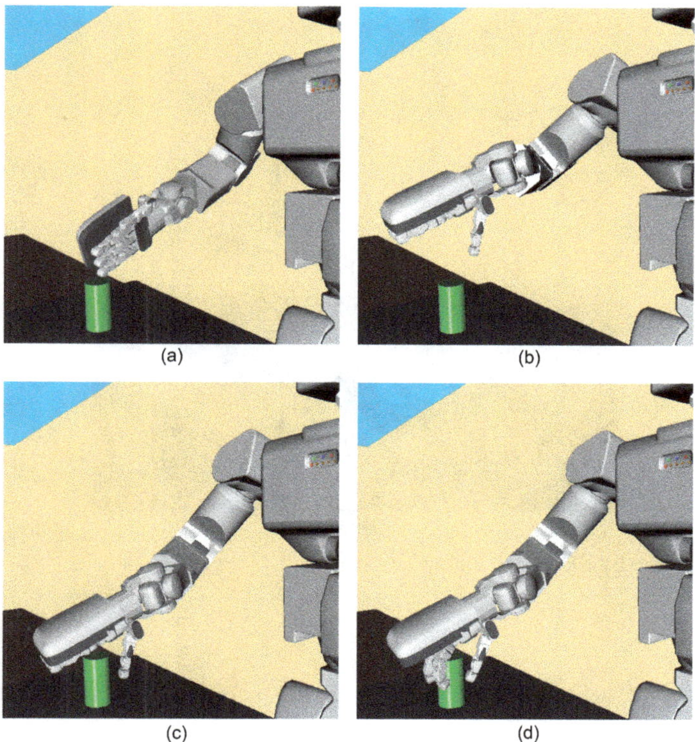

Fig. 13. Three-fingered fingertip grasp when grasping a can at waist hight

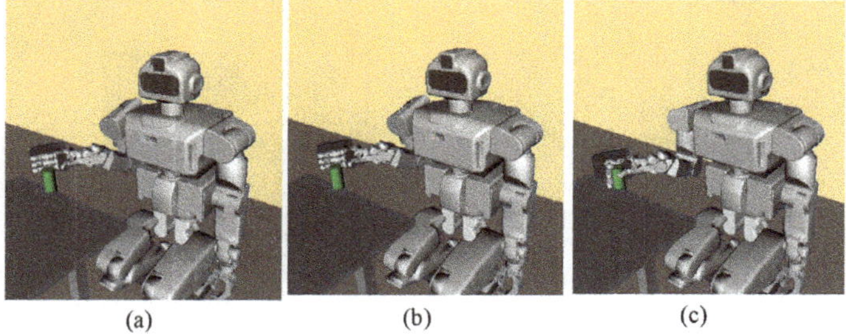

Fig. 14. Four-fingered enveloping grasp with squatting down

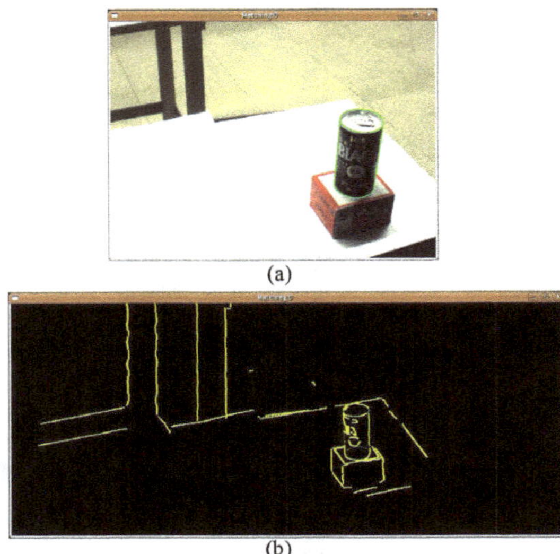

Fig. 15. Image taken by the stereo vision system

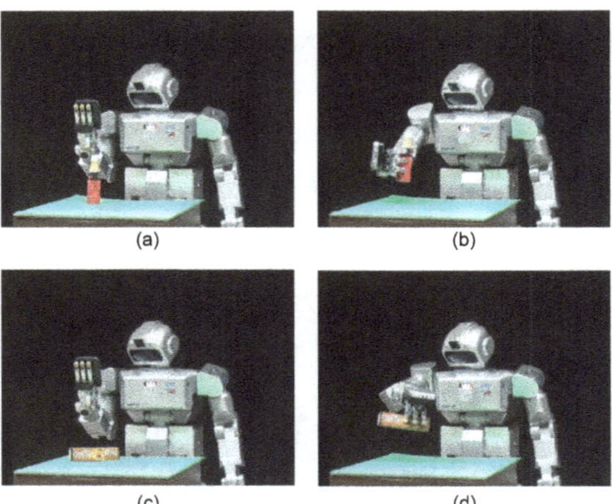

Fig. 16. Experimental result of grasping objects

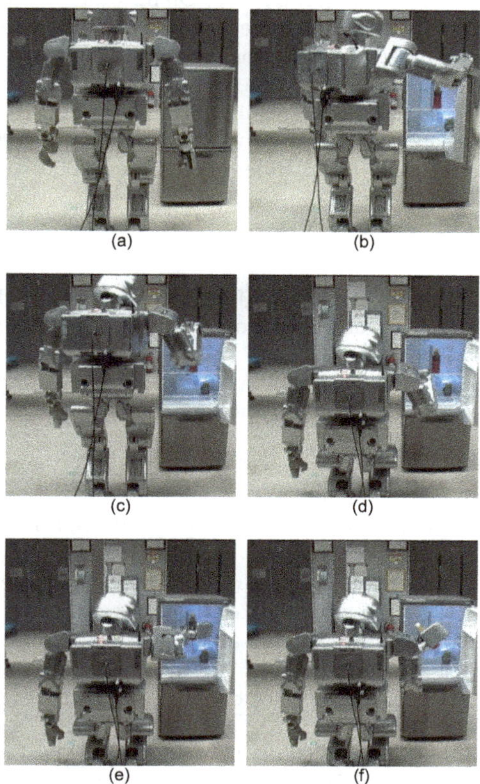

Fig. 17. Experimental result of taking out a pet bottle from a fridge

5. Conclusions and future works

In this chapter, we present algorithms of grasp planning for humanoid hand/robot system. By using the convex model for both hand and the object, the nominal palm position/orientation can be calculated easily. Also by using the ellipsoidal approximation of the friction cone, the grasp planning can be finished within a short period of time.
We improved the performance of the planner by considering the following things; First, depending on which part of the object to be grasped by the hand, our planner changes the grasping style. We numerically confirmed that, when grasping the side of the can, the robot grasps it by using the four-fingered fingertip grasp. On the other hand, when grasping the top of the can, the robot grasps it by using the three-fingered fingertip grasp. Also, if it is difficult to find the grasping posture only by using the arm/hand kinematics, the planner tries to use the whole body motion. We confirmed the effectiveness of our planner modularized as Choreonoid plugin by using experimental results.
Planner for the humanoid hand is necessary for future applications. Grasp planning would be improved on the dexterousness. The planner for multi-arm and handling objects has to be developed.

6. References

K. Akachi, K. Kaneko, N. kanehira, S. Ota, G. Miyamori, M. Hirata, S. Kajita, and F. Kanehiro, (2005). Development of Humanoid Robot HRP-3P, *IEEE-RAS/RSJ Int. Conference on Humanoid Robots*.

Ch. Borst, M. Fischer, G. Hirzinger, (1999). A fast and robust grasp planner for arbitrary 3D objects, *Proc. of the IEEE International Conf. on Robotics and Automation*, vol 3, pp.1890-1896.

C. Borst, M. Fischer, G. Hirzinger, (2003). Grasping the Dice by Dicing the Grasp, *IEEE/RSJ Int. Conf. on Intelligenr Robots and Systems*.

J.A. Coelho Jr. and R.A. Grupen, (1996). Online Grasp Synthesis, *IEEE Int. Conf. on Robotics and Automation*.

M.R. Cutkosky, (1989). On Grasp Choice, Grasp Models, and the Design of Hands for Manufacturing Tasks, *IEEE Trans. on Robitics and Automation*, vol.5, no.3.

S. Ekvall and D. Kragic, (2007). Learning and Evaluation of the Approach Vector for Automatic Grasp Generation and Planning, *IEEE Int. Conf on Robotics and Automation*.

C. Ferrari, and J. Canny, (1992). Planning optimal grasps, *IEEE Intl. Conf. on Robotics and Automation*, pp.2290-2295.

L. Han, J. C. Trinkle, and Z.X. Li, (2000). "Grasp Analysis as Linear Matrix Inequality Problems", *IEEE Trans. on Robotics and Automation*, vol. 16, no.6, pp. 663-674.

K. Harada, K. Kaneko, and F. Kanehiro (2008). Fast Grasp Planning for Hand/Arm Systems based on Convex Modes, *Procedings of IEEE Int. Conf. on Robitics and Automation*.

K.Kaneko, K. Harada, and F. Kanehiro, (2007). Development of Multi-fingered Hand for Life-size Humanoid Robots, *IEEE Int. Conf. on Robotics and Automation*, pp. 913-920.

S.B. Kang and K. Ikeuchi, (1995). Toward Automatic Robot Instruction from Perception-temporalsegmentation of Tasks from Human Hand Motion, *IEEE Trans. on Robotics and Automation*, vol. 11, no. 5.

I.A. Kapandji, (1985). Physiologie Architecture, *Maloine S.A. Editeur*.

J. Kerr and B. Roth, (1988). "Analysis of multifingered hands", *Int J Robot Res* vol. 4, no. 4, pp.3-17.

Y. H. Liu, (1993). "Qualitative test and force optimization of 3-D frictional form-closure grasps using linear programming", *IEEE Trans. Robot. Automat.*, vol. 15, no. 1, pp.163-173.

G.F. Liu, and Z.X. Li, (2004). "Real-time Grasping Force Optimization for Multifingered Manipulation: Theory and Experiments", *IEEE/ASME Transactions on Mechatronics*, vol.9, no.1, pp.65-77.

G.F. Liu, J.J. Xu, and Z.X. Li, (2004). "On Geometric Algorithms for Real-time Grasping Force Optimization", *IEEE Trans. on control System Technology*, vol.12, no.6, pp.843-859.

K.i Maruyama, Y. Kawai, F. Tomita, (2009). Model-based 3D Object Localization Using Occluding Contours, *Proc. 9th ACCV*, MP3-20.

A.T. Miller, S. Knoop, H.I. Christensen, and P.K. Allen, (2003). Automatic Grasp Planning using Shape Primitives, *IEEE Int. Conf. on Robotics and Automation.*

A.T. Miller and Peter K. Allen, (2000). GraspIt!: A Versatile Simulator for Grasp Analysis, *ASME Int. Mechanical Engineering Congress & Exposition.*

B. Mishra, J.T. Schwartz, and M. Sharir, (1987). "On the existence and synthesis of multifinger positive grips", *Algorithmica (Special Issue: Robotics)* vol. 2, no. 4, pp.541-558.

A. Morales, P.J. Sanz, A.P. del Pobil, and A.H. Fagg, (2006). Vision-based Three-finger Grasp Synthesis Constrained by Hand Geometry, *Robotics and Automous Systems*, no.54.

A. Morales, T. Asfour, and P. Azad, Integrated Grasp Planning and Visual Object Localization for a Humanoid Robot with Five-Fingered Hands, (2006). *IEEE/RSJ Int. Conf. on Intelligent Robots and Systems.*

Y. Nakamura, K. Nagai, and T. Yoshikawa, (1987). "Dynamics and stability in coordination of multiple robotic mechanisms", *Int J Robot Res* vol. 8, no. 2, pp.44-61.

S. Nakaoka, A. Nakazawa, K. Yokoi, H. Hirukawa, and K. Ikeuchi, (2003). Generating Whole Body Motions for a Biped Humanoid Robot from Captured Human Dances, *IEEE Int. Conf. on Robotics and Automation.*

V. Nguyen, (1988). "Constructing force closure grasps", *Int J Robot Res* vol. 7, no. 3, pp.3-16.

N. Niparnan and A. Sudsang, (2004). Fast Computation of 4-Fingered Force-Closure Grasps from Surface Points, *IEEE/RSJ Int. Conf. Intelligent Robots and Systems.*

N. Niparnan and A. Sudsang, (2007). "Positive Span of Force and Torque Components of Three-Fingered Three-Dimensional Force-Closure Grasps", *Proc. of the IEEE International Conf. on Robotics and Automation*, pp.4701-4706.

M.S. Ohwovoriole, (1980). *An extension of screw theory and its application to the automation of industrial assemblies*, Ph.D. dissertation, Department of Mechanical Engineering, Stanford University.

OpenRTM: http://www.is.aist.go.jp/rt/OpenRTM-aist/

Y.C. Park and G.P. Starr, (1992). "Grasp synthesis of polygonal objects using a three-fingered robot hand", *Int J Robot Res* vol. 11, no. 3, pp.163-184.

R. Pelossof, A. Miller, P. Allen, and T. Jebra, (2004). An SVM Learning Approach to Robotic Grasping, *IEEE Int. Conf. on Robotics and Automation.*

N.S. Pollard, (2004). Closure and Quality Equivalence for Efficient Synthesis of Grasps from Examples, *Int. J. of Robotics Research*, vol.23, no.6, pp.595-613.

J. Ponce and B. Faverjon, (1995). "On computing three-finger force-closure grasps of polygonal objects", *IEEE Trans Robot Automat* vol. 11, no. 6, pp.868-881.

J. Ponce, S. Sullivan, J.-D. Boissonnat, and J.-P. Merlet, (1993). On Characterizing and Computing Three- and Four-fingered Force Closure Grasps of Polygonal Objects, *IEEE Int. Conf. on Robotics and Automation.*

M. Prats, P.J. Sanz, and P. del Pobil, (2007). Task-Oriented Grasping using Hand Preshapes and Task Frames, *IEEE Int. Conf. on Robotics and Automation.*

qhull, (2010): http://www.qhull.org/

F. Reuleaux, (1876). *The kinematics of machinery*, Macmillan, New York.

J.K. Salisbury and B. Roth, (1982). "Kinematics and force analysis of articulated hands", *ASME J Mech Trans Automat Des*, 105, pp.33-41.

K.B. Shimoga, (1996). Robot Grasp Synthesis: A Survey. *Int. J. of Robotics Research*, vol.5, no.3.

T. Tsuji, K. Harada, and K. Kaneko, F. Kanehiro, Y. Kawai, (2008). Selecting a Suitable Grasp Motion for Humanoid Robots with a Multi-Fingered Hand *Proc. IEEE-RAS Int. Conf. on Humanoid Robots*, pp.54-60.

T. Tsuji, K. Harada, and K. Kaneko, (2009). Easy and Fast Evaluation of Grasp Stability by Using Ellipsoidal Approximation of Friction Cone, *Proc. IEEE/RSJ Int. Conf. on Intelligent Robots and Systems.*

T. Tsuji, K. Harada, K.Kaneko, F.Kanehiro, M.Morisawa, (2009). Grasp Planning for a Multifingered Hand with a Humanoid Robot, Int. J. of Robotics and Mechatronics.

M. Yashima, Y. Shiina and H. Yamaguchi, (2003). Randomized Manipulation Planning for A Multi-Fingered Hand by Switching Contact Modes, *IEEE Int. Conf. on Robotics and Automation.*

Y. Zheng, W.-H. Qian, (2005). "Simplification of the ray-shooting based algorithm for 3-D force-closure test", *IEEE Trans. Robot. Automat.*, vol. 21, no. 3, pp. 470-473.

5

Design of 5 D.O.F Robot Hand with an Artificial Skin for an Android Robot

Dongwoon Choi, Dong-Wook Lee, Woonghee Shon and Ho-Gil Lee
Department of Applied Robot Technology, Korea Institute of Industrial Technology
Republic of Korea

1. Introduction

There have been many researches of robot hands in robot fields and they have been considered one of the most complicated area. There are many reasons why researches of robotic hand are difficult, and these are from complicated structures and functions of hands. There are many types of robotic hands in robotics area, but they can be classified to major two categories. The one is a robotic hand for an operation in industrial area and the other is an experimental hand like human hand. The most of robotic hands in industrial area are 1 D.O.F or 2 D.O.F gripers and they are designed for precise, repetitive operations. In the other area, human like robotic hands, main concerns are how the shape of robotic hands resembles human hands and how the robotic hands can operate like human hands. Most human like robotic hands have 3 ~ 5 fingers like human hands and their shape, size and functions are designed based on human hand. For a long time, major area in researches of robotic hand has been an industrial area, but the importance of human like robotic hands are getting more and more increasing, because the needs of robots will be changed from industrial fields to human friendly environment such as home, office, hospital, school and so on. In brief, the mainstream of robotics will be changed from industrial robots to service robots. One of the important factors for service robots in human friendly environment is their appearance. In general, most of humans are feeling friendly and comfortably to similar appearance like them, so the appearance of service robots should resemble human and their hands should imitate human hands, too. For this reason, there have been many researches for human like robotic hands.

Haruhisa Kawasaki developed Gifu hand 2 which has 5 independent fingers. It has 16 D.O.F and 20 joints, so it is one of the most complicated hands. It can operate all joint of each fingers and with attached tactile sensors, delicate grip can be operated. However, its size is big to install to the human size robot (Haruhisa Kawasaki. et al., 2002). F. Lotti used spring joint and tendon to make UBH 3. It has 5 fingers and human like skin and like Gifu hand, each finger has independent joint. The characteristics of this hand is using a spring to its joint and this make its structure simple, but this hand uses too many motors and they are located in other place, so this hand is not good to humanoid robot (F. Lotti. et al., 2005). Kenji KANEKO developed human size multi fingered hand. It has 13 D.O.F complicated fingers and all devices are located in hand but it has 4 fingers and the back of the hand is too big like glove. In this reason, the shape of this hand is a little bit different to human hand, so

it can be used to humanoid robot not android robot (Kenji KANEKO. et al., 2007). The HONDA ASIMO is the most well-known humanoid robot in the world and its hand shape is like human. Its appearance and size is like human and all devices are included in hand but it has only 1 D.O.F, so it is impossible to express variable gestures (K. hirai. et al., 1998) (H. Hirose. et al., 2001). N.Dechev used spring, links and ball nut joint to make multi fingered prosthetic hand. This hand can passive adaptive grasp by using only one actuator and it has human like shape, 5 fingers but it can't express variable motions because there is only one active joint. This hand shows possibility that prosthetic hands can be applied to human like robot (N. Dechev. et al., 2000). Many of exist hands used motors as actuators but Feifei Zhao used pneumatic type actuators. This hand was made of rubber tube, so there are no links or wires and the structure is so simple. However, this hand can't make gesture of human finger and its shape is different to human hand (Feifei Zhao. et al., 2006).
The many existing researches of human like hands can be applied to humanoid robots but not matched to android robots. An android robot is one of humanoid robots but its appearance is more similar to human appearance than appearance of humanoid robot. When the android robot is designed, uncanny-vally must be concerned, in particular (Shimada M. et al., 2009). In this reason, the hand of an android required the nearest appearance to human hand than existing variable human like hands and the design is more difficult than one of humanoid hand. There are three required factors to the design of an android robot hands. Firstly, the size is very important, because it must be matched to whole body in proportion. It can be hard to make small size hand due to its complexity, but if the proportion is contrary to human body, it can look ugly. Secondly, even if the size of hand is satisfied, shape must be concerned. Most of humanoid robot hands satisfy the size policy, but the shape is not curved surface like human hand. Mechanical parts are generally angulate, so the space which can be used is narrow and this makes the design of an android robot hand hard. Thirdly, android robot hands need an artificial skin. The artificial skin is major difference between humanoid robot hands and android robot hands. The artificial skins which is used to android robot hands need human like touch, color and flexibility for driving.
In this research, an android robot hand for an android robot EveR 3 is presented. The EveR 3 is an android robot for stage performances as an actress, so the hand of EveR 3 was made for variable gestures as acting not grasping. This hand has five fingers with 5 D.O.F and artificial skin. DC motors and screw-nut are used as actuators and the hand is driven by links. The shape is based on 3D model which was made by scanning data from real human. The artificial skin was made by silicon complex and its shape was also based on 3D model, so its appearance is very similar to human hand.

2. Design policies

2.1 Applied android robot

The presented android robot hand is made for an android robot EveR 3. EveR 3 is the latest model of EveR 1 which was the first android robot in Korea. It was designed for stage performances as an actress. Because this android robot was designed as an actress, it has an elegant appearance and delicate body which can express human like motions. To make human like motions, it has 41 D.O.F (9 D.O.F in face, 22 D.O.F in body, 10 D.O.F in both hands) and its height is 165cm, weight is 55kg. It can move on the stage by two wheeled mobile lower body and was controlled by wireless LAN. Fig.1 shows an android robot EveR 3 with skin, dress and mechanical drawing. Why this robot was made as a woman is to hide the mobile lower body. As an actress, moving on the stage is necessity, but its appearance is different to humans one, so we can hide this parts with long skirt and to wear a skirt, the woman robot is chosen not a man. There is already an android robot can walk, but it is not enough to fast, stable moving (Shin'ichiro Nakaoka. et al., 2009) (Kenji KANEKO. et al., 2009). To make a woman android robot is more difficult than a man android robot, because it has more curved body, narrow space and especially it is more sensitive in appearance. This android robot EveR 3 is based on Korean young woman. To play emotional acting, it has face which has 9 D.O.F and an artificial skin to make expressions. The full body was made by 3D data which is from real human scanning data and this data is also used to make hand, skin. The hand is designed by using this data, so the size and shape are decided from this reference. The artificial skin was also made by this data but little bit different, because the skin design needs making mold. To make mold for an artificial skin, RP (Rapid prototype) mock up was made from 3D data and the mold was made by using mock up. Silicon complex is used as an artificial skin. There are many materials which are reviewed for an artificial skin, but why silicon complex is selected is its characteristics are most like to human skin. The appearance is the most important factor to android robots, so these are design policies for hands.

Fig. 1. The android robot EveR 3 and its mechanical drawing.

2.2 Design policies

The goal of this design is to make a human like hand as possible. To achieve this goal, the design policies are established and those are about size, shape and skin. The priority of these policies focused on appearance but not performance.

2.2.1 Size

The size of the hand is based on an android robot which was already mentioned, EveR 3. The model of an android robot EveR 3 is Korean young woman whose height is 165cm and weight is 50kg. When humanoid robots which use real human as model are designed, they use only size like height, length of the limb etc. in general. However, android robots need more strict observation of size not only height, length of joint but also circumference of joint, each part. To make exact reference of model, the original model was scanned by 3D scanner and the scanned data was handled by 3D MAX. From this process, the data which can be used CAD program (Solid works) was earned. This data contained all information of size of original model and this can measure exactly by CAD. The size of hand from original model can be measured exactly, too. The required size data which is needed to realize same size of original model is length of each finger, length of each joint of finger, circumference of each finger, thickness of palm and length of palm. Of course, it is hard to realize exact size of each factor of original model, but this process can help to make a robot hand which is the nearest to human hand than any other existing hands. Fig. 2 shows the real size of robotic hand especially mechanical structure and comparison with real human hand whose is the designer of this hand (168cm, Korean man). Even if this man's hand is small, the robotic hand is smaller than his one. Table.1 shows parameters of main factor which are considered to design.

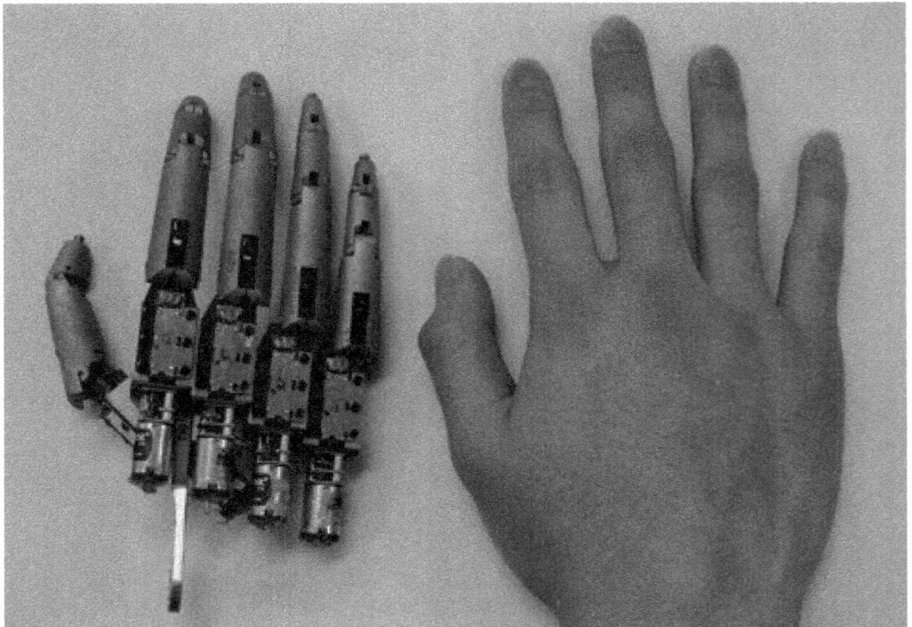

Fig. 2. Mechanical part of robotic hand and comparison with real human hand.

Total length	168 mm (from fingertip to wrist joint)
Width	80 mm (when thumb finger is bended)
Thickness (Palm)	27 mm
Weight	450 g
Index finger	37/23/ 18 mm (1st, 2nd, 3rd joint)
Middle finger	40/28/18 mm (1st, 2nd, 3rd joint)
Ring finger	36/25/18 mm (1st, 2nd, 3rd joint)
Little finger	26/18/15 mm (1st, 2nd, 3rd joint)
Thumb finger	32/18 mm (1st, 2nd, 3rd joint)

Table 1. The pamameters of developed android hand.

2.2.2 Shape

The consideration of shape of an android robot hand is as important as one of its size. There are many humanoid robot hands whose size satisfy human like size, because humanoid robots are designed as real human size. However, most of their designs are not like real human hand especially shape. Even if humanoid robots have similar appearance to human, its design is based on robot basically, so shape of human is curved but one of humanoid robots is angulate (AKACHI K. et al., 2005) (Ill-Woo Park. et al., 2005) (Jun-Ho Oh. et al., 2006). The android robot hand should have curved design and it make hard to design, because most of mechanical parts are angulate, so the space which can be used is very narrow. To design human like curved shape, 3D scanned hand data which was already mentioned was used. There is consideration to use that data. The shape of this data is full straightened, so it is hard to know the shape when the finger bended. Some processing was applied to original data to make bended shape by 3D graphics program (Maya). Fig. 3 shows variable shape of finger to check. From these processes, the nearest reference 3D data to real human hand can be obtained.

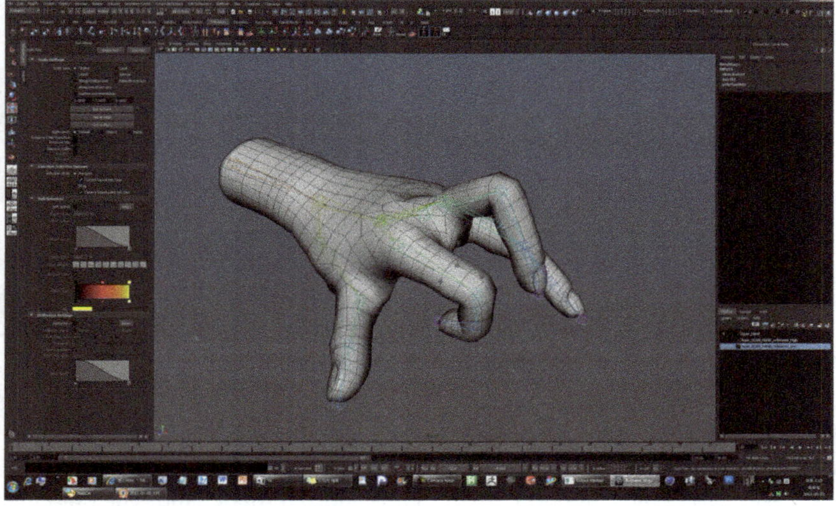

Fig. 3. The handled 3D data to check shape by graphics program.

2.2.3 Skin

The skin is the most important factor to make human like hand, because skin is the part which is shown intuitively. There are some considerations for an artificial skin. These are texture, color and details like wrinkle, so it needs not only technology but also arts. The silicon composite was selected as an artificial skin, because it is the closest material to human skin than any other materials and it is easy to handle. The artificial skin also used 3D data as a reference, but there are more steps to use. To make a silicon complex skin, the mold is needed, because the process of making silicon complex is like one of making plasters. The 3D data was used to make mock up model. To make exactly same model to original model, RP (Rapid prototype) was used to make mock up model. After making mock up, mold was made by using this mock up model, and a silicon complex skin can be made. The artificial skin which is made by silicon complex is very similar to human skin, but to raise the similarity, make up was taken in final step. Though the mechanical part was made by consideration of shape, there are gaps between parts and skin. This gaps can make wrinkles when the finger bended. To solve this problem, art clay was attached to gaps between mechanical parts and skin. Fig. 4 shows a silicon composite artificial skin for hand.

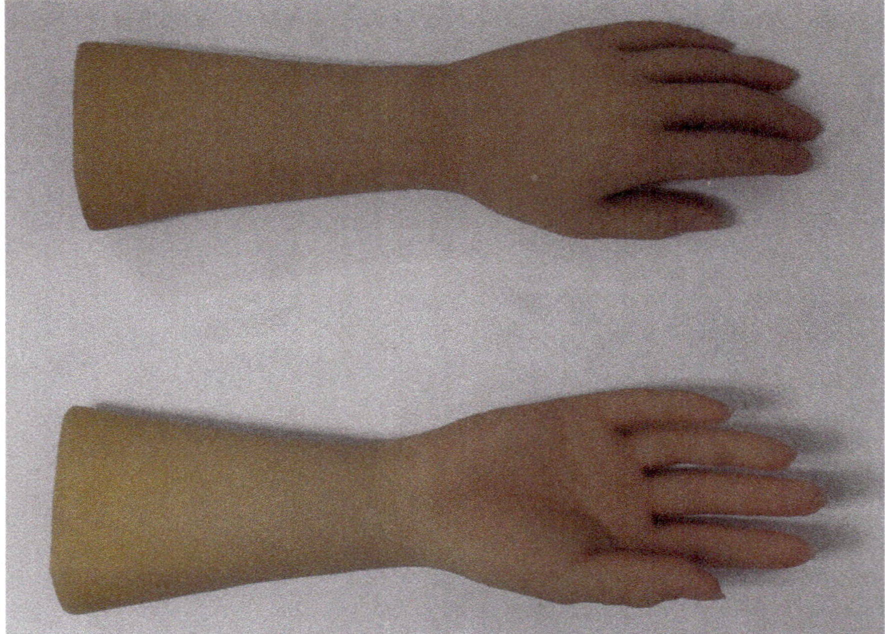

Fig. 4. A silicon composite artificial skin for an android robot hand.

3. Hardware design

3.1 Finger design

The finger design is the most important part in this android robot hand. The presented hand has 5 fingers like human hand and total 5 D.O.F with 1 D.O.F each finger. Actually, human finger has 3 D.O.F at proximal joint, middle joint and distal joint to bend or stretch and 1 D.O.F to spread, so there are 4 D.O.F in each finger basically. In case of thumb finger, there is no middle joint, so it has 3 D.O.F (proximal, distal and 1 D.O.F for abduction and adduction). There are at least 25 D.O.F in real hand, so it is very hard to realize by robotic hands, because there are not enough spaces to install actuators in hand. For this reason, most of humanoid or android hands have smaller number of D.O.F than one of human. Of course, there are some hands which have many D.O.F like one of human, but their size are bigger than human hands (H. Liu. et al., 2008) or they have other large spaces which have a lot of actuators instead (Yuichi Kurita. et al., 2009). For these reasons, the hands for humanoid robots or android robots should have less D.O.F than one of human hand. In this research, the 5 D.O.F hand is designed to express fundamental motion of human hand like straightening - bending of each finger. The best feature of this hand is that the each finger was made as a modular structure independently. Most of robotic hands have their actuators like motors in palm or forearm and their finger is connected to palm, but the finger of this hand has an actuator (DC motor), sensor and gears as its own components. The combination of these components becomes independent finger module. Why this finger was designed as a module is for easy maintenance. The android robot EveR 3 which this hand is applied to was designed for stage performance especially commercial performances not just using in the laboratory for researches, so reliability and maintenance is one of the most important factors. The modular structure can give fast and easy maintenance of hardware. For example, when the middle finger breaks down, replacing of the middle finger module is the only maintenance of all. It is fast, easy and low cost. Fig. 5 shows the disassembled hand by each module.

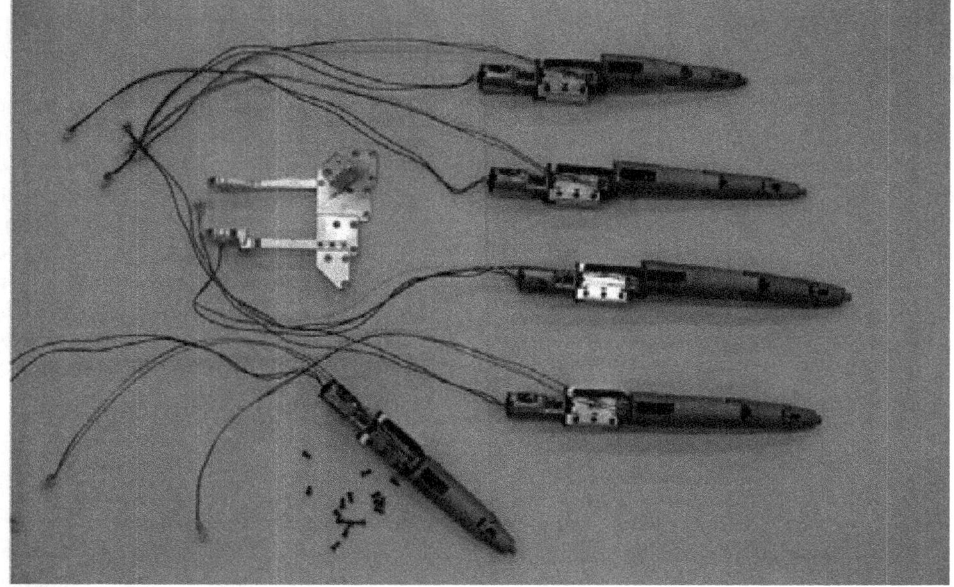

Fig. 5. The disassembled hand by each module (fingers and palm).

The organization of a finger module consists of a geared DC motor as an actuator, a screw-nut, a linear potentiometer and joint links. The joint links are composed of three phalanges as a proximal joint, middle and a distal like human finger. In case of a thumb finger, it is little different to other fingers. There are only a proximal and a distal phalange. These phalanges are frame of finger and these joints are connected inner links and the inner links make subordination of driving of three joints. Fig. 6 shows the composition of a finger module.

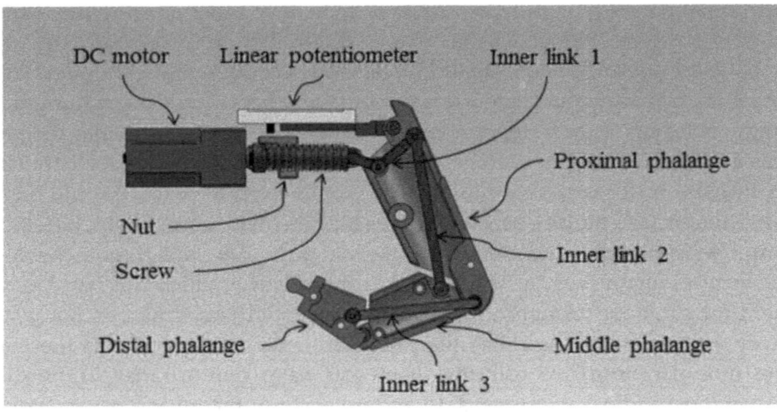

Fig. 6. The composition of a finger module.

The power of the used DC motor is 6mNm and it uses 5V (GM-12F, Motor Bank). The dimension of motor is 36x10x12mm (including gear head) and its weight is 10g. The power of motor is not enough to drive the finger with skin, a gear head whose ratio is 1/50 was attached to the motor. The screw-nut is used to drive links of finger for saving spaces. If the motor is placed to rotation of links horizontally, it is hard to arrange the motors, because all joints of proximal phalanges are placed in same axis. The advantage of this type to arrange motors is easy to design and disadvantage is using of space ineffectively. To place motor vertically to rotational axis of links, worm gear, bevel gear or screw-nut can be considered which connect motor to joint. Why the screw-nut is selected is that it is easy to make small size and a linear potentiometer can be attached to nut. The nut is connected to a linear potentiometer and an inner link 1. The linear potentiometer (RDC1047, ALPS) is used to feedback control of motor and initialize. The motor is small, so it is hard to attach an encoder to it, so the linear potentiometer is used by calculating of gear ratio and lead of screw-nut, the RPM of motor can be obtained. The linear potentiometer uses 5V and its linearity is +/- 5%. The most of parts are made by aluminum (6061alloy) for light weight, a screw-nut is made by brass for low friction and inner links are made by steel (sus304).

3.2 Mechanism of motion

The motion of the finger is occurred by rotation of inner link 1 and the one of left links are subordinated by kinematic relation. Fig. 7 describes the motion of each links when the finger bended and straightened. The motion of finger can be described by following. The proximal phalange is connected to housing of actuating parts and middle phalange, distal phalange are connected in serial order. Three inner links are connected among three phalanges and these inner links make subordinate relation of each phalange. The inner link 1 is connected to nut and its center of rotation is located under of proximal phalange and their centers of rotations are different, so when the inner link 1 is pulled by nut; the proximal phalange rotates by difference of their center of rotation. The shape of inner link 1 is like a boomerang and each tail is connected to nut and inner link 2, so when the inner link 1 is pulled, it starts rotate and it pulls inner link 2. The mechanism of rotation of middle phalange is same to one of proximal phalange. The inner link 2 is connected to middle phalange and the point is different to the center of rotation of the middle phalange, so when the inner link 2 is pulled by inner link 1, the middle phalange rotates. The inner link 3 is connected to proximal phalange and distal phalange. The connected point of inner link 3 in proximal phalange is different to the center of rotation of the middle phalange and the connected point in distal phalange is different to the center of rotation of distal phalange. When the inner link 3 is pulled by rotation of the middle phalange, the distal phalange rotates as same to the middle phalange.

The angle of rotation of each joint is decided by the length and center of rotation of each inner links. The length of proximal phalange, middle phalange and distal phalange is fixed value which is based on the original human model, so the values of inner links (length, position of center of rotation) are design factors in this hand. This robotic hand was made to gesture but not grasp. The gesture which is purposed is natural fist, so the design factors of the inner links are decided when fully bended shape of hand becomes fist. The angles of proximal joint, middle joint and distal joint are 60 degree, 90degree and 45 degree when the shape of hand is fist. The structure of fingers is same except the thumb finger but the sizes are different. In case of human hand, the middle finger is the longest, the index finger and ring finger are similar, and little finger is the smallest size, so this hand wad followed that order.

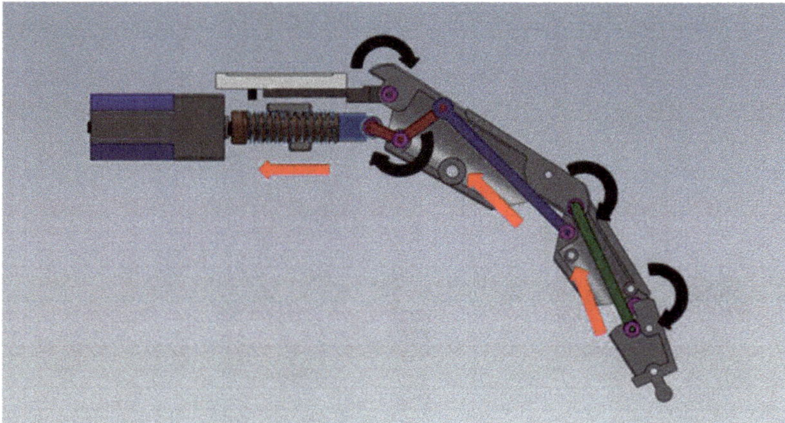

Fig. 7. The mechanism of bending.

3.3 Palm and wrist design

The design of palm is simple as compared with the one of finger. The presented hand is just combination of finger modules, so the palm part is the connecter of finger modules. Even if the role is simple, there are some considerations in palm design, because it can decide the shape of hand by arrangement of finger modules. The most of robot fingers are arranged in same axis in front, top view but in case of human, they are different. The design of palm considered this difference, so the height and distance of each fingers is adjusted by arrangement of attached points. Fig. 8 shows arrangement of finger modules. The one more consideration of palm design is the attachment of thumb finger. The human thumb finger has 3 D.O.F especially including abduction-adduction and some of humanoid robot has same D.O.F. In case of robot hand which is designed for grasping, the D.O.F for abduction-adduction motion is very important, but in this research, there is only 1 D.O.F by out of space. Because the only 1 D.O.F is for bending, the attachment to palm is important to decide the shape of hand. The attachment position is considered natural shape of hand when the hand is fully straightened and fist. The attachment angle is decided 81 degree in top view and 15 degree in front views. This arrangement is shown in Fig. 8. The angle and position are decided by one part and the shape can be changed easily by replacement of this part, thumb connecter. The palm part consists of a palm connecter which connects 4 finger modules, thumb connecter and wrist connecter, so it is very simple and it is easy to change the shape.

The wrist design is based on 3D model like hand design. The human wrist has 2 D.O.F but the wrist of this robot has 1 D.O.F by out of space. The wrist design is included to forearm design and this forearm has wrist joint, forearm yaw joint and controllers for hand and forearm, so there are space problems. To solve this problems, smallest harmonic drive (CSF-5-100-2XH, Harmonic drive), BLBC motors (RE20Flat 5W, Maxon) and self-developed controller was used for getting over narrow space problem. The controller for hand used DSP and it has 6 analog input ports for sensors (Linear potentiometer) and it can control 6 DC motors. Though this hand has 5 D.O.F, this controller has 6 ports for abduction-adduction motion of thumb finger in future works. The BLBC controller was also self-developed to satisfy small spaces. It can control 2 BLDC motors and there are 4 ports for encoders, proximity sensors and it used DSP, too. The proximity sensor (GL-6H, Sunx) was used for initializing. From these efforts, slim size wrist design which can match to reference model can be made. Fig. 9 shows the wrist design and controllers.

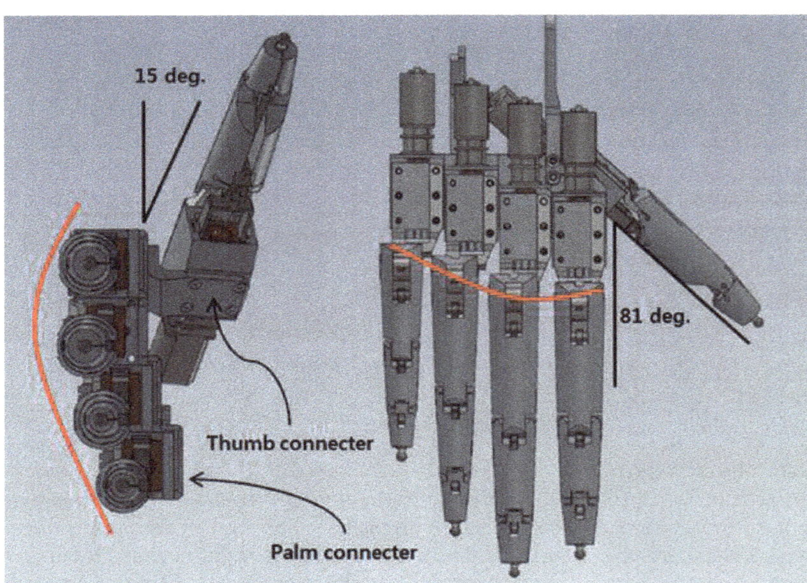

Fig. 8. The shape of palm, thumb joint and attachment position.

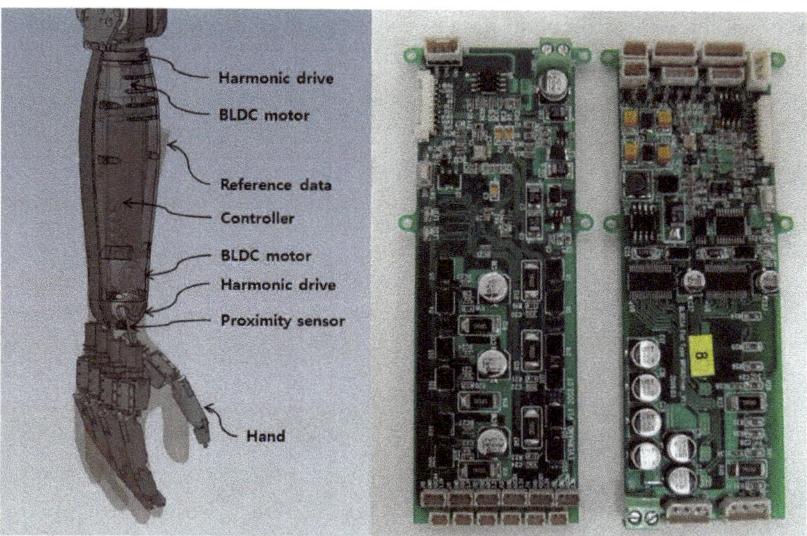

Fig. 9. The wrist design and controllers (hand controller, forearm controller).

3.4 Skin design

The skin part, especially an artificial skin, is very important factor of hand for an android robot, because it is the only shown part optically in outside. The skin design is basically similar to one of hand, but there are more steps than hand design. There are three steps to make skin part. These are mock up, mold and material making. Firstly, the mock up is made by original 3D model. This mock up is used to make mold for skin, so it is needed to be like original model as possible. To realize original model exactly, the mold was made by RP (Rapid prototype) and the exact same model to original can be made. Secondly, the mold was used to cast skin and this is made by mock up. This process is similar to make a plaster caster. The mold was made by silicon, but this is not tough to make skin several times. The CNC based metal mold was replaced to the silicon mold and this mold was tough to use several times, but it was expensive to make. Thirdly, the artificial skin is made by mock-up and mold. There are some materials as an artificial skin like urethane, latex, rubber and silicon. The silicon complex was selected the material of artificial skin, because it has the closet texture to human skin and it is easy to handle. In addition, the characteristic of silicon can be changed easily by the mixture ratio of an emulsion. This mixture ratio between silicon and an emulsion decides the durableness of the skin. It is very important the durableness of material to make an artificial skin, because if the durableness of the skin is hard, it can be stiff resistance to move but soft, it tears easily. Through many times of experiments, the optimal mixture ratio between silicon and an emulsion can be obtained. The pigments were used to make human like color and they were made by blending of some colors and added with an emulsion. After completion of skin, the wrinkles are added to skin by make-up to raise reality. Even if this artificial skin of hand can realize the human hand and mechanical parts are designed by 3D data from human, there are some gaps between skin and mechanical parts. The kind of art clay was used to fill these gaps. The art clay is easy to handle and build shape. From these processes, the artificial skin for robot hand which has the closest shape to human can be produced. Fig. 10 shows the mock-up, mold and the artificial skin.

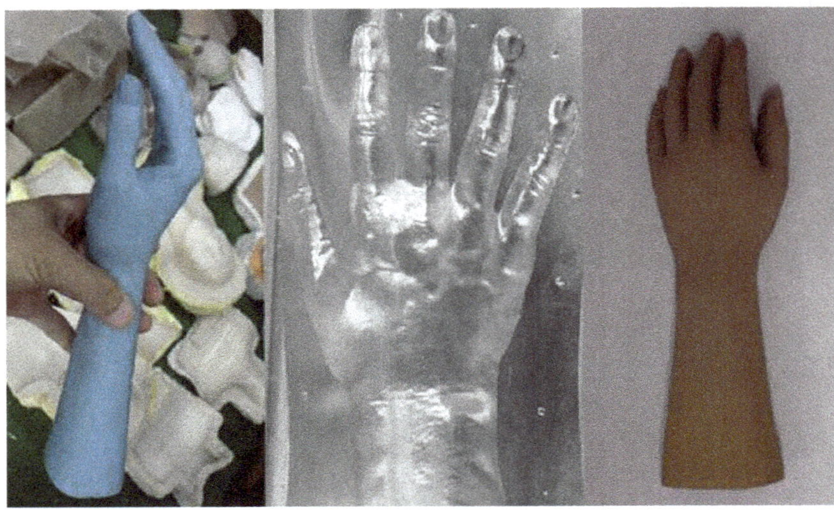

Fig. 10. The mock-up, mold and finished artificial hand skin

4. Kinematics of finger modules

The kinematics of this hand is complicated, because all links are subordinated by 1 D.O.F and the motion is driven by linear and rotary movement. All parameters for kinematics are shown in Fig. 11 and parameters are described in Table. 2. The most important part in kinematics is to know θ_1 from input from screw-nut $U(x_d, y_d)$. Because this hand has only 1 D.O.F, the angle which is controlled is only θ_1. The left angles, θ_2, θ_3 are decided by θ_1 through kinematics. The purpose of this hand is not grasping but gesture, so the fully bended angles are decided to make fist shape. From this, the initial angles are $\theta_1 = 0$, $\theta_2 = 0$, $\theta_3 = 0$ and final angels are $\theta_1 = 60^o$, $\theta_2 = 90^o$, $\theta_3 = 45^o$ when maximum distance of nut is 6mm and these angles are used as the boundary condition to solve equations. The length of each phalange is already determined from original human hand and the length of inner links is design factor. From boundary condition and simple model by CAD, the length of inner links can be earned without kinematics. The kinematics was used to check these approximate values and the relations between angles. The geometric method was used to solve and the Matlab was used to organize and solve equations.

The equation of θ_1 from input $U(x_d, y_d)$ is given as follows formula (1) by geometric method. The const. is 6mm from design and the angle of P_1 is 120°. When it is used to hardware, the formula (1) is used only, because θ_1, θ_2 are not needed to control and full kinematics is complicated and this can calculate slowly by computer. To know θ_2, the position of P_3 should be calculated and the length of $\overline{J_2P_4}$ which is the distance between the center of rotation of middle phalange J_2 and P_4. From these factors, the θ_2 can be solved in formula (2).

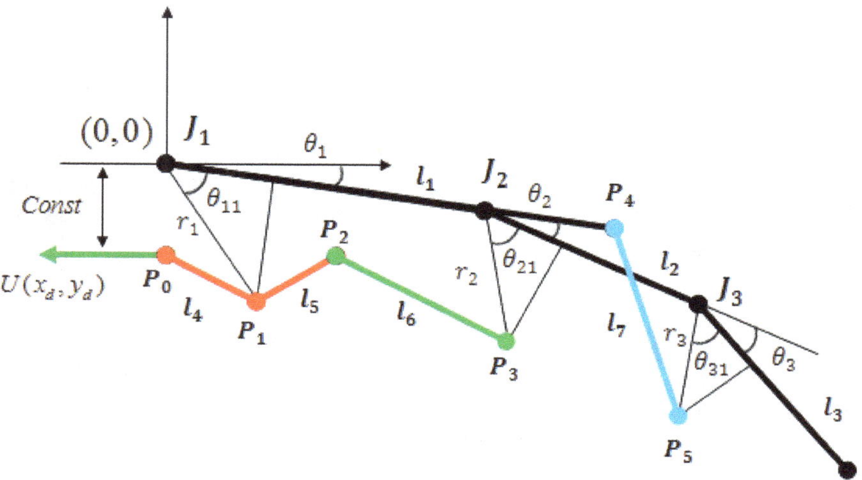

Fig. 11. The simplified image of finger and parameters

J_1	Proximal joint
J_2	Middle joint
J_3	Distal joint
l_1	Proximal phalange
l_2	Middle phalange
l_3	Distal phalange
l_4	Inner link 1
l_5	Inner link 1 (other limb)
l_6	Inner link 2
l_7	Inner link 3
$U(x_d, y_d)$	Distance by screw-nut (input)

Table 2. Parameters for kinematics.

$$\theta_1 = \cos^{-1}\left(\frac{u_1}{\sqrt{x_d^2 + y_d^2}}\right) + \sin\left(\frac{y_d}{\sqrt{x_d^2 + y_d^2}}\right) - \theta_{11}$$

where (1)

$$u_1 = \frac{x_d^2 + y_d^2 + r_1^2 - l_4^2}{2r_1}, \quad (x_d, y_d) = U(x, y)$$

$$\theta_2 = \cos^{-1}\left(\frac{r_2^2 + \overline{J_2P_4}^2 - \overline{P_3P_4}^2}{2r_2 \cdot \overline{J_2P_4}}\right) - \theta_{21} \quad (\theta_{21} \text{ is from design}) \tag{2}$$

The formula (2) needs $\overline{P_3P_4}$ and this can be known formula (3) which are the relations between P_2 and P_3, because the P_4 can be earned from $\overline{J_2P_4}$.

$$\begin{cases} l_6 = \sqrt{(P_{3x} - P_{2x})^2 + (P_{3y} - P_{2y})^2} \\ P_3 = f(J_2) \end{cases} \tag{3}$$

The l_6 is the designed value and P_3 can be known by forward kinematics from J_2. These formulas are also solved by Matlab, because they are too complicated to solve by hand.
The structure of distal joint is simple 4 bar linkage, so the θ_3 can be solved easier than θ_1, θ_2. The θ_3 is known by solving simultaneous equations (4).

$$\begin{cases} l_7 = \sqrt{(P_{4x} - P_{3x})^2 + (P_{4y} - P_{3y})^2} \\ P_5 = f(J_3) \end{cases} \tag{4}$$

How to get unknown values is same to get unknown in θ_2.
Even if the complicated kinematics was solved, it was not used to operate hand except for calculating θ_1.

5. Experiment and discussions

This hand was made for gestures like human hand not grasping, so experiments are very simple. From this purpose, the experiments were taken to check how this hand can realize gestures like the one of human similarly and verify the torque of hand is enough or not to move under the resistance of skin. Even if some experiments were taken to know the resistance of skin by pieces of silicon, there are great differences between pieces of skin and hand shaped skin. The exact experiments can be taken by material engineering area, so it was not efficient way to check the resistance of skin, tests to the real model was taken. The performed experiments to gestures are basically to express rock-paper-scissors posture. The 6 postures are performed including rock-paper-scissors and during these postures, how the shapes are natural like the shape of hand of human in same posture was checked. The tearing and wrinkle at the surface, adequateness of torque are also checked. Fig. 12 shows 6 postures with the completed hand. The gear ratio was 1/30 at first, but it is not enough to bend the finger fully, so the gear ratio was changed to 1/50 and it worked well. In addition to change of the gear ratio, the thickness of skin was thinner and this made problem. Because the transmittance becomes high, the inner mechanical part can be shown even if there is make-up to the skin. The mixture ratio among silicon, pigments and an emulsion should be more researched. There are no barometers for a point of similarity between presented hand and human hand, so the evaluation of this hand is subjectively. The esthetic valuation basis should be needed as future works. There are some needs of improvement after experiments. Firstly, the exact measure of the resistance of skin is needed. It is not easy to know the resistance of the real shape, but trial-error ways are not efficient. Secondly, the cost of hand is too expensive. This hand is not just for researches, the cost is one of the most important factors. The small and exquisite parts caused high cost, so the more simple structure and parts should be designed. Thirdly, the 1 D.O.F thumb finger and no spread of fingers are not enough to gesture like human. Even if there are not enough spaces in hand, the spread fingers and abduction-adduction motion of thumb finger should be added to make natural gestures like human.

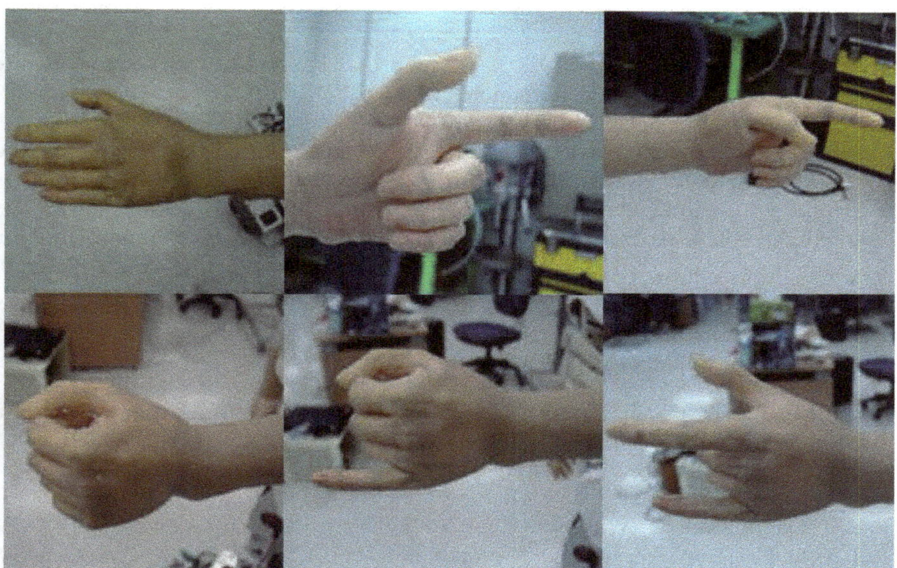

Fig. 12. The 6 postures with completed hand.

6. Conclusion

In this paper, 5 D.O.F hand for an android robot with an artificial skin was presented. The hand of an android robot required human like appearance because of an android robot is the nearest robot to human. The presented robotic hand has 5 D.O.F fingers and its shape and size are based on Korean young woman. There are three design policies which are size, shape and skin to make this hand and these are for the closest realization of human hand. The finger used D.C motor, screw-nut and liner potentiometer and linkage structure as a power transmission. The characteristic of mechanical design is a modular structure. Each finger module has its own power and sensor independently, so this design can bring easy maintenance by changing modules. The finger module was designed to suit the shape which is based on 3D data from human hand. The hand is just combination of each finger module and palm part. The palm part decides the shape of hand by arrangement of finger modules and it is also designed by 3D data based on human hand. The artificial skin was made of silicon complex which was selected as the nearest material to human skin. To make this silicon complex, mock-up and mold and mixture of materials are needed. After these processes, the closest android hand to real human hand can be produced. This hand is not for grasping but gesture, so experiments are for evaluation how it has similar appearance to human hand and can make variable gestures. In experiments, this hand can express variable postures include rock-paper-scissors and its appearance is similar to human hand. There are some improvements to this hand. The 5 D.O.F is not enough to realize variable gestures of human hand especially spread of fingers and adduction-abduction. In addition, the exact valuation standard of similarity in appearance should be researched. These should be the future works in this research.

7. References

Akachi K., Kaneko K., Kanehira N., Ota S., Miyamori G., Hirata M., Kajita S. & Kanehiro F. (2005). Development of humanoid robot HRP-3P, *In Proceedings of the 2005 IEEE/RAS International Conference on Humanoid Robots*, pp. 50-55.

Feifei Zhao, Shujiro Dohta, Tetsuya Akagi & Hisashi Matsushita. (2006). Development of a Bending Actuator using a Rubber Artificial Muscle and its Application to a Robot Hand, *SICE-ICASE, 2006. International Joint Conference*, pp.381-384.

Hirai K., Hirose M., Haikawa Y. & Takenaka T. (1998). The development of Honda humanoid robot, *In Proceedings of the International Conference on Robotics & Automation (ICRA 1998)*, pp. 1321-1326.

Hirose H., Haikawa Y., Takenaka T. & Hirai K. (2001). Development of Humanoid robot ASIMO, *In Proceedings of the 2001 IEEE/RSJ International Conference on Intelligent Robotics & Systems*

Ill-Woo Park, Jung-Yup Kim, Jungho Lee & Jun-Ho Oh. (2005). Mechanical design of humanoid robot platform KHR-3 (KAIST Humanoid Robot 3: HUBO), *In Proceedings of the 2005 IEEE/RAS International Conference on Humanoid Robots*, pp. 321-326.

Jun-Ho Oh, David Hanson, Won-Sup Kim, Young Han, Jung-Yup Kim & Ill-Woo Park. (2006). Design of Android type Humanoid Robot Albert HUBO, *In Proceedings of the 2006 IEEE/RAS International Conference on Intelligent Robots and Systems*, pp. 1428-1433.

Kaneko K., Harada K. & Kanehiro F. (2007). Development of Multi-fingered Hand for Life-size Humanoid Robots, *In Proceedings of the 2007 International Conference on Robotics & Automation (ICRA 2007)*, pp. 913-920.

Kaneko K., Kanehiro F., Morisawa M., Miura K., Nakaoka S. & Kajita S. (2009). Cybernetic human HRP-4C, *In Proceedings of the 2009 IEEE/RAS International Conference on Humanoids*, pp. 7-14.

Kawasaki H., Komatsu T., Uchiyama K. & Kurimoto T. (2002). Dexterous anthropomorphic robot hand with distributed tactile sensor: Gifu hand II, Mechatronics, IEEE/ASME Transactions on Vol. 7, Issue 3, pp. 296-303.

Kurita Y., Ono Y., Ikeda A. & Ogasawara T. (2009). NAIST Hand 2: Human-sized Anthropomorphic Robot Hand with Detachable Mechanism at the Wrist, *In Proceedings of the 2009 IEEE/RSJ International Conference on Intelligent Robots and Systems(IROS 2009)*, pp. 2271-2276.

Lotti F., Tiezzi P., Vassura G., Biagiotti L., Palli G. & Melchiorri C. (2005). Development of UB Hand 3: Early Results, *In Proceedings of the International Conference on Robotics & Automation (ICRA 2005)*, pp. 4488-4493.

N. Dechev, W. L. Cleghorn & S. Naumann. (2000) Thumb Design of an Experimental Prosthetic Hand, *In Proceedings of the International Symposium on Robotics and Automation(ISRA 2000)*

Nakaoka S., Kanehiro F., Miura K., Morisawa M., Fujiwara K., Kaneko K., Kajita S. & Hirukawa H. (2009). Creating facial motions of Cybernetic Human HRP-4C, *In Proceedings of the 2009 IEEE/RAS International Conference on Humanoids*, pp. 561-567.

Shimada M., Minato T., Itakura S. & Ishiguro H. (2007). Uncanny Valley of Androids and Its Lateral Inhibition Hypothesis, *The16th IEEE International Symposium on Robot and Human interactive Communication (ROMAN 2007)*, pp. 374-379

6

Development of Multi-Fingered Universal Robot Hand with Torque Limiter Mechanism

Wataru Fukui, Futoshi Kobayashi and Fumio Kojima
Kobe University
Japan

1. Introduction

Today, various industrial robots are developed and used all over the world. However, these industrial robots are specialized in particular operations. In fact, one industrial robot is not able to be designed for operating various tasks. One of the causes is that general-purpose and multifunctional robot hands substituted human manual-handling task are not brought to realization. If these robot hands like human hands are consummated, the applicable field of industrial robots is extended, and the utilization efficiency is improved very much.

A human hand has mechanical handling function such as grab, grip, pinch, push and pull. In addition, it can sense the feeling such as configuration, hard, flexible, smoothness and asperity. In other words, a human hand is a multifunctional and a universal end effector. Many research works on robot hand have been studied all over the world in order to imitate human hand and achieve the similar function to human hand. However, it is not attained that a robot hand system has the coordinative function to human hand's one yet. For resolving this problem, it is necessity that software system processes various sensors' information effectively. Additionally, it is also necessity that hardware system has drive mechanism, multi Degrees Of Freedom (DOFs) linkage mechanism and some sensors that allocated in limited spatial restrictions. Consequently, we produced the Universal Robot Hand I as shown in Fig.1. The robot hand system has tactile sensors, joint torque sensors, joint angle sensors and the similar structure to human hand's one. We have studied on the robot hand's mechanism, the sensory information processing and the kinematic control.

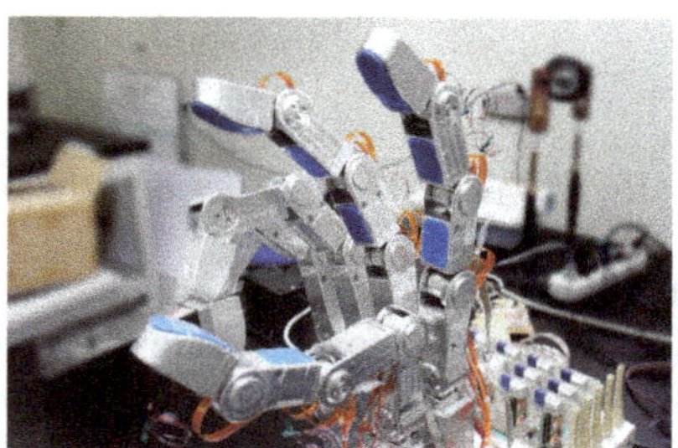

Fig. 1. Universal robot hand I

In this paper, a new robot hand is developed on resulting knowledge for advancing our study. Universal Robot Hand II has actuators, transmission gears, reduction gears, and Torque Limiter Mechanisms in the fingers. Using the Torque Limiter Mechanisms, the fingers can sustain overload not by the gears but by the structure. This is the imitative behavior of a human finger. This paper describes that new small robot hand mechanism has five fingers at the first. At the second, this Torque Limiter Mechanism is introduced. At the third, the effects of Torque Limiter Mechanism are verified in experiments. At the last, results of these experiments are summarized and concluded.

2. Specifications of developed robot hand

2.1 Basic design

This section describes the basic design of developed robot hand. Fig. 2 (left) shows "Universal Robot Hand II". The height between the lower limit of the palm and the upper limit of the middle finger is 290mm. The width of the robot hand opened up between the thumb and the little finger is 416mm. The size of this robot hand is a little larger than human hand. Thus, this size is enough to imitate human hand workings. The weight of the robot thumb is 0.262kg, the weight of the every other finger is 0.250kg, and the total weight of the robot hand without the pedestal is 1.323kg.

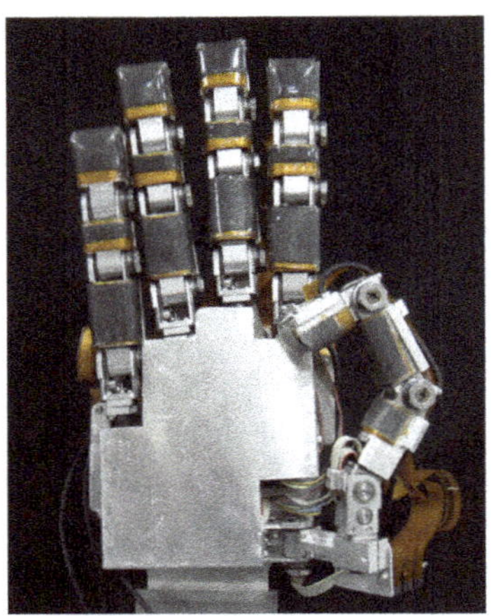

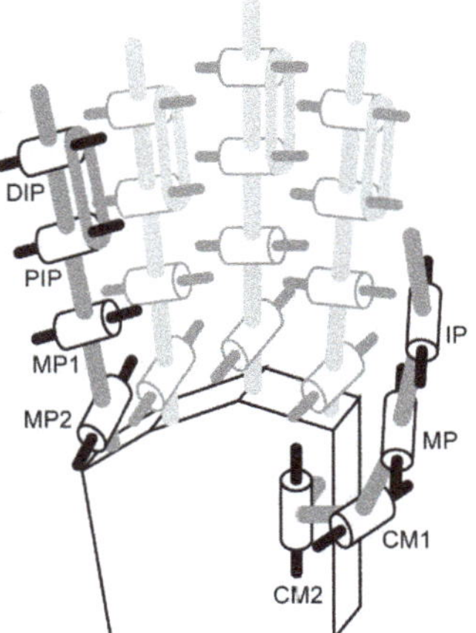

Fig. 2. Universal robot hand II and configuration of DOFs

This robot hand has 16 DOFs. Thumb has four DOFs (the IP, the MP, the CM1 and the CM2 joints), and the other fingers have three DOFs (the PIP, the MP1 and the MP2 joint). Every DIP joint is interlocked with the PIP joint. These DOFs and the movable directions of joints are shown in Fig. 2 (right).

This robot hand has the multi-axis force/torque sensors in every fingertip and tactile sensors on every finger pad. The multi-axis force/torque sensor is able to measure the force and torque at fingertip. This sensor in every fingertip is as shown in Fig.3 (upper right). Tactile sensor is able to measure the pressure distribution on the finger pad. This sensor on every finger pad is as shown in Fig. 3 (lower right).

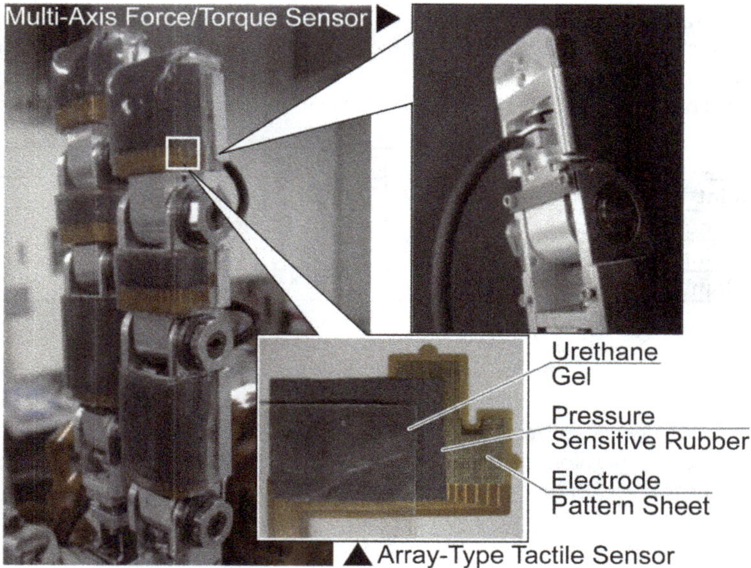

Fig. 3. Multi-axis force/torque sensor (BL AUTOTEC, LTD.) and array-type tactile sensor

The overview of the control system for this robot hand is shown in Fig. 4. This control computer gets the pulse from the encoders in every motor, the value from multi-axis force/torque sensors in every fingertip and the pressure distribution from the array-type tactile sensors on every finger pad. The fingers are controlled through driver circuits according to these data.

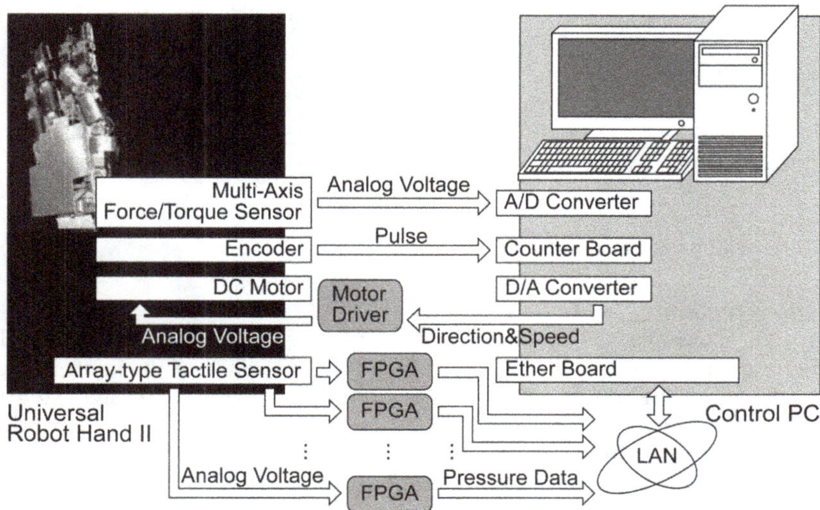

Fig. 4. Control system for universal robot hand

2.2 Basic performance

It is shown that the basic performance of the developed robot hand. The movable range of joints is as shown in Table 1. 0 [deg.] is extended position and the flexion direction is the plus direction. This movable range of robot hand is similar or over the human's one.

Thumb (deg.)		Others (deg.)	
IP Joint	0 -110	DIP Joint	0 - 95
MP Joint	0 -110	PIP Joint	0 - 95
CM1 Joint	0 -110	MP1 Joint	0 -110
CM2 Joint	0 -110	MP2 Joint	0 -110

Table 1. Movable range of each joint

The step responses of every finger are shown in Fig. 5. From this figure, DIP & PIP operates slower than the other joints. As the PIP joint is operated with the DIP joint, the load of DIP & PIP is about twice larger than the other joints' one. Thus, DIP & PIP operates with about half the angular velocity. Therefore this Universal Robot Hand II has enough response velocity for our future study.

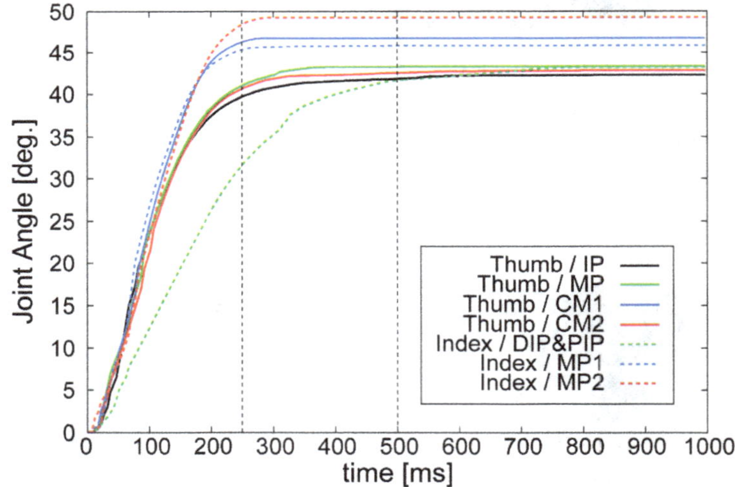

Fig. 5. Result of step response experiments

2.3 Superior function

Typically, the robot finger is classified into a hard finger and an elastic finger. In the hard finger, the rotation of the actuator responds plainly to the angle of the joint. In the elastic finger, the fingertip can be moved with elastic members depending on the external force. However, a human finger acts as both a hard finger and an elastic finger depending on a situation. Thus, Torque Limiter Mechanism is fitted into the joint of this Universal Robot Hand II. With this mechanism, driving mechanism in joints is started to skid from setup skidding torque. By implementation with this mechanism, the driving mechanism can be protected against overload, and the robot hand may grasp objects flexibly.

3. Torque limiter mechanism

3.1 Mechanism

Torque Limiter Mechanism is constructed by a fixed plate, a rotating plate and rollers held between these plates as shown in Fig. 6. These rollers are tilted on an angle of α degrees. The skidding torque T is expressed in (1).

$$T = \mu r P \sin\alpha \tag{1}$$

where, μ is the coefficient of the friction between rollers, plates. r is the radius of rollers, and P is the pressure by the adjustment nut. Every 20 joints of this robot hand have this mechanism as shown in Fig. 7.

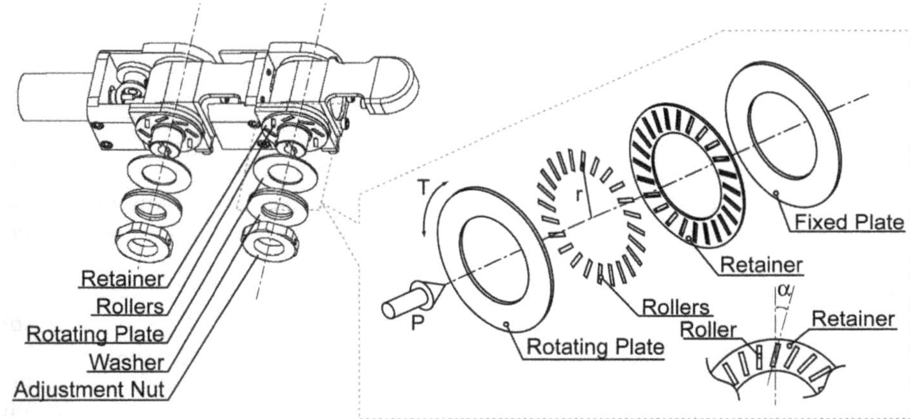

Fig. 6. Inner structure of finger joint with torque limiter mechanism

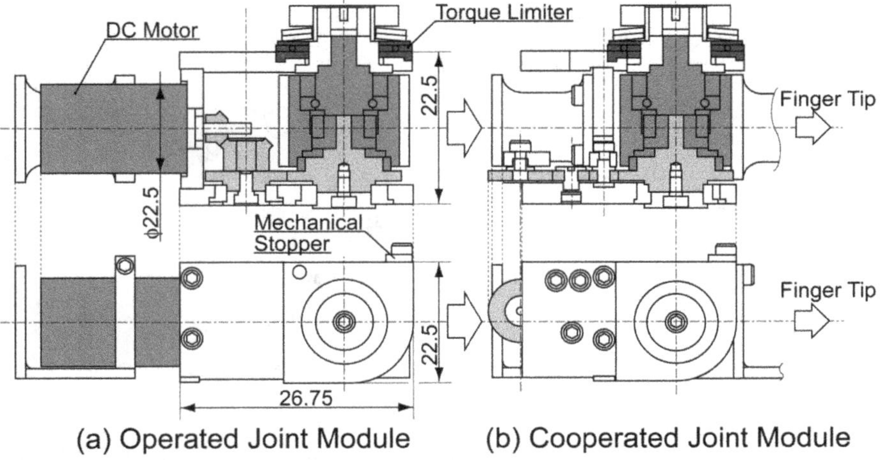

Fig. 7. Cross-section view and side view of torque limiter mechanism

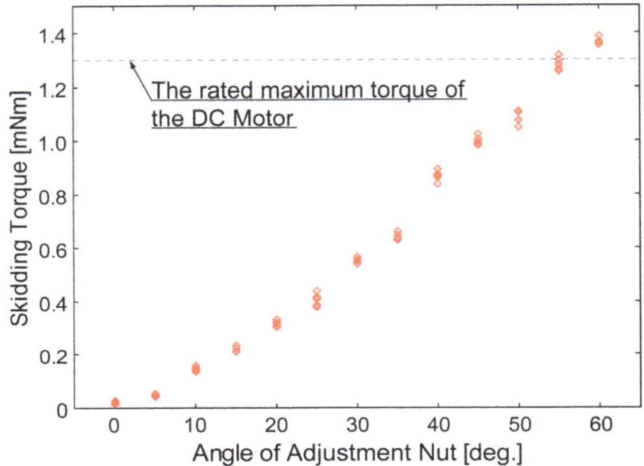

Fig. 8. Angle of adjustment nut vs skidding torque

Fig. 8 shows the relation between the clenched angle of an adjustment nut and the skidding torque. From Fig. 8, the skidding torque is adjustable from maximum to minimum of motor torque. In other word, this finger is able to be adjusted as a hard finger or a passive finger.

3.2 Advantages in finger behavior

Torque Limiter Mechanism has some advantages on the operation of the Universal Robot Hand II. The behavior of the finger with the mechanism is shown in Fig. 9 and Fig. 10. The PIP joint and the DIP joint are normally located as shown in Fig. 9 (a). The DIP joint is flexed in conjunction with PIP joint as same degrees. It is thought that excess overload is operated at the distal phalanx. As shown in Fig. 9 (b). If torque by the external force exceeds setup skidding torque, modules of drive gearing are turned over toward the direction of fending off the force to the mechanical stopper as one structure. This mechanism doesn't only make the actuators to be protected against the overload, but also support the external force mechanistically without output power of actuators by the position of particularity.

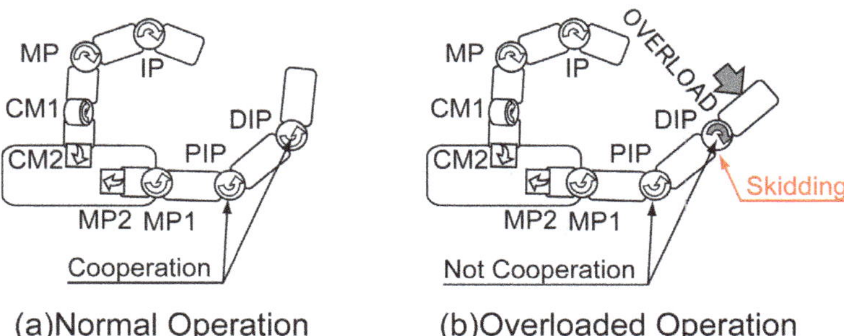

Fig. 9. Overloaded operation with torque limiter mechanism

Flexible grasp with Torque Limiter Mechanism is as shown by Fig. 10. Typically, a robot finger takes the form in Fig. 10 (a) in case of grasping a thin object. This is because that the commonly-used robot hand has engaged DIP and PIP joints in imitation of human joints. On the other hand, in case of grasping a thin object with human fingers, these fingertips are collimated, and increases area of contact between these finger pads. Thus, developed robot hand operates skidding mechanism in the DIP joint. Robot fingertips are collimated, and increases area of contact in Fig. 10 (b). Herewith, developed robot hand doesn't pinch a thin object with a point contact but with plane contact.

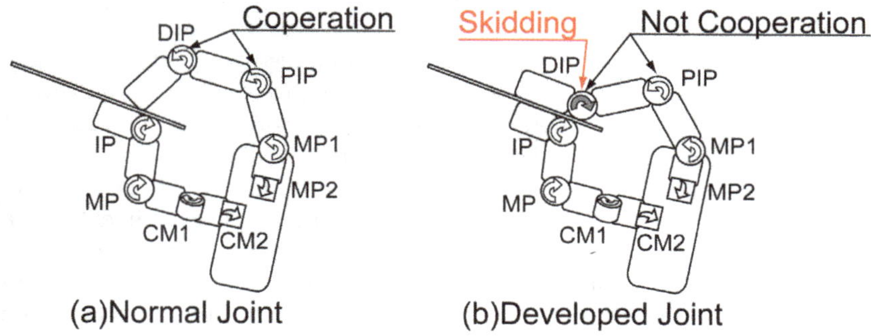

Fig. 10. Clip operation with thin object

4. Experiments

4.1 Alleviation of impact force

4.1.1 Experimental setup

In this experiment, the protection of drive train is verified. The outline of this experiment is shown in Fig. 11. The DIP and the MP2 are fixed with a bump against the mechanical stopper. The PIP is flexed at a tilt, 45 [deg.], and the skidding mechanism is tried and enabled to operate with tuning the adjustment nut. The adjustment nut of MP1 is clenched up to the disabled angle. Drive the MP1 and contact hardly the fingertip to a rigid object. Keep sliding the PIP to a bump against the stopper of the PIP joint and fastening the fingertip on the object for a few seconds. Extend the MP1 to a bump against stopper of the MP1 joint. Values of the fingertip force and the encoder (MP1) are measured during this experiment.

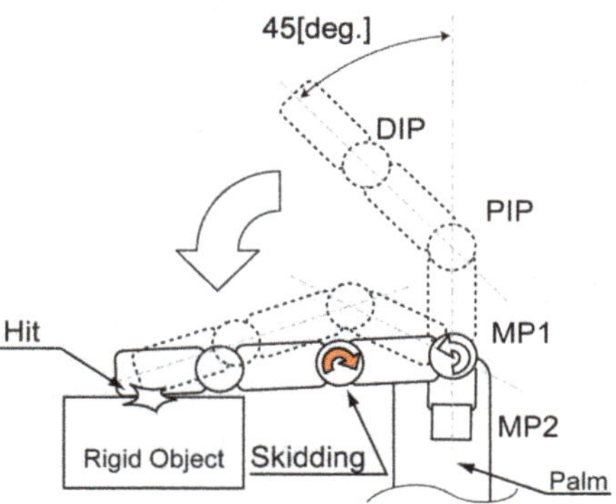

Fig. 11. Impact force experiment

In addition, the adjustment nut of PIP is clenched up and the similar experiment without Torque Limiter Mechanism is conducted for comparison.

4.1.2 Experimental results

The result of this experiment is shown in Fig. 12. The blue line represents the fingertip force in the case of "Torque Limiter Mechanism is active", and the black line represents the inactive case. The red line is the value of encoder in the MP1. At 130[step], the fingertip force increases drastically. There is strong evidence that the fingertip touched on a rigid object. At 600[step], the fingertip force decreases precipitously. The fingertip pulled away from the object at this time. As shown by this graph, the case with active skidding mechanism has the lower impact force than the case with inactive one. After that, the fingertip force is kept low during the joint is skidding. The fingertip force in active case converges to the force in inactive case in accordance with the joint is skidded to a bump against the stopper. The fingertip force during this period is lower than converged value. By the result of multiple experiments, the average peak of the force is 0.33 [kgf] in inactive case and 0.20 [kgf] in active case. The peak in active case is drop by about 40% from in inactive case. As identified above, this Torque Limiter Mechanism protects the finger against the accidental overload. Meanwhile, a transition at 700 [step] is impact force by a bump against the stopper at the MP1 joint.

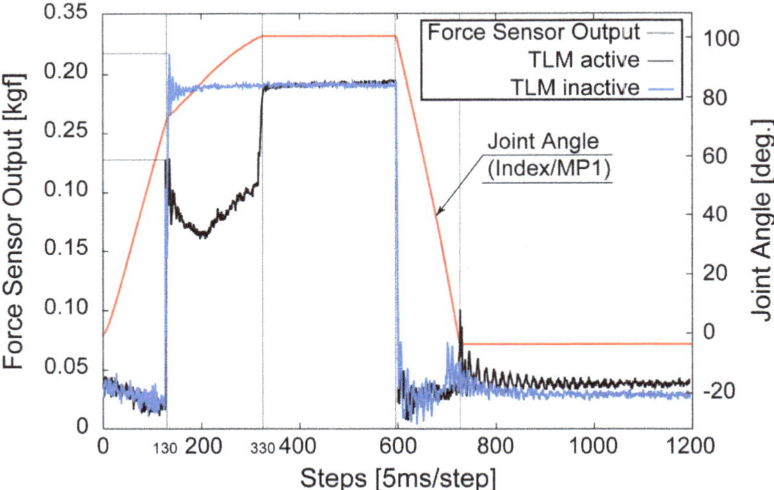

Fig. 12. Transition of encoder contacted against rigid object

4.2 Mismatch between joint angle and counted pulse

This skidding mechanism protects the finger against the accidental overload by the experiment in Section 4.1. However, this skidding mechanism has one problem. In the case of the joint is driving and skidding, this problem must be considerable. In this case, the joint angle recognized by the encoder is different from the real joint angle. The encoder is set in every motor, and the joint angle is recognized indirectly by the number of rotations. The motor drives the joint through the skidding mechanism, and the recognized angle has a gap with the real angle in the skidding case. Thus, in this section, compensating method for this gap is validated the evidence.

4.2.1 Experimental setup

In this section, this gap is compensated in the following equation.

$$\theta + = \begin{cases} 0 & \begin{pmatrix} f > F_{threshold} \\ \& t > T_{threshold} \end{pmatrix} \\ \theta_i - \theta_{i-1} & otherwise \end{cases} \tag{2}$$

where, f is the fingertip force and $F_{threshold}$ is the threshold of the fingertip force. t is the motor torque and $T_{threshold}$ is the threshold of the torque. θ is the angle of the joint and i is the control step. The fingertip force f is over the constant value $F_{threshold}$, in other words, the fingertip contacts an object. In addition, the motor torque t is over the constant value $T_{threshold}$, in other words, the torque is able to operate the skidding mechanism. In this instance, as operating the skidding mechanism, the angle is not counted up (down). In other instance, the angle is counted up (down).

As shown in Fig. 13, the finger hits a rigid object three times from the position of 0 [deg.] using the above method. Meanwhile, the rigid object is set at 91 [deg.], and the finger is extended back to 12.5 [deg.] by measuring the experimental movie. The finger is controlled by the time-control method.

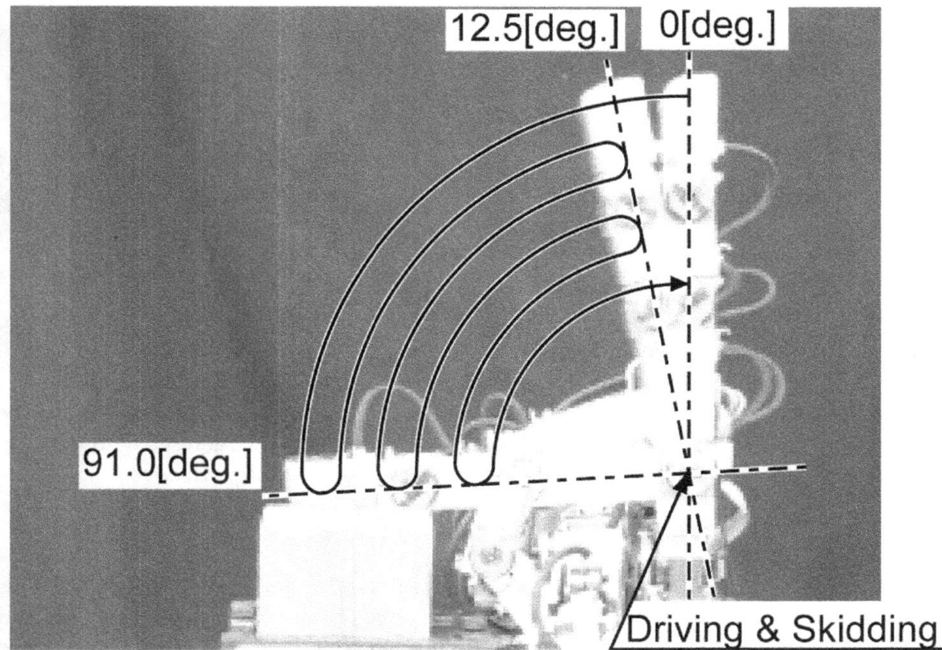

Fig. 13. Outline of experiment about mismatch between joint angle and counted pulse

4.2.2 Experimental results

The result of this experiment is shown by Fig. 14. The red line is the fingertip force. The black line and the blue line is the value of the encoder in MP1. The black line is the compensated data, and the blue line is the raw data. As shown in Fig. 14, the impact force is measured. This has a reason that the DIP and PIP joints of the finger extended to the stopper and the skidding mechanism of the DIP and PIP joints are inactive. Thus, the impact force is not alleviated.

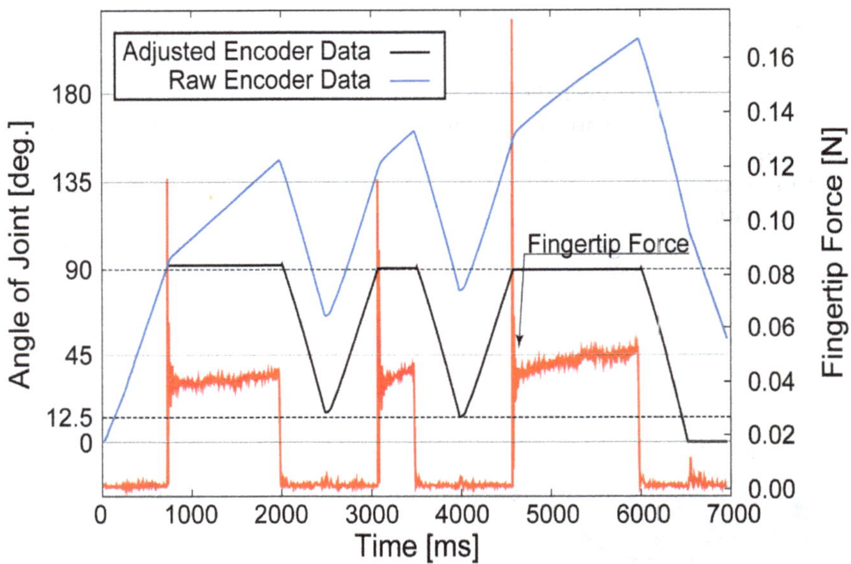

Fig. 14. Transition of fingertip force contacted against rigid object

The angular difference is 1.5 [deg.] between the first tap and the second tap in the compensated data. The angular difference is 0.5[deg.] between the second and the third. After multi-cycle experiments, the angular difference is 2 [deg.] at a maximum by one tap. This compensating method has cumulative difference but is practical. Depending on the desired accuracy of control system, the values of encoders should be reset with a bump against the stopper. The results show that this compensating method is effective.

5. Conclusions

In this paper, it is declared that the multi-fingered universal robot hand is developed. This robot hand is named "Universal Robot Hand." This robot hand has 5 fingers, 20 joints and 16 DOFs. This robot hand is a little bigger than a human hand. Every DOF is driven by the DC motor in the finger. Every joint has Torque Limiter Mechanism. This mechanism is the clutch brake system. The drive mechanism in the joint can be protected against overload by using the skidding mechanism. At the skidding time, the joint angle recognized by the encoder is different from the real joint angle. However the difference can be corrected by the software method.

6. Acknowledgment

This work was supported by the robot study group in the Advanced Materials Processing Institute Kinki Japan. The following researchers participate this study group: Hiroyuki Nakamoto (Hyogo Prefectural Institute of Technology), Tadashi Maeda (Maeda Precision Manufacturing Limited Kobe), Nobuaki Imamura (Hiroshima International University), Kazuhiro Sasabe (The Kansai Electric Power Co., Inc.), Hidenori Shirasawa (The Advanced Materials Processing Institute Kinki Japan).

7. References

Fukui, W.; Nakamoto, H.; Kobayashi, F.; Kojima, F.; Imamura, N.; Maeda, T.; Sasabe, K.; & Shirasawa, H. (2008). Development of Multi-Fingered Universal Robot Hand, *Proceedings of the 35th Annual Conference of the IEEE Industrial Electronics Society*, pp.2225-2230, ISBN 978-1424446490, Porto, Portugal, November 3-5, 2009

Imamura, N.; Nakamura, Y.; Yamaoka, S.; Shirasawa, H. & Nakamoto, H. (2007). Development of an Articulated Mechanical Hand with Enveloping Grasp Capability, *Journal of Robotics and Mechatronics*, Vol.19, No.3, pp.308-314, ISSN 1883-8049

Jacobsen, S. C.; Iversen, E. K.; Knutti, D. F.; Johnson, R. T.; & Biggers, K. B. (1986). Design of the Utah/MIT Dexterous Hand, *Proceedings of the 1986 IEEE International Conference on Robotics and Automation*, pp.1520-1532, San Francisco, California, USA, April 7-11, 1986

Lovchic, C. S. & Diftler, M. A. (1999). The Robonaut Hand : A Dexterous Robot Hand for Space, *Proceedings of the 1999 IEEE International Conference on Robotics and Automation*, pp. 907-912, ISBN 978-0780351806, Detroit, Michigan, USA, May 10-15, 1999

Morita, T.; Iwata, H. & Sugano. S. (2000). Human Symbiotic Robot Design based on Division and Unification of Functional Requirements, *Proceedings of the 2000 IEEE International Conference on Robotics and Automation*, pp.2229-2234, ISBN 978-0780358867, San Francisco, California, USA, April 24-28, 2000

Mouri, T.; Kawasaki, H.; Nishimoto, Y.; Aoki, T.; Ishigure, Y. & Tanahashi. M. (2008). Robot Hand Imitating Disabled Person for Education/Training of Rehabilitation, *Journal of Robotics and Mechatronics*, Vol.20 No.2, pp.280-288, ISSN 1883-8049

Nakamoto, H.; Kobayashi, F.; Imamura, N. & Shirasawa, H. (2006). Universal Robot Hand Equipped with Tactile and Joint Torque Sensors -Development and Experiments on Stiffness Control and Object Recognition-, *Proceedings of The 10th World Multi-Conference on Systemics, Cybernetics and Informatics*, pp.347-352, ISBN 978-1934272015, Orelando, Floroda, USA, July 16-19, 2006

Yamano, I.; Takemura, K. & Maeno. T. (2003). Development of a Robot Finger for Five-fingered Hand using Ultrasonic Motors, *Proceedings of 2003 IEEE/RSJ International Conference on Intelligent Robots and Systems*, pp. 2648-2653, ISBN 978-0780378612, Las Vegas, Nevada, USA, Octover 27-31, 2003.

Part 3

Interactive Applications of Humanoid Robots

7

Affective Human-Humanoid Interaction Through Cognitive Architecture

Ignazio Infantino
Istituto di Calcolo e Reti ad Alte Prestazioni, Consiglio Nazionale delle Ricerche
Italy

1. Introduction

Development of humanoid robots has to address two vital aspects, namely physical appearance and gestures, that will allow the machines to closely resemble humans. Other aspects such as "social" and "emotional" will enable human-machine interaction to be as natural as possible. The field of robotics has long been investigating how effective interaction between humans and autonomous and intelligent mechanical system can be possible (Goodrich & Schultz., 2007). Several distinctive features have been determined depending on whether a robot that acts as an assistant (for example, in the course of a business) or as a companion is required. In the case of humanoid robots, the human appearance and behavior may be very closely linked and integrated if you adopt a cognitive architecture that can take advantage of the natural mechanisms for exchange of information with a human. The robot that cooperates in the execution of an activity would benefit from the execution of its tasks if it had a mechanism that is capable of recognizing and understanding human activity and intention (Kelley et al., 2010), with perhaps the possibility of developing imitation learning by observation mechanisms.

On the other hand, if we consider the robot as a partner, then it plays an important role in sharing the emotional aspects: it is not essential to equip the robot with emotions, but it is important that it can "detect" human emotional states (Malatesta et al. 2009).

The cognitive architectures allow software to deal with problems that require contributions from both the cognitive sciences and robotics, in order to achieve social behavior typical of the human being, which would otherwise be difficult to integrate into traditional systems of artificial intelligence. Several cognitive models of the human mind can find common ground and experimental validation using humanoid agents. For example, if we approach the study of actions and social interactions involving "embodied" agents, the concept of motor resonance investigated in humans may play an important role (Chaminade & Cheng, 2009) to achieve sophisticated, yet simple to implement, imitative behaviors, learning by demonstration, and understanding of the real scene.

In recent years, there is often talk of mirror neurons, which are evidence of the physiological motor resonance at the cellular level with regard to action, action understanding and imitation. But the resonance is applicable in other contexts such as cognitive emotions, the sensations of physical pain, and in various components of the actions of agents interacting socially (Barakova & Lourens, 2009; Fogassi, 2011).

Cognitive models proposed would make the humanoid robot capable of overcoming the so-called "Uncanny Valley of eeriness" (Saygin et al., 2011), by allowing the humanoid is perceived by human beings not as artificial machine but as a credible social interacting entity. In this sense, the recent experimental data have confirmed the importance of "natural" movements (Saygin et al., 2011) that are expected from the observation of a robot with human features, even if it is a process yet to fully understand, that continually generate predictions about the environment, and compares them with internal states and models. Mirror neurons are assumed to be the neural basis for understanding the goals and intentions (Fogassi 2011), allowing for the prediction of the actions of the individual who is observed, and its intentions. Various studies indicate that they are also involved in the system of empathy, and emotional contagion (Hatfield et al. 1994), explaining that the human tendency to automatically mimic and synchronize facial expressions, vocalizations, postures, and movements with those of another person.

The classical approach in robotics based on the perception-reasoning-action loop has evolved towards models that unify perception and action, such as the various cognitive theories arising from the Theory of Event Coding (Hommel et al., 2001). Similarly, the objectives are integrated with the intentions, and emotions with reasoning and planning.

An approach that considers the human-robot interaction based on affective computing and cognitive architectures, can address the analysis and reproduction of social processes (and not only) that normally occur between humans, so that a social structure can be created which includes the active presence of a humanoid. Just as a person or group influences the emotions and the behavior of another person or group (Barsade, 2002), the humanoid could play a similar role in owning their own emotional states and behavioral attitudes, and by understanding the affective states to humans to be in "resonance" with them.

The purpose of this chapter is to consider the two aspects, intentions and emotions, simultaneously: discussing and proposing solutions based on cognitive architectures (such as in Infantino et al., 2008) and comparing them against recent literature including areas such as conversational agents (Cerezo et al., 2008).

The structure of the chapter is as follows: firstly, an introduction on the objectives and purposes of the cognitive architectures in robotics will be presented; then a second part on the state of the art methods of detection and recognition of human actions, highlighting those more suitably integrated into an architecture cognitive; a third part on detecting and understanding emotions, and a general overview of effective computing issues; and finally the last part presents an example of architecture that extends on the one presented in (Infantino et al., 2008), and a discussion about possible future developments.

2. Cognitive architectures

To achieve the advanced objective of human-robot interaction, many researchers have developed cognitive systems that consider sensory, motor and learning aspects in a unified manner. Dynamic and adaptive knowledge also needs to be incorporated, basing it on the internal representations that are able to take into account variables contexts, complex actions, goals that may change over time, and capabilities that can extend or enrich themselves through observation and learning. The cognitive architectures represent the infrastructure of an intelligent system that manages, through appropriate knowledge representation, perception, and in general the processes of recognition and categorization, reasoning, planning and decision-making (Langley et al., 2009). In order for the cognitive architecture to be capable of generating behaviors similar to humans, it is important to consider the role of emotions. In this way, reasoning and planning may be influenced by emotional processes and representations as happens in humans. Ideally, this could be thought of as a representation of emotional states that, in addition to influencing behavior, also helps to manage the detection and recognition of human emotions. Similarly, human intentions may somehow be linked to the expectations and beliefs of the intelligence system. In a wider perspective, the mental capabilities (Vernon et al. 2007) of artificial computational agents can be introduced directly into a cognitive architecture or emerge from the interaction of its components. The approaches presented in the literature are numerous, and range from cognitive testing of theoretical models of the human mind, to robotic architectures based on perceptual-motor components and purely reactive behaviors (see Comparative Table of Cognitive Architectures[1]).

Currently, cognitive architectures have had little impact on real-world applications, and a limited influence in robotics, and the humanoid. The aim and long-term goal is the detailed definition of the Artificial General Intelligence (AGI) (Goertzel, 2007), i.e. the construction of artificial systems that have a skill level equal to that of humans in generic scenarios, or greater than that of the human in certain fields. To understand the potential of existing cognitive architectures and indicate their limits, you must first begin to classify the various proposals presented in the literature. For this purpose, it is useful a taxonomy of cognitive architectures (Vernon et al. 2007; Chong et al., 2007) that identifies three main classes, for example obtained by characteristics such as memory and learning (Duch et al., 2008) . In this classification are distinguished symbolic architectures, emerging architectures, and hybrid architectures. In the following, only some architectures are discussed and briefly described, indicating some significant issues that may affect humanoids, and affective-based interactions. At present, there are no cognitive architectures that are strongly oriented to the implementation of embodied social agents, nor even were coded mechanisms to emulate the so-called social intelligence. The representation of the other, the determination of the self, including intentions, desires, emotional states, and social interactions, have not yet had the necessary consideration and have not been investigated approaches that consider them in a unified manner.

The symbolic architectures (referring to a cognitivist approach) are based on an analytical approach of high-level symbols or declarative knowledge. SOAR (State, Operator And Result) is a classic example of an expert rule-based cognitive architecture (Laird et al., 1987). The classic version of SOAR is based on a single long-term memory (storing production-rules), and a single short-term memory (with a symbolic graph structure). In an extended version of the architecture (Laird 2008), in addition to changes on short and long-term memories, was added a module that implements a specific appraisal theory. The intensity of individual appraisals (express either as categorical or numeric values) becomes the intrinsic rewards for reinforcement learning, which significantly speeds learning. (Marinier et al., 2009) presents a unified computational model that combines an abstract cognitive theory of behavior control (PEACTIDM) and a detailed theory of emotion (based on an appraisal theory), integrated in the SOAR cognitive architecture. Existing models that integrate emotion and cognition generally do not fully specify why cognition needs emotion and conversely why emotion needs cognition. Looking ahead, we aim to explore how emotion can be used productively with long-term memory, decision making module, and interactions.

The interaction is a very important aspect that makes possible a direct exchange of information, and may be relevant both for learning to perform intelligent actions. For

[1]Biologically Inspired Cognitive Architectures Society -Toward a Comparative Repository of Cognitive Architectures, Models, Tasks and Data. http://bicasociety.org/cogarch/

example, under the notion of embodied cognition (Anderson, 2004), an agent acquires its intelligence through interaction with the environment. Among the cognitive architectures, EPIC (Executive Process Control Interactive) focuses his attention on human-machine interaction, aiming to capture the activities of perception, cognition and motion. Through interconnected processors working in parallel are defined patterns of interaction for practical purposes (Kieras & Meyer, 1997).

Finally, among the symbolic architecture, physical agents are relevant in ICARUS (Langley & Choy, 2006), integrated in a cognitive model that manages knowledge that specify the reactive abilities, reactions depending on goals and classes of problems. The architecture consists of several modules that bind in the direction of bottom-up concepts and percepts, and in a top-down manner the goals and abilities. The conceptual memory contains the definition of generic classes of objects and their relationships, and the skill memory stores how to do things.

The emergent architectures are based on networks of processing units that exploit mechanisms of self-organizations and associations. The idea behind this architecture is based on connectionism approach, which provides elementary processing units (processing element PE) arranged in a network that changes its internal state as a result of an interaction. From these interactions, relevant properties emerge, and arise from the memory considered globally or locally organized. Biologically inspired cognitive architectures distribute processing by copying the working of the human brain, and identify functional and anatomical areas correspond to human ones such as the posterior cortex (PC), the frontal cortex (FC), hippocampus (HC). Among these types of architecture, one that is widely used is based on adaptive resonance theory ART (Grossberg, 1987). The ART unifies a number of network designs supporting a myriad of interaction based learning paradigms, and address problems such as pattern recognition and prediction. ART-CogEM models use cognitive-emotional resonances to focus attention on valued goals.

Among the emerging architectures, are also considered models of dynamic systems (Beer 2000, van Gelder & Port, 1996) and models of enactive systems. The first might be more suitable for the development of high-level cognitive functions as intentionality and learning. These dynamic models are derived from the concept that considers the nervous system, body and environment as dynamic models, closely interacting and therefore to be examined simultaneously. This concept also inspired models of enactive systems, but emphasize the principle of self-production and self-development. An example is the architecture of the robot iCub (Sandini et al., 2007), that also includes principles Global Workspace Cognitive Architecture (Shanahan, 2006) and Dynamic Neural Field Architecture (Erlhagen and Bicho, 2006). The underlying assumption is that cognitive processes are entwined with the physical structure of the body and its interaction with the environment, and the cognitive learning is an anticipative skill construction rather than knowledge acquisition.

Hybrid architectures are approaches that combine methods of the previous two classes. The best known of these architectures is ACT-R (Adaptive Components of Rational-thought), which is based on perceptual-motor modules, memory modules, buffers, and pattern matchers. ACT-R (Anderson et al., 2004) process two kinds of representations: declarative and procedural: declarative knowledge is represented in form of chunks, i.e. vector representations of individual properties, each of them accessible from a labeled slot; procedural knowledge is represented in form of productions. Other popular hybrid architectures are: CLARION- The Connectionism Learning Adaptive rule Induction ON-Line (Sun, 2006), LIDA-The Learning Intelligent Distribution Agent (Franklin & Patterson, 2006).

More interesting for the purposes of this chapter is the PSI model (Bartl & Dorner, 1998; Bach et al., 2006) and its architecture that involves explicitly the concepts of emotion and motivation in cognitive processes. MicroPsi (Bach et al., 2006) is an integrative architecture based on PSI model, has been tested on some practical control applications, and also on simulating artificial agents in a simple virtual world. Similar to LIDA, MicroPsi currently focuses on the lower level aspects of cognitive process, not yet directly handling advanced capabilities like language and abstraction. A variant of MicroPsi framework is included also in CogPrime (Goertzel, B. 2008). This is a multi-representational system, based on a hyper-graphs with uncertain logical relationships and associative relations operating together. Procedures are stored as functional programs; episodes are stored in part as "movies" in a simulation engine.

3. Recognition of human activities and intentions

In the wider context of capturing and understanding human behavior (Pantic et al., 2006), it is important to perceive (detect) signals such as facial expressions, body posture, and movements while being able to identify objects and interactions with other components of the environment. The techniques of computer vision and machine learning methodologies enable the gathering and processing of such data in an increasingly accurate and robust way (Kelley et al., 2010). If the system captures the temporal extent of these signals, then it can make predictions and create expectations of their evolution. In this sense, we speak of detecting human intentions, and in a simplified manner, they are related to elementary actions of a human agent (Kelley et al., 2008).

Over the last few years has changed the approach pursued in the field of HCI, shifting the focus on human-centered design for HCI, namely the creation of systems of interaction made for humans and based on models of human behavior (Pantic et al., 2006). The Human-centered design, however, requires thorough analysis and correct processing of all that flows into man-machine communication: the linguistic message, non-linguistic signals of conversation, emotions, attitudes, modes by which information are transmitted, i.e. facial expressions, head movements, non-linguistic vocalizations, movements of hands and body posture, and finally must recognize the context in which information is transmitted. In general, the modeling of human behavior is a challenging task and is based on the various behavioral signals: affective and attitudinal states (e.g. fear, joy, inattention, stress); manipulative behavior (actions used to act on objects environment or self-manipulative actions like biting lips), culture-specific symbols (conventional signs as a wink or a thumbs-up); illustrators actions accompanying the speech, regulators and conversational mediators as who nods the head and smiles.

Systems for the automatic analysis of human behavior should treat all human interaction channels (audio, visual, and tactile), and should analyze both verbal and non verbal signals (words, body gestures, facial expressions and voice, and also physiological reactions). In fact, the human behavioral signals are closely related to affective states, which are conducted by both physiological and using expressions. Due to physiological mechanisms, emotional arousal affects somatic properties such as the size of the pupil, heart rate, sweating, body temperature, respiration rate. These parameters can be easily detected and are objective measures, but often require that the person wearing specific sensors. Such devices in future may be low-cost and miniaturized, distributed in clothing and environment, but which are now unusable on a large scale and in non structured situations.

The visual channel that takes into account facial expressions and gestures of the body seems to be relatively more important to human judgment that recognizes and classifies behavioral states. The human judgment on the observed behavior seems to be more accurate if you consider the face and body as elements of analysis.

A given set of behavioral signals usually does not transmit only one type of message, but can transmit different depending on the context. The context can be completely defined if you find the answers to the following questions: Who, Where, What, How, When and Why (Pantic et al., 2006). These responses disambiguating the situation in which there are both artificial agent that observes and the human being observed.

In the case of human-robot interaction, one of the most important aspects to be explored in the detection of human behavior is the recognition of the intent (Kelley et al., 2008): the problem is to predict the intentions of a person by direct observation of his actions and behaviors. In practice we try to infer the result of a goal-directed mental activity that is not observable, and characterizing precisely the intent. Humans recognize, or otherwise seek to predict the intentions of others, using the result of an innate mechanism to represent, interpret and predict the actions of the other. This mechanism probably is based on taking the perspective of others (Gopnick & Moore, 1994), allowing you to watch and think with eyes and mind of the other.

The interpretation of intentions can anticipate the evolution of the action, and thus capture its temporal dynamic evolution. An approach widely used in statistical classification of systems that evolve over time, is what uses Hidden Markov Model (Duda et al., 2000). The use of HMM in the recognition of intent (emphasizing the prediction) has been suggested in (Tavakkoli et al., 2007), that draws a link between the HMM approach and the theory of the mind.

The recognition of the intent intersects with the recognition of human activity and human behavior. It differs from the recognition of the activity as a predictive component: determining the intentions of an agent, we can actually give an opinion on what we believe are the most likely actions that the agent will perform in the immediate future. The intent can also be clarified or better defined if we recognize the behavior. Again the context is important and how it may serve to disambiguate (Kelley et al., 2008). There are a pairs of actions that may appear identical in every aspect but have different explanations depending on their underlying intentions and the context in which they occur.

Both to understand the behaviors and the intentions, some of the tools necessary to address these problems are developed for the analysis of video sequences and images (Turaga et al., 2008). The aspects of security, monitoring, indexing of archives, led the development of algorithms oriented to the recognition of human activities that can form the basis for the recognition of intentions and behaviors. Starting from the bottom level of processing, the first step is to identify the movements in the scene, to distinguish the background from the rest, to limit the objects of interest, and to monitor changes in time and space. We use then, techniques based on optical flow, segmentation, blob detection, and application of space-time filters on certain features extracted from the scene.

When viewing a scene, the man is able to distinguish the background from the rest, that is, instant by instant, automatically rejects unnecessary information. In this context, a model of attention is necessary to select the relevant parts of the scene correctly. One problem may be, however, that in these regions labeled as background is contained the information that allows for example the recognition of context that allows the disambiguation. Moreover, considering a temporal evolution, what is considered as background in a given instant, may be at the center of attention in successive time instants.

Identified objects in the scene, as well as being associated with a certain spatial location (either 2D, 2D and 1/2, or 3D) and an area or volume of interest, have relations between them and with the background. So the analysis of the temporal evolution of the scene, should be accompanied with a recognition of relationships (spatial, and semantic) between the various entities involved (the robot itself, humans, actions, objects of interest, components of the background) for the correct interpretation of the context of action. But defining the context in this way, how can we bind the contexts and intentions? There are two possible approaches: the intentions are aware of the contexts, or vice versa the intentions are aware of the contexts (Kelley et al., 2008). In the first case, we ranked every intention carries with it all possible contexts in which it applies, and real-time scenario is not applicable. The second approach, given a context, we should define all the intentions that it may have held (or in a deterministic or probabilistic way). The same kind of reasoning can be done with the behaviors and habits, so think of binding (in the sense of action or sequence of actions to be carried out prototype) with the behaviors.

A model of intention should be composed of two parts (Kelley et al, 2008): a model of activity, which is given for example by a particular HMM, and an associated label. This is the minimum amount of information required to enable a robot to perform disambiguation of context. One could better define the intent, noting a particular sequence of hidden states from the model of activity, and specifying an action to be taken in response. A context model, at a minimum, shall consist of a name or other identifier to distinguish it from other possible contexts in the system, as well as a method to discriminate between intentions. This method may take the form of a set of deterministic rules, or may be a discrete probability distribution defined on the intentions which the context is aware.

There are many sources of contextual information that may be useful to infer the intentions, and perhaps one of the most attractive is to consider the so-called affordances of the object, indicating the actions you can perform on it. It is possible then builds a representation from probabilities of all actions that can be performed on that object. For example, you can use an approach based on natural language (Kelley et al., 2008), building a graph whose vertices are words and a label is the weighed connecting arc indicating the existence of some kind of grammatical relationship. The label indicates the nature of the relationship, and the weight can be proportional to the frequency with which the pair of words exists in that particular relationship. From such a graph, we can calculate the probability to determine the necessary context to interpret an activity. Natural language is a very effective vehicle for expressing the facts of the world, including the affordances of the objects.

If the scene is complex, performance and accuracy can be very poor when you consider all the entities involved. then, can be introduced for example the abstraction of the interaction space, where each agent or object in the scene is represented as a point in a space with a defined distance on it related to the degree of interaction (Kelley et al, 2008). In this case, then consider the physical artificial agent (in our case the humanoid) and its relationship with the space around it, giving more importance to neighboring entities to it and ignore those far away.

4. Detection of human emotions

Detection of human emotions plays many important roles in facilitating healthy and normal human behavior, such as in planning and deciding what further actions to take, both in interpersonal and social interactions. Currently in the field of human-machine interfaces, systems and devices are now being designed that can recognize, process, or even generate emotions (Cerezo et al., 2008). The "affect recognition" often requires a multidisciplinary and multimodal approach (Zeng et al., 2009), but an important channel that is rich with information is facial expressiveness (Malatesta et al., 2009). In this context, the problem of expression detection is supported by robust artificial vision techniques. Recognition has proven critical in several aspects: such as in defining basic emotions and expressions, the subjective and cultural variability, and so on.

Consideration must be given the more general context of affect, for which research in psychology has identified three possible models: categorical, dimensional and appraisal-based approach (Grandjean et al., 2008). The first approach is based on the definition of a reduced set of basic emotions, innate and universally recognized. This model is widely used in automatic recognition of emotions, but as well as for human actions and intentions, can be considered more complex models that address a continuous range of affective and emotional states (Gunes et al., 2011). Dimensional models are described by geometric spaces that can use the basic emotions, but represented by a continuous dynamic dimensions such as arousal, valence, expectation, intensity. The appraisal-based approach requires that the emotions are generated by a continuous and recursive evaluation and comparison of an internal state and the state of the outside world (in terms of concerns, needs, and so on). Of course this model is the most complex to achieve the recognition, but is used for the synthesis of virtual agents (Cerezo et al., 2008).

As mentioned previously, most research efforts on the recognition and classification of human emotions (Pantic et al., 2006) focused on a small set of prototype expressions of basic emotions related to analyzing images or video, and analyzing speech. Results reported in the literature indicate that typically performances reach an accuracy from 64% to 98%, but detecting a limited number of basic emotions and involving small groups of human subjects. It is appropriate to identify the limitations of this simplification. For example, if we consider the dimensional models, it becomes important to distinguish the behavior of the various channels of communication of emotion: the visual channel is used to interpret the valence, and arousal seems to be better defined by analyzing audio signals. By introducing a multi-sensory evaluation of emotion, you may have problems of consistency and masking, i.e. that the various communication channels indicate different emotions (Gunes et al., 2011).

Often the emotion recognition systems have aimed to the classification of emotional expressions deduced from static and deliberate, while a challenge is on the recognition of spontaneous emotional expressions (Bartlett, et al. 2005; Bartlett, et. Al. 2006 ; Valstar et al., 2006), i.e. those found in normal social interactions in a continuous manner (surely dependent on context and past history), capable of giving more accurate information about affective state of human involved in a real communication (Zeng et al., 2009).

While the automatic detection of the six basic emotions (including happiness, sadness, fear, anger, disgust and surprise) can be done with reasonably high accuracy, as they are based on universal characteristics which transcend languages and cultures (Ekman, 1994), spontaneous expressions are extremely variable and are produced - by mechanisms not yet fully known - by the person who manifests a behavior (emotional, but also social and cultural), and underlies intentions (conscious or not).

If you look at human communication, some information is related to affective speech, and in particular to the content. Some affective mechanisms of transmission are clear and directly related to linguistics, and other implicit (paralinguistic) signals that may affect especially the way in which words are spoken. You can then use some of the dictionaries that can link the word to the affective content and provide the lexical affinity (Whissell, 1989). In addition, you can analyze the semantic context of the speech to determine more emotional content, or endorse those already detected. The affective messages transmitted through paralinguistic signals, are primarily affected by prosody (Juslin & Scherer, 2005), which may be indicative

of complex states such as anxiety, boredom, and so on. Finally, are also relevant non-linguistic vocalizations such as laughing, crying, sighing, and yawning (Russell et al., 2003).
Considering instead the channel visual, emotions arise from the following aspects: facial expressions, movements (actions) facial movements and postures of the body (which may be less susceptible to masking and inconsistency).
Most of the work on the analysis and recognition of emotions is based on the detection of facial expressions, addressing two main approaches (Cohn, 2006; Pantic & Bartlett, 2007): the recognition based on elementary units of facial muscles action (AU), that are part of the coding system of facial expression called the Facial Action coding - FACS (Ekman & Friesen 1977), and recognition based on spatial and temporal characteristics of the face.
FACS is a system used for measuring all visually distinguishable facial movements in terms of atomic actions called Facial Action Unit (AU). The AU is independent of the interpretation, and can be used for any high-level decision-making process, including the recognition of basic emotions (Emotional FACS - EMFACS[2]), the recognition of various emotional states (FACSAID - Facial Action Coding System Affect Interpretation Dictionary[2]), and the recognition of complex psychological states such as pain, depression, etc.. The fact of having a coding, has originated a growing number of studies on spontaneous behavior of the human face based on AU (e.g., Valstar et al., 2006).
The facial expression can also be detected using various pattern recognition approaches based on spatial and temporal characteristics of the face. The features extracted from the face can be geometric shapes such as parts of the face (eyes, mouth, etc.), or location of salient points (the corners of the eyes, mouth, etc.), or facial characteristics based on global appearance and some particular structures, such as wrinkles, bulges, and furrows. Typical examples of geometric feature-based methods are those that face models described as set of reference points (Chang et al., 2006), or characteristic points of the face around the mouth, eyes, eyebrows, nose and chin (Pantic & Patras, 2006), or grids that cover the whole region of the face (Kotsia & Pitas, 2007). The combination of approaches based on geometric features and appearance is likely (eg Tian et al., 2005) the best solution for the design of systems for automatic recognition of facial expressions (Pantic & Patras, 2006). The approaches based on 2D images of course suffer from the problem of the point of view, which can be overcome by considering 3D models of the human face (eg, Hao & Huang, 2008; Soyel & Demirel, 2008, Tsalakanidou & Malassiotis, 2010).

5. Integration of a humanoid vision agent in PSI cognitive architecture

SeARCH-In (Sensing-Acting-Reasoning: Computer understands Human Intentions) is an intentional vision framework scheme oriented towards human-humanoid interactions (see figure 1). It extends on the system presented in the previous work (Infantino et al., 2008), improving vision agent and expressiveness of the ontology. Such a system will be able to recognize user faces, to recognize and track human postures by visual perception. The described framework is organized on two modules mapped on the corresponding outputs to obtain intentional perception of faces and intentional perception of human body movements. Moreover a possible integration of an intentional vision agent in the PSI (Bart & Dorner, 1998; Bach et al., 2006) cognitive architecture is proposed, and knowledge management and reasoning is allowed by a suitable OWL-DL ontology.

[2]http://face-and-emotion.com/dataface/general/homepage.jsp

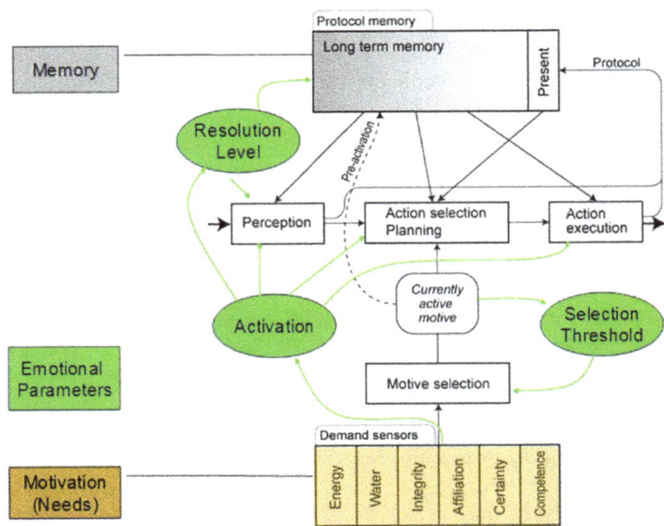

Fig. 1. Cognitive-emotional-motivational schema of the PSI cognitive architecture[3].

In particular, the ontological knowledge approach is employed for human behavior and expression comprehension, while stored user habits are used for building a semantically meaningful structure for perceiving human wills. A semantic description of user wills is formulated in terms of the symbolic features produced by the intentional vision system. The sequences of symbolic features belonging to a domain specific ontology are employed to infer human wills and to perform suitable actions.

Considering the architecture of PSI (see Figure 1) and the intentional vision agent created by the SEARCH-In framework, you can make some considerations on the perception of the intentions of a human being, the recognition of his identity, the mechanism that triggers of sociality, how memory is used, the symbolic representation of actions and habits, and finally the relationship of the robot's inner emotions and those observed.

The perception that regards the agent is generated from the observation of a human being who acts in an unstructured environment: human face, body, actions, and appearance are the object of humanoid in order to interact with him. The interaction is intended to be based on emotional and affective aspects, on the prediction of intents recalled from the memory and observed previously. Furthermore, the perception concerns, in a secondary way for the moment, the voice and the objects involved in the observed action.

The face and body are the elements analyzed to infer the affective state of the human, and for the recognition of identity. The face is identified in the scene observed by the cameras of the robot using the algorithm of Viola-Jones (Viola & Jones, 2004), and its OpenCV[4] implementation. The implementation of this algorithm is widely used in commercial devices since it is robust, efficient, and allows real-time use. The human body is detected by the Microsoft Kinect device, which is at the moment is external to humanoid, but the data are accessible via the network. From humanoid point of view, the Kinect[5] device is in effect one

[3]Figure available at the link www.macs.hw.ac.uk/~ruth/psi-refs.html (author: Ruth Aylett)
[4]http://opencv.willowgarage.com/wiki/
[5]http://en.wikipedia.org/wiki/Kinect

of its sensor, and the software architecture integrated it as the other sensors. Again, you are using a device that is widely used, and that ensures accurate perceptive results in real time. This sensor produces both a color image of the scene, and a depth map, and the two representations are aligned (registered), allowing you to associate each pixel with the depth estimated by IR laser emitter-detector pair. Through software libraries, it is possible to reconstruct the human posture, through a reconstructed skeleton defined as a set of points in three dimensional space corresponding to the major joints of the human body.

In the region of the image containing the detected face, are run simultaneously two sets of algorithms. The first group is used to capture the facial expression, identifying the position and shape of the main characteristic features of the face: eyes, eyebrows, mouth, and so on. The recognition of the expression is done using the 3D reconstruction (Hao Tang & Huang, 2008), and identifying the differences from a prototype of a neutral expression (see figure 3). The second group, allows the recognition, looking for a match in a database of faces, and using the implementation (API NaoQi, ALFaceDetection module) already available to the NAO humanoid robot (see figure 2).

The PSI model requires that the internal emotional states modulate the perception of the robot, and are conceived as intrinsic aspects of the cognitive model. The emotions of the robots are seen as an emergent property of the process of modulation of the perceptions, behavior, and global cognitive process. In particular, emotions are encoded as configuration settings of cognitive modulators, which influence the pleasure/distress dimension, and on the assessment of the cognitive urges.

The idea of social interaction based on affect recognition and intentions, that is the main theme of this chapter, simply leads to a first practical application of cognitive theory PSI. The detection and recognition of a face meets the need for social interaction that drives the humanoid robot, consistent with the reference theory which deals with social urges or drives, or affiliation. The designed agent includes discrete levels of pleasure/distress: the greatest pleasure is associated with the fact that the robot has recognized an individual, and has in memory the patterns of habitual action (through representations of measured movement parameters, normalized in time and in space, and associated with a label); the lowest level when it detects a not identified face, showing a negative affective state, and a lack of recognition and classification of the observed action. It is possible to implement a simple mechanism of emotional contagion, which executes the recognition of human affective state (limited to an identified human), and tends to set the humanoid on the same mood (positive, neutral, negative). The Nao may indicate his emotional state through the coloring of some leds placed in eyes and ears, and communicates its mood changes by default vocal messages to make the human aware of its status (red is associated with a state of stress, green with neutral state, yellow with euphoria, blue with calm).

The symbolic explicit representation provided by the PSI model requires that the objects, situations, plans are described by a formalism of executable semantic networks, i.e. semantic networks that can change their behaviors via messages, procedures, or changes to the graph. In previous work (Infantino et al., 2008), it has been defined a reference ontology (see figure 3) for the intentional vision agent which together with the semantic network allows for two levels of knowledge representation, increasing the communicative and expressive capabilities.

The working memory, in our example of emotional interaction, simply looks for and identifies human faces, and contains actions for random walk and head movements to allow it to explore space in its vicinity until it finds a human agent to interact with. There is not a world model to compare with the one perceived, even if the reconstructed 3D scene by depth sensor could be used, and compare it with a similar internal model in order to plane exploration through anticipation in the cognitive architecture. The long-term memory is represented by the collection of usual actions (habits), associated with a certain identity and emotional state, and in relation to certain objects. Again, you might think to introduce simple mechanisms affordances of objects, or introduce a motivational relevance related to the recognition of actions and intentions.

Fig. 2. NAO robot is the humanoid employed to test the agent that integrates SeARCH-In framework and PSI cognitive model.

Fig. 3. Example of face and features extraction, 3D reconstruction for expression recognition (on the left), and 3D skeleton of human (on the right).

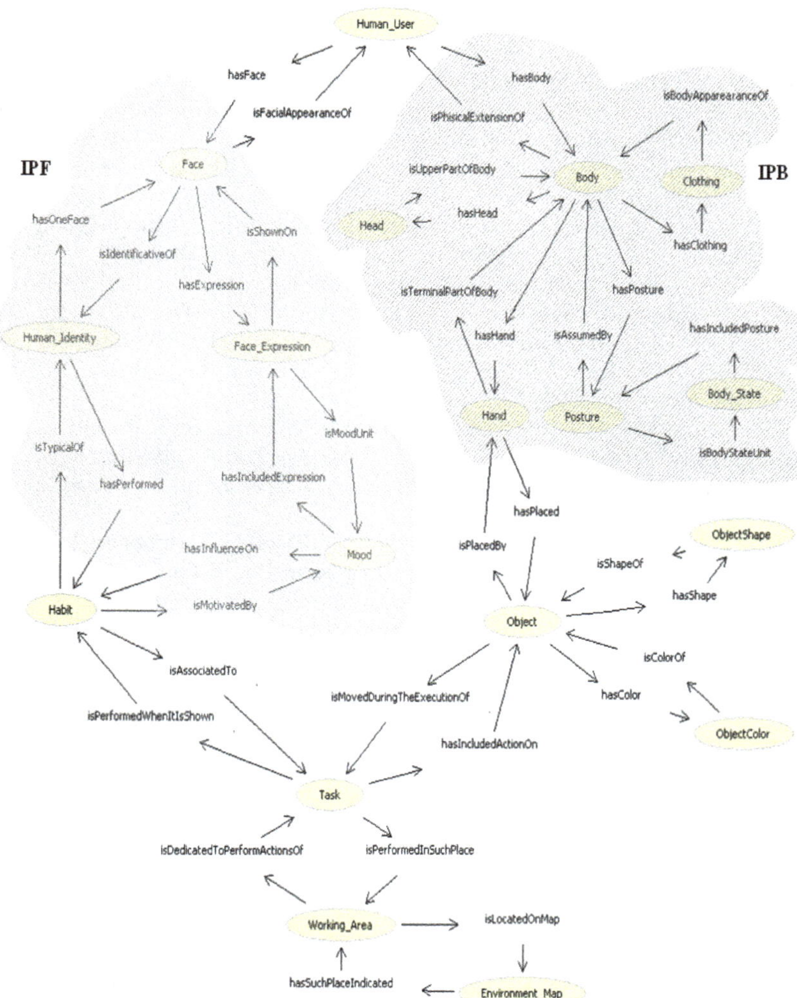

Fig. 4. SearchIn Ontology (see Infantino et al., 2008). Gray areas indicate Intentional Perception of Faces module (IPF) and Intentional Perception of Body module (IPB).

6. References

Anderson, J. R.; Bothell, D.; Byrne, M. D.; Douglass, S.; Lebiere, C. & Qin, Y . (2004). An integrated theory of the mind. *Psychological Review,* vol. 111, n.4, pp. 1036-1060.

Bach, J. (2003). The MicroPsi Agent Architecture. Proceedings of ICCM5 International Conference on Cognitive Modeling Bamberg Germany (Vol. 1, pp. 15-20). Universitäts-Verlag. Retrieved from citeseer.ist.psu.edu/bach03micropsi.html

Bach, J.; Dörner, D., & Vuine, R. (2006). Psi and MicroPsi. A Novel Approach to Modeling Emotion and Cognition in a Cognitive Architecture. Tutorial at ICCM 2006 available at http://www.macs.hw.ac.uk/EcircusWeb/webContent/.

Bachorowski, J., & Fernandez-Dols, J.. (2003). Facial and Vocal Expressions of Emotion. Ann. Rev. Psychology, vol. 54, pp. 329-349.

Barakova, E. I. & Lourens, T. (2009). Mirror neuron framework yields representations for robot interaction, *Neurocomputing*, vol. 72, n. 4-6, Brain Inspired Cognitive Systems (BICS 2006) / Interplay Between Natural and Artificial Computation (IWINAC 2007), January 2009, pp. 895-900. doi: 10.1016/j.neucom.2008.04.057.

Barsade S.G., (2002). The Ripple Effect: Emotional Contagion and Its Influence on Group Behavior, Administrative Science Quarterly, vol. 47, n. 4. (Dec., 2002), pp. 644-675.

Bartl, C., & Dörner, D. (1998). PSI: A theory of the integration of cognition, emotion and motivation. F. E. Ritter, & R. M. Young (Eds.), Proceedings of the 2nd European Conference on Cognitive Modelling, pp. 66-73.

Bartlett, M.S. ; Littlewort, G. ; Frank, M.; Lainscsek, C.; Fasel, I., & Movellan, J.. (2005). Recognizing Facial Expression: Machine Learning and Application to Spontaneous Behavior, Proc. IEEE Int'l Conf. Computer Vision and Pattern Recognition (CVPR '05), pp. 568-573.

Bartlett, M.S. ; Littlewort, G. ; Frank, M.; Lainscsek, C.; Fasel, I., & Movellan, J.. (2006). Fully Automatic Facial Action Recognition in Spontaneous Behavior. In Proc. IEEE Int'l Conf. Automatic Face and Gesture Recognition (AFGR '06), pp. 223-230.

Beer, R. D. (2000) Dynamical approaches to cognitive science, *Trends in Cognitive Sciences*, vol. 4, no. 3, 1 March 2000, pp. 91-99. doi: 10.1016/S1364-6613(99)01440-0.

Birlo, M & Tapus, A.. (2011) The crucial role of robot self-awareness in HRI, in Proc. of the 6th international conference on Human-robot interaction HRI'11, ACM. doi>10.1145/1957656.1957688

Cerezo, E.; Baldassarri, S; Hupont, I. & Seron, F. J. (2008). Affective Embodied Conversational Agents for Natural Interaction, *Affective Computing*, Jimmy Or (Ed.), ISBN: 978-3-902613-23-3, I-Tech Education and Publishing, Available from: http://www.intechopen.com/articles/show/title/affective_embodied_conversational_agents_for_natural_interaction.

Chaminade, T. & Cheng, G. (2009). Social cognitive neuroscience and humanoid robotics, *Journal of Physiology–Paris*, vol.103, pp. 286-295. doi:10.1016/j.jphysparis.2009.08.011

Chaminade, T.; Zecca, M.; Blakemore, S.-J.; Takanishi, A.; Frith, C.D. & al. (2010) Brain Response to a Humanoid Robot in Areas Implicated in the Perception of Human Emotional Gestures. *PLoS ONE* 5(7): e11577. doi:10.1371/journal.pone.0011577.

Chang, Y.; Hu, C.; Feris, R., & Turk, M.. (2006). Manifold Based Analysisof Facial Expression. *Journal of Image and Vision Computing*, vol. 24, no. 6, pp. 605-614.Chella, A.; Dindo, H. & Infantino I. (2008). Cognitive approach to goal-level imitation, *Interaction Studies: Social Behaviour and Communication in Biological and Artificial Systems*, vol. 9, n. 2, pp. 301-318.

Chong, H. ; Tan, A. & Ng, G.W.. (2007) Integrated cognitive architectures: a survey. Artificial Intelligence Review, pp. 103-130

Cohn, J.F.. (2006). Foundations of Human Computing: Facial Expression and Emotion. In Proc. Eighth ACM Int'l Conf. Multimodal Interfaces (ICMI '06), pp. 233-238.

Duch, W. ; Oentaryo, R. J. & Pasquier, M.. (2008). Cognitive Architectures: Where do we go from here?. In Proceeding of the 2008 conference on Artificial General Intelligence 2008: Proceedings of the First AGI Conference, Pei Wang, Ben Goertzel, and Stan Franklin (Eds.). IOS Press, Amsterdam, The Netherlands, The Netherlands, 122-136.

Duda, R.; Hart, P. & Stork, D. (2000). Pattern Classification, Wiley-Interscience

Ekman, P. (1994). Strong Evidence for Universals in Facial Expressions: A Reply to Russell's Mistaken Critique," Psychological Bull., vol. 115, no. 2, pp. 268-287.

Ekman, P. & Friesen, W.. (1977). Facial Action Coding System. Consulting Psychologists Press, 1977.

Ekman, P.; Friesen, W.V., & Hager, J.C.. (2002). Facial Action Coding System, A Human Face, 2002.

Epstein, S. L.. (1994) For the Right Reasons: The FORR Architecture for Learning in a Skill Domain. *Cognitive Science*, vol. 18, no. 3, pp. 479-511.

Erlhagen, W. & Bicho, E.. (2006) The dynamic neural field approach to cognitive robotics. *Journal of Neural Engineering*, vol. 3, pp. 36–54.

Fogassi, L. (2011). The mirror neuron system: How cognitive functions emerge from motor organization, *Journal of Economic Behavior & Organization*, vol. 77, n. 1, Special Issue: Emotions, Natural Selection and Rationality, January 2011, pp. 66-75, doi: 10.1016/ j.jebo.2010.04.009.

Franklin, S. & F. G. Patterson, Jr. (2006). The Lida Architecture: Adding New Modes of Learning to an Intelligent, Autonomous, Software Agent. Integrated Design and Process Technology, IDPT-2006, San Diego, CA, Society for Design and Process Science.

Gazzola, V. ; Rizzolatti , G.; B. Wicker, B. & C. Keysers, C. (2007). The anthropomorphic brain: The mirror neuron system responds to human and robotic actions, NeuroImage, vol. 35, n 4, 1 May 2007, Pages 1674-1684, ISSN 1053-8119, doi: 10.1016/j.neuroimage.2007.02.003.

Goertzel, B. & Pennachin, C.. (2007). *Artificial General Intelligence*. Springer-Verlag, Berlin, Heidelberg.

Goertzel, B. (2008). OpenCog Prime: Design for a Thinking Machine. Online wikibook, at http://opencog.org/wiki/OpenCogPrime.

Goodrich, M. A. & Schultz, A (2007). *Human-Robot Interaction : A Survey,* Foundations and Trands in Human-Computer interaction, vol.1, No. 3, 2007, pp. 203-275.

Gopnick, A. & Moore, A. (1994). "Changing your views: How understanding visual perception can lead to a new theory of mind," in Children's Early Understanding of Mind, eds. C. Lewis and P. Mitchell, 157-181. Lawrence Erlbaum

Grandjean, D.; Sander, D., & Scherer, K. R.. (2008). Conscious emotional experience emerges as a function of multilevel, appraisal-driven response synchronization. Consciousness & Cognition, vol. 17, no. 2, pp. 484–495, 2008, Social Cognition, Emotion, & Self-Consciousness.

Grossberg, S. (1987). Competitive learning: From interactive activation to adaptive resonance. *Cognitive Science,* vol. 11, pp. 23-63.

Gunes, H. & Piccardi, M. (2009). Automatic Temporal Segment Detection and Affect Recognition from Face and Body Display. IEEE Transactions on Systems, Man, and Cybernetics Part B, Special Issue on Human Computing 39(1), 64–84.

Gunes, H. & Piccardi,M. (2006). A Bimodal Face and Body Gesture Database for Automatic Analysis of Human Nonverbal Affective Behavior. In: Proc. of the 18th Intl. Conference on Pattern Recognition (ICPR 2006), vol. 1, pp. 1148–1153 (2006).

Gunes, H.; Schuller, B.; Pantic, M., & Cowie, R.. (2011). Emotion representation, analysis and synthesis in continuous space: A survey," in Proc. of IEEE Int. Conf. on Face and Gesture Recognition - 2011.

Hao, T., & Huang, T.S. (2008). 3D facial expression recognition based on automatically selected features. Computer Vision and Pattern Recognition Workshops, 2008. CVPRW '08. IEEE Comp. Society Conf. pp.1-8. doi: 10.1109/CVPRW.2008.4563052

Hatfield, E; Cacioppo, J. T. & Rapson, R. L. (1994). Emotional contagion. Current Directions, *Psychological Science,* vol. 2, pp. 96-99, Cambridge University Press. doi: 10.1111/j.1467-8721.1993.tb00114.x

Hommel, B.; Müsseler, J; Aschersleben, G. & Prinz W. (2001). The Theory of Event Coding (TEC): A framework for perception and action planning. *Behavioral and Brain Sciences,* vol. 24, pp. 849-878. doi:10.1017/S0140525X01000103.

Infantino, I.; Lodato, C.; Lopes, S. & Vella, F. (2008). Human-humanoid interaction by an intentional system, In proc. of 8th IEEE-RAS International Conference on Humanoids 2008, pp.573-578, 1-3 Dec. 2008. doi: 10.1109/ICHR.2008.4756007.

Jerritta, S.; Murugappan, M.; Nagarajan, R., & Wan, K.. (2011). Physiological signals based human emotion Recognition: a review. Signal Processing and its Applications (CSPA), IEEE 7th Intl. Colloquium on, pp. 410-415, 4-6 March 2011. doi: 10.1109/CSPA.2011.5759912

Juslin, P.N., & Scherer, K.R.. (2005). Vocal Expression of Affect. The New Handbook of Methods in Nonverbal Behavior Research, J. Harrigan, R. Rosenthal, and K. Scherer, eds., Oxford Univ. Press.

Kelley, R.; Tavakkoli, A.; King, C.; Nicolescu, M.; Nicolescu, M. & Bebis, G.. (2008). Understanding human intentions via hidden markov models in autonomous mobile robots. In Proceedings of the 3rd ACM/IEEE international conference on Human robot interaction (HRI '08). ACM, New York, NY, USA, 367-374. DOI=10.1145/1349822.1349870

Kelley, R.; Tavakkoli, A.; King, C.; Nicolescu, M. & Nicolescu, M. (2010). Understanding Activities and Intentions for Human-Robot Interaction, *Human-Robot Interaction,* Daisuke Chugo (Ed.), ISBN: 978-953-307-051-3, InTech, Available from: http://www.intechopen.com/articles/show/title/understanding-activities-and-intentions-for-human-robot-interaction.

Kieras, D. & Meyer, D.E. (1997). An overview of the EPIC architecture for cognition and performance with application to human-computer interaction. *Human-Computer Interaction,* vol. 12, pp. 391-438*MacDorman, K.F. & Ishiguro, H. (2006). The uncanny advantage of using androids in cognitive and social science research. *Interaction Studies,* vol. 7, n. 3, pp. 297-337. doi: 10.1075/is.7.3.03mac.

Kipp, M., & Martin, J.-C. (2009). Gesture and Emotion: Can basic gestural form features discriminate emotions? In proc. of the Intl. Conf. on Affective Computing and Intelligent Interaction 2009, pp.1-8.

Knox, J. (2010). Response to 'Emotions in action through the looking glass', *Journal of Analytical Psychology,* vol. 55, n. 1, pp. 30-34. doi:10.1111/j.1468-5922.2009.01822.x.

Kotsia I., & Pitas, I.. (2007). Facial Expression Recognition in Image Sequences Using Geometric Deformation Features and Support Vector Machines. *IEEE Trans. Image Processing*, vol. 16, no. 1, pp. 172-187.

Laird, J. E. (2008). Extending the Soar Cognitive Architecture. In proc. of the 2008 conference on Artificial General Intelligence 2008: Proceedings of the First AGI Conference, Pei Wang, Ben Goertzel, and Stan Franklin (Eds.). IOS Press, Amsterdam, The Netherlands, The Netherlands, pp. 224-235.

Laird, J. E.; Newell, A. & Rosenbloom, P.S., (1987). "Soar: An architecture for general intelligence," Artificial Intelligence, vol. 33, pp. 1-64.

Langley, P., & Choi, D. (2006). A unified cognitive architecture for physical agents. In proc. of the 21st national conference on Artificial intelligence (AAAI'06), vol. 2, pp. 1469-1474.

Langley, P.; Laird, J. E.; & Rogers, S. (2009). Cognitive architectures: Research issues and challenges. *Cognitive Systems Research*, 10, 141-160.

Malatesta, L.; Murray, J.; Raouzaiou, A.; Hiolle, A; Cañamero, L. & and Karpouzis, K. (2009). Emotion Modelling and Facial Affect Recognition in Human-Computer and Human-Robot Interaction, *State of the Art in Face Recognition*, Julio Ponce and Adem Karahoca (Ed.), ISBN: 978-3-902613-42-4, I-Tech Education and Publishing, Available from: http://www.intechopen.com/articles/show/title/emotion_modelling_and_facial_affect_recognition_in_human-computer_and_human-robot_interaction .

Marinier R. P. III; Laird, J.E & Lewis, R. L.. (2009). A computational unification of cognitive behavior and emotion, *Cognitive Systems Research*, vol. 10, n. 1, Modeling the Cognitive Antecedents and Consequences of Emotion, March 2009, pp. 48-69, doi: 10.1016/j.cogsys.2008.03.004.

Pantic, M. & Bartlett, M.S.. (2007). Machine Analysis of Facial Expressions. Face Recognition, K. Delac and M. Grgic, eds., pp. 377-416, I-Tech Education and Publishing Russell, J.A..

Pantic, M., & Patras, I.. (2006). Dynamics of Facial Expression: Recognition of Facial Actions and Their Temporal Segments Form Face Profile Image Sequences. *IEEE Trans. Systems, Man, and Cybernetics Part B*, vol. 36, no. 2, pp. 433-449.

Pantic, M.; Pentland, A.; Nijholt, A. & Huang, T.S. (2006). Human Computing and Machine Understanding of Human Behavior: A Survey, in proc. Of Eighth ACM Int'l Conf. Multimodal Interfaces (ICMI '06), pp. 239-248.

Russell, J. A.; Bachorowski, J., & Fernández-Dols, J. M. (2003). Facial and Vocal Expressions of Emotion, Annual Review of Psychology, vol. 54, pp. 329-349, doi:10.1146/annurev.psych.54.101601.145102

Salvucci, D. D., & Taatgen, N. A. (2011). Toward a unified view of cognitive control. Topics in Cognitive Science, 3, 227–230.

Sandini, G.; Metta, G. & Vernon, D. (2007). The iCub Cognitive Humanoid Robot: An Open-System Research Platform for Enactive Cognition, in *50 Years of AI*, M. Lungarella et al. (Eds.), Festschrift, LNAI 4850, pp. 359–370, 2007, Springer-Verlag, Heidelberg

Saygin, A.P.; Chaminade, T.; Ishiguro, H.; Driver, J. & Frith C. (2011). The thing that should not be: predictive coding and the uncanny valley in perceiving human and humanoid robot actions. *Soc Cogn Affect Neurosci*, first published online April 22, 2011. doi:10.1093/scan/nsr025

Shanahan, M.. (2006) A cognitive architecture that combines internal simulation with a global workspace, *Consciousness and Cognition*, vol. 15, no. 2, June 2006, pp. 433-449. doi: 10.1016/j.concog.2005.11.005.

Soyel, H., & Demirel, H.. (2008). 3D facial expression recognition with geometrically localized facial features, Computer and Information Sciences, 2008. ISCIS '08. 23rd Intl. Symposium on , pp.1-4. doi: 10.1109/ISCIS.2008.4717898

Sun, R. (2006). The CLARION cognitive architecture: Extending cognitive modeling to social simulation. In: Ron Sun (ed.), *Cognition and Multi-Agent Interaction*. Cambridge University Press, New York.

Tavakkoli, A.; Kelley, R.; King, C.; Nicolescu, M.; Nicolescu, M. & Bebis, G. (2007). "A Vision-Based Architecture for Intent Recognition," Proc. of the International Symposium on Visual Computing, pp. 173-182

Tax, D., Duin, R. (2004). "Support Vector Data Description." Machine Learning 54. pp. 45-66.

Tian, Y.L.; Kanade, T., & Cohn, J. F.. (2005). Facial Expression Analysis, *Handbook of Face Recognition*, S.Z. Li and A.K. Jain, eds., pp. 247-276, Springer.

Tistarelli, M. & Grosso, E. (2010). Human Face Analysis: From Identity to Emotion and Intention Recognition, Ethics and Policy of Biometrics, Lecture Notes in Computer Science, vol. 6005, pp. 76-88. doi: 10.1007/978-3-642-12595-9_11.

Trafton, J. G. ; Cassimatis, N. L. ; Bugajska, M. D. ; Brock, D. P.; Mintz, F. E. & Schultz, A. C.. (2005) Enabling effective human-robot interaction using perspective-taking in robots. IEEE Transactions on Systems, Man, and Cybernetics, vol. 35, n.4, pp. 460-470.

Tsalakanidou, F. & Malassiotis, S.. (2010). Real-time 2D+3D facial action and expression recognition. *Pattern Recognition*, vol. 43, no. 5, pp. 1763-1775. doi: 10.1016/j.patcog. 2009.12.009.

Turaga, P.; Chellappa, R.; Subrahmanian, V. S. & Udrea, O.. (2008). Machine Recognition of Human Activities: A Survey. IEEE Treans. On Circuits and Systems for video Technology, vol18,no. 11, pp. 1473-1488.

Valstar, M.; Pantic, M; Ambadar, Z. & Cohn, J. F.. (2006). Spontaneous versus Posed Facial Behavior: Automatic Analysis of Brow Actions, in Proc. of Eight Int'l Conf. Multimodal Interfaces (ICMI '06), pp. 162-170.

van Gelder, T. & Port, R. F.. (1996). It's about time: an overview of the dynamical approach to cognition. In *Mind as motion*, Robert F. Port and Timothy van Gelder (Eds.). Massachusetts Institute of Technology, Cambridge, MA, USA, pp. 1-43.

Vernon, D.; Metta, G. & Sandini G.. (2007) A Survey of Artificial Cognitive Systems: Implications for the Autonomous Development of Mental Capabilities in Computational Agents. *IEEE Transaction on Evolutionary Computation*, vol. 11, n. 2, pp. 151-180.

Viola, P., & Jones, M. J. (2004) Robust Real-Time Face Detection. *International Journal of Computer Vision*, vol. 57, no 2, pp. 137-154.

Whissell, C. M.,. (1989). The Dictionary of Affect in Language, Emotion: Theory, Research and Experience. The Measurement of Emotions, R. Plutchik and H. Kellerman, eds., vol. 4, pp. 113-131, Academic Press.

Zeng, Z.; Pantic, M.; Roisman, G.I. &; Huang, T.S. (2009). A Survey of Affect Recognition Methods: Audio, Visual, and Spontaneous Expressions. *IEEE Transactions on Pattern Analysis and Machine Intelligence*, vol.31, no.1, pp.39-58. doi: 10.1109/TPAMI.2008.52

8

Speech Communication with Humanoids: How People React and How We Can Build the System

Yosuke Matsusaka
National Institute of Advanced Industrial Science and Technology (AIST)
Japan

1. Introduction

Robots are expected to help increase the quality of our life. Robots are already widely used in industry to liberate humans from repetitive labour. In recent years, entertainment is getting more momentum as an application in which robots can be used to increase peoples quality of life (Moon, 2001)(Wada et al., 2002).

We have been developing the robot TAIZO as a demonstrator of human health exercises. TAIZO encourages the human audience to engage in the health exercise by demonstrating (Matsusaka et al, 2009). When demonstrating, TAIZO, the robot and the human demonstrator will stand in front of the human audience and demonstrate together. For this to work, the human demonstrator has to control the robot while they themselves are demonstrating the exercise to the audience.

A quick and easy method for controlling the robot is required. Furthermore, in human-robot collaborative demonstration, the method of communication used between the human and robot can be used to affect the audience.

In this chapter, we will introduce the robot TAIZO, and it's various functions. We also evaluated the effect of using voice commands compared to keypad input during the robot-human collaborative demonstration. In Section 2 we explain the social background behind the development of TAIZO. In Section 2.5 we will discuss about effects of using voice commands compared to key input in human-robot collaborative demonstration. In section 2.6 we present an overview of the system used in TAIZO. In Section 2.7 is the evaluation and discussion of the results from experiments in which weãĂĂmeasured the effect of using voice commands through simulated demonstrations. Finally, in Section 2.10 and Section 2.11 we will discuss about the benefits and problems of using humanoid robot to this application.

In latter part of the chapter, we will discuss how to develop the communication function for the humanoid robot.

Recently, "behavior-based" scripting method is applied in many practical robotic systems. The application presented by Brooks (1991) used hieratical structure model, the recent applications (Kaelbling, 1991) (Yartsev et al, 2005) uses state transition model (finite state automata) to model the situation of the system. The developer incrementally develop the script by adding each behaviors which fits to each small situations. Diverse situation understanding ability can be realized as a result of long-term incremental development.

The behavior-based scripting method can also be applied to communication robots by incorporating speech input with the situation model. Application of the behavior-based scripting method to the communication robot is first presented by Kanda et al (2002) in 2002. In their work, they not only proposed an incremental development framework, but also implemented an on-line development environment which can realize automated control of the robot. They have confirmed through 25 days field study that with help of the development environment, the conversation ability of the robot was incremented on-line and succeed to decrease operation time of the human operator (Kanda et al, 2009).

However, in the existing behavior-based scripting methods for communication robot, there is an inefficiency in terms of reusing the script to develop different types of robots (this problem is described in Section 3.2.2). This chapter, we present the extension-by-unification method in order to push forward the behavior-based scripting approach to develop communication robots.

In Section 3.1, a component based architecture we have developed that can develop a robotic dialog system only by connecting a components will be shown.

In Section 3.2, a formal discussion of an incremental development methods for the state-transition model is presented. Here, we introduce the formalization of the proposed incremental development method and clarify its characteristics by comparing it to the previous method.

2. TAIZO robot and robot-human health exercise demonstration

2.1 Background: Aging society in Japan

The aging society is becoming a serious problem in Japan. Like many developed countries, the decrease in birth rate and advance in life expectancy is proceeding very steeply. Due to the post war baby boom, the population of the elderly (over 65 year olds), has exceeded 20 percent of the whole population.

In the past, social welfare service has been focused on giving good medical care to the elderly. But recently, attention is shifting towards minimizing the needs of medical care itself.

Minimization of medical care not only has financial benefits, but is also important for increasing the individuals quality of life. By keeping good health and not needing medical care, an elderly person can continue to make their own decisions. This is rarely possible if they are hospitalized. Health exercises are considered to be an effective way to keep their health and minimize medication.

There is a lot of activity going on for spreading the health exercises. In the Ibaraki prefecture, the local government and a local university has co-developed a program for teaching health exercises. One characteristic of the system developed in the Ibaraki prefecture is that the trainer of the exercise is also an elderly person. The prefecture gives certificates to elderly people who have mastered the specified teaching program. This in turn qualifies them to participate as volunteers to teach the exercise to other elderly people. Due to this elderly-to-elderly teaching system, the number of health exercise demonstrators has increased to over 2300.

TAIZO is developed to assist spreading the health exercise.

2.2 Humanoid robot as an exercise demonstration medium

We use a humanoid robot as a medium to demonstrate the exercise, because the following points make it an effective demonstrator.

Similarity of the shape of body: Because a humanoid robot is designed to imitate the shape of a human, a person can easily observe and imitate the demonstrative body motion expressed by the robot.

Attraction: Meeting face to face with a humanoid robot is not yet a common occurrence. It is easy to spark the curiosity of someone who is not familiar with humanoid robots and grab their attention with an artificial being which moves and talks.
Embodiment in 3D space: The robot has a real body which has an actual volume in 3D space. Compared to the virtual agents which only appear in 2D display, it can be observed from very wide view points (Matsusaka, 2008) (Ganeshan, 2006). From this characteristic, even the user at the back of the robot can observe the robots motions. A user to the side can also observe other users interactions with the robot.
Similarity of the body shape assists precise communication of the body movement to the trainee and useful to enhance the effectiveness of the exercise. Attraction also becomes a good incentive for the trainee to engage in the exercise. Embodiment in 3D space assists the social communication between the robot and the trainees. A trainee can watch other trainees while training. By looking at other trainees, they can see how well they communicate with the robot and how eager they engage in the exercise. In most of our demonstration experiments, this inter-audience peer-to-peer effect gives positive feedback to enhance the individuals eagerness to engage in the exercise (explained and discussed in section 2.9).

2.3 Demonstration setup and scenario

Health exercise is intended to strengthen the body of elderly people. It consists of several kind of exercise that involve stretching and muscle training. Health exercise is recently increasing in importance and is gaining attention as an effective way to reduce the number of elderly people who become bedridden and need nursing care and increase their quality of life. Despite these good points, there are still some difficulty in applying this illness prevention activity. One of the biggest problems is the difficulty to encourage people to engage in these health exercise to begin with.
People are not wary of their health while they are healthy. They notice after they realize they have a serious illness. Although, we have statistics on the percentage of people who become seriously ill, it is still difficult to estimate how healthy we are ourselves. Moreover, it is much more difficult to understand the effect of health exercise, since we cannot compare what would have happened if we didn't (or did) engage in the exercise.
Because of above, we usually require a special incentive to engage in the health exercise.
TAIZO (Figure 1) is designed to help demonstrate health exercise. It will stand to the side of the demonstrator and assist the trainer in demonstrating. Demonstration is usually done in front of 5 to 15 trainees. TAIZO is used to demonstrate at events with 40 to 80 trainees. The robot is used as an eye-catcher to capture the attention of people who don't know the exercise, but could be a potential regular trainee. By using the robot as a demonstrator, human demonstrator can get more interest from a larger variety of people compared to a demonstration done by humans alone. This leads to more people having a chance to engage in health exercises.

2.4 Role of the human demonstrator and the robot

Both human demonstrator and the robot stand in front of the audience (Figure 2). Human demonstrator leads the training program and the robot follows. Both human and the robot show the demonstration to the audience.
To follow the human demonstrators lead, the robot has to accept commands given by the human demonstrator. In addition, although the human demonstrator takes the lead in most situations, the robot has to collaborative with demonstration activity in order to make it more

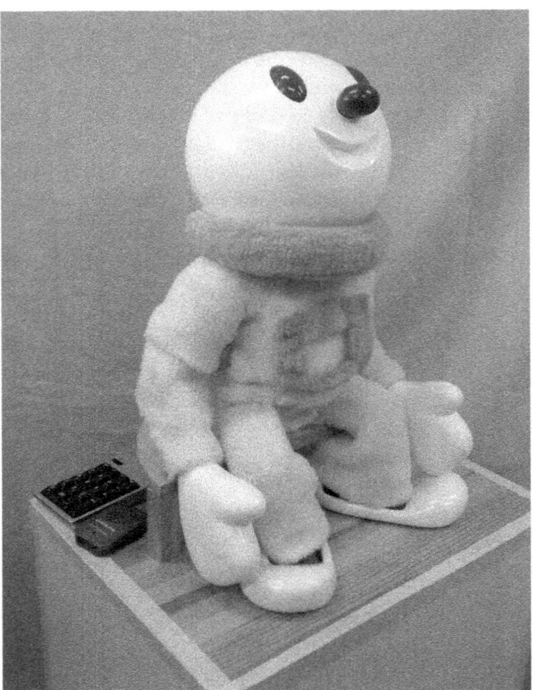

Fig. 1. TAIZO robot

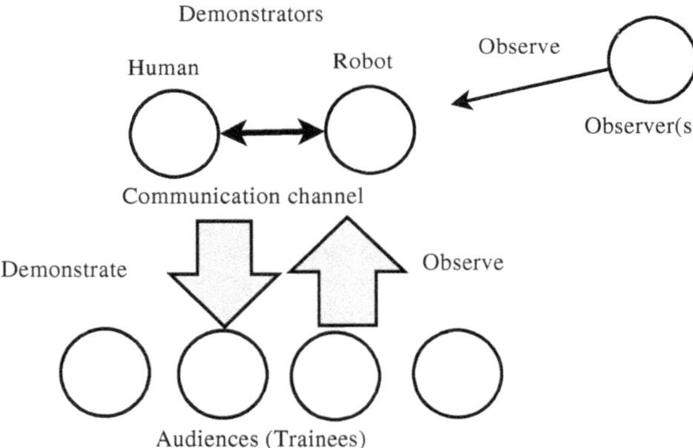

Fig. 2. Demonstrative setup

attractive. The collaboration activity itself could sometimes be an attention catcher to increase attendance of the demonstration. This chapter focuses on this communication effect in the human-robot collaborative demonstration.

2.5 Difference of communication channels in robot-human demonstration

2.5.1 In-demonstration conversation

When demonstrating the exercise there is a dialog between the trainer and the supporting robot. We call this dialog "in-demonstration conversation".

There exists research which handles this in-demonstration conversation. Katagiri et al (2001) has realized and evaluated the effect of demonstration through virtual agents which uses in-demonstration conversation. In Japan, this dialog has been developed into a two-man stand-up comedy called manzai. There several pairs of robots which have been used for manzai to entertain a human audience (e.g. Hayashi et al (2008)).

However, in most of previous research, the demonstration setup consists of only two robots. If one of the members becomes human, there will be a communication issues between the human and robot. In this chapter, we introduce our health exercise robot which has handled the communication issue specific to a human-robot setup.

2.5.2 Speech or keyboard : Discussion from both sides

In human-robot collaborative demonstration, what is the most appropriate method to give commands to the robot? Here, we compare two input methods, key input and vocal input.

When we compare the two input methods from accuracy, key input is more precise than the speech input.

When we compare two the input methods from their characteristics, key input uses a private channel, while vocal commands are public. In other words, vocal commands can be heard from the audience, but key input cannot be heard or observed by most of the audience.

From the demonstrator's side of view, a more precise input channel may be preferable, because they don't want to make errors. But when we consider the audience's point of view, a more transparent communication may be more preferable. From this point of view, vocal commands may be preferred, because it will allow the audience to observe the flow of the dialog between the human demonstrator and the robot. Interaction using an audio medium is publicly observable, and from a users standpoint, could overcame keypad input despite it's inaccuracy.

In this chapter, we will not only evaluate the effectiveness of each input method from a demonstrator's point of view, but will also focus on the demonstrative effect to the audience. In Section 2.7 evaluation design is described.

2.6 Architecture of TAIZO robot

2.6.1 Speech input system

Figure 3 shows the overall architecture of TAIZO robot.

Speech input system consists of a speaker-phone device (Figure 4) to capture and emit sound. The speech recognition system Julius (Kawahara et al, 2000) inputs the captured sound and outputs the recognized phrase. The recognized phrase is matched with a phrase database in the phrase matcher. The phrase database contains a set of phrases written in a form of script which associates phrases and commands. Details of the speech I/O system will be shown in Section 3.1.

2.6.2 Key input system

Key input system consists of a keypad device (Figure 4). The human demonstrator types a number using this keypad.

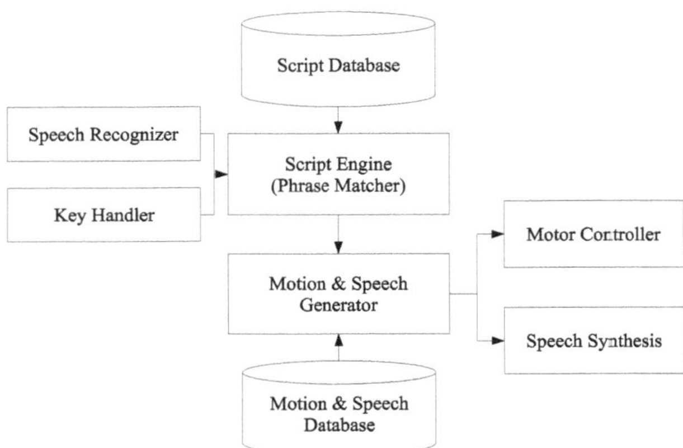

Fig. 3. System architecture

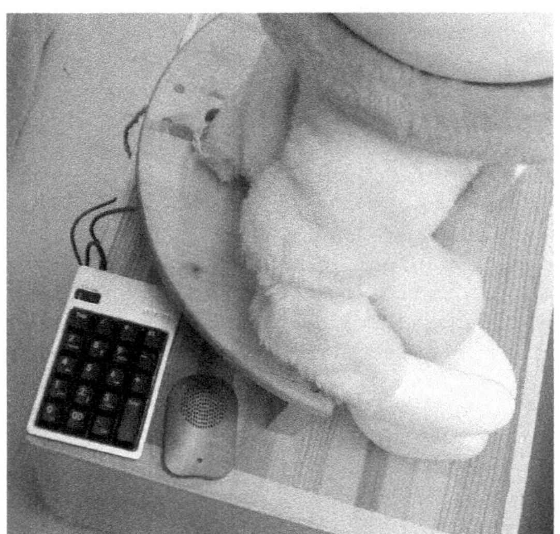

Fig. 4. Keypad (left) and speaker-phone (right)

The keypad is used for both key input mode and speech input mode. In speech input mode, the keypad is used to control the volume of the microphone device. Input to the microphone device switches on when the key is pressed, and turns off when the key is released (push-to-talk).

2.6.3 Motion and speech database

The motion database contains sequences of values specified to each joint of the robot. These sequences are designed to express the exercise motion of the robot.

In this chapter we have designed 17 motions (see Figure 5). When designing the motion, we first create an abstract design of the motion using a robot simulator (Hirukawa et al, 2003). Final adjustment is done by playing back the sequence on the real robot.

Fig. 5. Motions designed for health exercise.

We play back prerecorded speech for the robot. Patterns of prerecorded speech consists of exercise related phrases (e.g. "Raise your arm behind your head", "Twist your waist"), question answer type phrases ("Yes", "Okay", "No") and greetings ("Hello", "Good Bye"). Exercise related phrases are recorded by the script developer, each time a new exercise is designed. The script developer can also write a script for interaction by using prerecorded QA type phrases.

Type	Text (followed by "Do you ...")	When
Expectation	think the demonstration by the robot effective?	Before
Enjoyableness	think the demonstration was enjoyable?	Each
Easiness	think using robot is easy?	Each
Fulfillness	think the demonstration was fulfilling?	Each
Preference	prefer speech input or key input?	After
Effectiveness	think the demonstration by the robot effective?	After
Willingness	want to use the robot in real demonstration?	After

Table 1. Items in question sheet (each question allows free commenting).

2.6.4 Motion and speech generation system

During motion generation, motion data in the database is transmitted to the motor controller. Motion data is transmitted sequentially at a specific rate using the internal clock of the robot. Speech recordings are synchronized in parallel with the motion at specified timings.

2.6.5 Script engine and script database

The scenario database contains scripts which define speech phrases and key numbers of the keypad commands. It also includes information which associates the commands to motion data.

The script engine is based on a state transition model. For this specific experiment, we use a flat (one-state) structure model with 17 commands. Details will be shown in Section 3.2.

2.7 Evaluation of command input methods

2.7.1 Experiment condition

We have designed 4 experiments controlled by two conditions. Condition 1 is the input which be either keypad input or vocal command. Condition 2 is defined by the subjects role which is demonstrator or audience. The conditions regarding the role of the demonstrator switches when the same subject experiences the same demonstration as a demonstrator or as a passive participant. Each subject is asked to attend all of these 4 (2×2) experiments.

Experiment subjects are 60-80 year old, and consists of 1 male and 5 females. All subjects have experience in the health exercise training program and are certificatified to instruct. Although they are knowledgeable in the health exercise, when participating as passive audience, we asked them to answer the questions (described in next section) as if they are a novice trainee.

2.7.2 Experiment sequence

The six subjects were divided into two groups of three. There are six sessions altogether for each group. The first three sessions used the keypad and the latter three sessions used vocal commands. In each of the three sessions, three subjects in turn take the role of demonstrator. Each subject is asked to fill in a questionnaire sheet after each session. Table 1 shows the questions used in this experiment. Questions consist of asking the ease of use, whether the experience was enjoyable and whether the demonstration was fulfilling.

Before and after the experiment, the subject is asked to fill the question sheet. In the questionnaire handed to the subject before the experiment, expectation for the robot is asked. In the post experiment questionnaire, the subjects preference to keypad or vocal commands and willingness to continue using the robot in a real demonstration is asked.

2.8 Result

Figure 6 shows the error rate of the key input and the speech input.

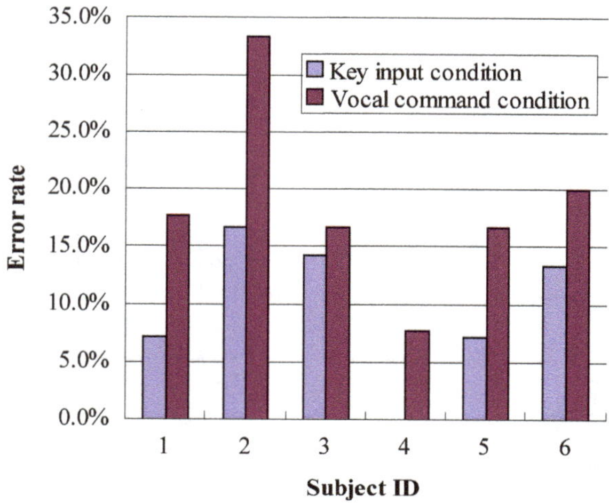

Fig. 6. Error rate of each input methods.

Although the key input is a precise input method, there are some errors due to mistyping. Mistyping can be classified into two types. One is typing error, which happens when the demonstrator is in a hurry to type the keypad in the middle of the demonstration. The other is memory error. Memory error happens because the keypad only accepts numeric input and the demonstrator has to remember the mapping between the number and the training pattern. They often forget the mapping and type the wrong number. Because of these errors, actual precision of the key input is not as high as we expected.

Most of the errors in vocal commands happen due to speech recognition errors. Some demonstrators had problems in pronunciation and other demonstrators spoke superfluous words to enhance the demonstration. In the case of vocal input, there was less memory error, because the demonstrator only has to pronounce the name of training pattern and there is no need to remember the mapping to the numbers.

Figure 7 shows the impression of the demonstration asked after each session. Demonstrator feels using the vocal command is easier than the key input. This can be understood from the memory error as we discussed above. When vocal commands were used, the audience both enjoyed the demonstration and found it as fulfilling as the keypad input demonstration, despite happenings due to inaccuracy. This could be one of the effects of the observability of speech in a human-robot collaborative demonstration.

Figure 8 shows the impression of the robot assisted demonstration before and after the demonstration. Almost all the subjects have answered that they are willing to use the robot as an instructor again.

2.9 Feasibility test in the real demonstration

We have already started applying this robot in real demonstration events. Figure 10 shows photos from the "Nenrin-pic" event. Nenrin-pic event is one of the event hosted by Ibaraki prefecture intended to encourage sports for elderly people.

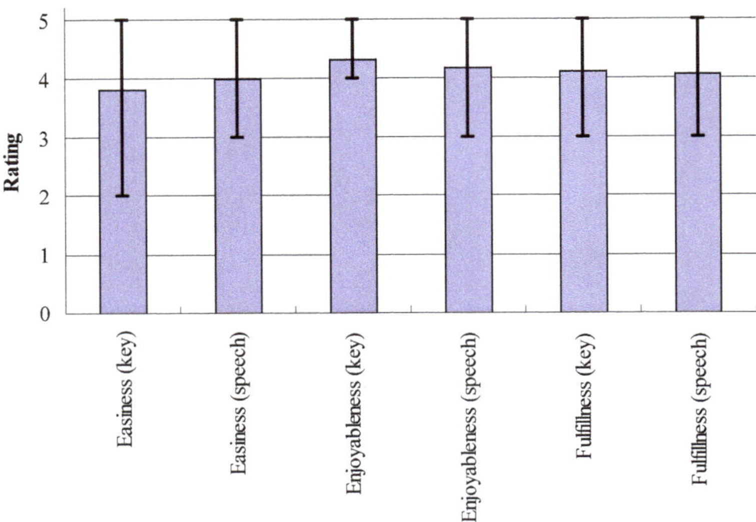

Fig. 7. Impression of the demonstration asked after each session. The error bar indicates maximum and minimum rating of the subjects. Easiness is asked to the demonstrator. Enjoyableness and fulfillness are asked to the audience.

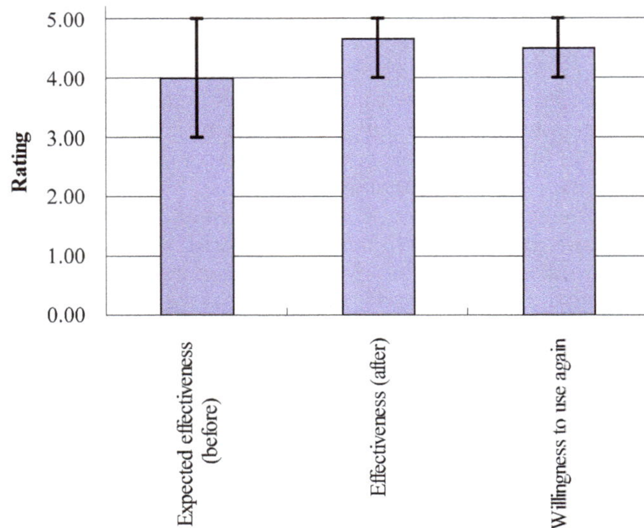

Fig. 8. Impression of the robot before and after the demonstration. The error bar indicates maximum and minimum rating of 6 subjects.

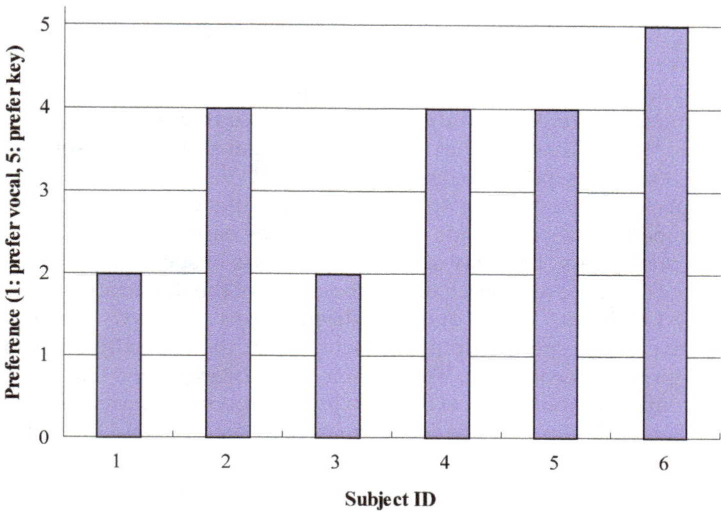

Fig. 9. Preference of using speech input or key input asked to each subject.

Fig. 10. Photos from a real demonstration at nenrin-pic event.

During the 3 day event, we have demonstrated 10 times a day using TAIZO robot. More than 600 peoples has joined the training experience. As we can see from the photos, almost all the audiences were eagerly followed the demonstration given by the human-robot demonstrators. One of the unexpected effects of using TAIZO was that we were able to catch the attention of a wide variety of ages. TAIZO was intended to catch attention of the elderly, but during the nenrin-pic event, many young adults and their children accompanying their parents or grand-parents were drawn to the demonstration. The attraction of the TAIZO robot was strong enough to catch also those accompanying persons. It seems to have a good effect for the elderly, because they can enjoy their exercise by participating together with their families.

2.10 Summary

In Section 2.7 we have tried to evaluate effect of using key and speech inputs especially focused to human-robot collaborative demonstration setup.

As we discussed in Section 2.5, vocal communication increases the transparency of the human-robot communication to the audience. We could see from the experiment results that the enjoyment and the fulfillment of the demonstration from audience perspective was not let down by the imprecision in speech recognition.

One of the unexpected effect of using speech is, because it is very intuitive, it decreases the burden for remembering the commands. Error rate of key based commands is unexpectedly high, despite it's preciseness. This may also support the use of speech input.

Despite supportive evidence for speech input, about half of the demonstrators answered that they prefer using key input. In the question sheet, subject can leave comments. Subjects who are supportive to the speech input commented that they actually enjoyed reacting to mistakes made by speech recognition. This comment can be explained that by increasing the transparency of communication channel between human-robot demonstrators, the subject can observe what is happening in the situation (human demonstrator said the right thing, robot mistakes) and feel the mistaken response of the robot funny. On the other hand, subjects who are supportive to the key input commented that they want to demonstrate the exercise in a more precise manner.

We are currently preparing to do an evaluation which is more focused on measuring these effects and searching for a way to realize an appropriate interface for both people who prefers enjoyment or precision.

2.11 Left problems

As we have looked and discussed in this section, humanoid robot has different character than the other artifacts. *It has human shape* which can attract human to join in the activity. This character also gives some effect to *gain expectation to use natural communication method (voice)* as the human do. *It has physical body and exist in same world* which can give effect also to the observers. These characters are especially useful for applications such as exercise demonstration.

However, this character sometimes gives negative effect to the usefulness. Because the human gain too much expectation to the robot to use natural communication method, human tends to use colloquial expression towered the robot, which is difficult for the robot to understand. In Section 2.7, we have seen the command acceptance rate using voice recognition evaluated by elderly users. In this experiment the command acceptance rate is low, not because voice recognition rate is low, but mostly because conversation patterns programmed to the dialog manager was not enough to understand the all varieties of colloquial expressions given by the users.

Because the robot has physical body and exist in physical world, the voice recognition system of the robot have to work under noisy condition of the real environment.

Although for ideal benefits of using humanoid robots, above practical problems are need to be solved beforehand to enhance the usefulness of the robots.

We are not only developing the applications for humanoid robots, but also developing a support tools for assist development of the communication functions for humanoid robots. From the next section, we will introduce our development tools.

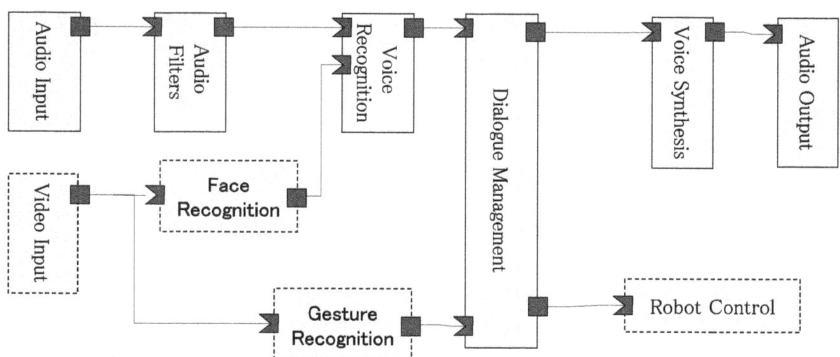

Fig. 11. Architecture of the OpenHRI software suite.

3. Reduce the difficulties of building the communication system

In this section, we introduce our efforts to reduce development difficulties.
We are currently taking two approaches to reduce the development difficulties. The one is component based system design and the other is incremental script development.

3.1 Component based system design

At present, we are developing a set of software called Open Source Software Suite for Human Robot Interaction (OpenHRI). Using OpenHRI, we aim to solve the above problems and enable the development of communication functions for robots. For this purpose, we employ the following approach.

Introduce a uniform component model: We construct our set of software on RT-Middleware, an object management group (OMG)-compliant robot technology middleware specification (Ando et al, 2005). The RT-Middleware specification can be used to connect all the components without requiring implementation issues to be taken into account. Further, because it is a standard architecture for building robotic systems, individual components developed in different institutes can easily be connected.

Provide the required functions in a reconfigurable manner: We implement various functions from audio signal processing to dialog management in a uniform and reconfigurable manner. The developer can develop the entire system at a comparatively less development cost. In addition, the system can easily be adapted to different environments for realizing accurate recognition.

Figure 11 illustrates the overall architecture of the components provided in OpenHRI. The software covers all the functions for the development of the communication system and also incorporates an interface for establishing connections with other components that can provide multi-modal information.

The component architecture of our software is based on RT-Middleware. RT-Middleware is a middleware architecture for robotic applications that has been standardized by the OMG.

In RT-Middleware, each function of the robot is implemented as a "node." An application system can be developed by selecting the required components and connecting them to each other (the connections are called "links"). Figure 12 shows the "RT-SystemEditor" development tool to edit the links between the components.

In the specification of RT-Middleware, a "data port" is defined as a connection point of a link that realizes the transmission of a data stream. A "service port" is defined as an entry point of

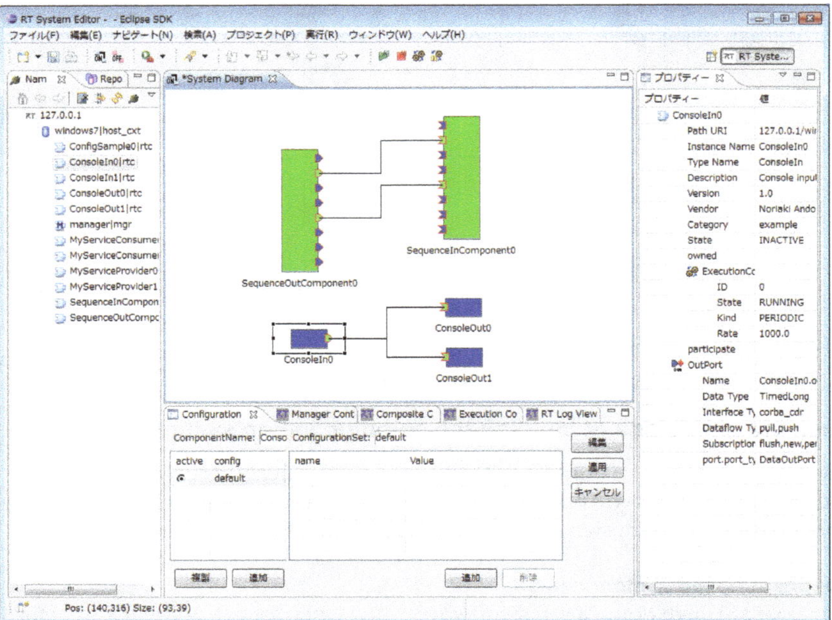

Fig. 12. Screenshot of RT-SystemEditor.

each function call for a service function. "Configuration parameters" are defined to configure each component.

OpenRTM-aist is an implementation of the RT-Component specification that supports C++, JAVA, and Python languages. It runs on various platforms such as Windows, Linux, Mac OS, and FreeBSD.

3.1.1 Audio input/output components

Audio input components accept the audio information from the sound device as input, convert it to a OpenRTM-aist data stream, and pass on the output to other linked components. These components also accept audio data streams from other components as input and pass on the output to the sound device.

We use portaudio, a cross platform audio input/output library, to implement both the audio input and output components. The components support both Windows and Linux platforms from monoral input to multichannel inputs.

3.1.2 Audio filter components

Audio filter components contain input and output ports.

The "sample rate conversion component" converts the sample rate of the audio stream by using an up/down sampling algorithm. The "echo cancel component" has two input streams; it subtracts the input of stream 1 from that of stream 2 by finding a maximum correlation. The "emphasis component" applies a signal processing algorithm to enhance or de-enhance the magnitude of the specified frequency in the data stream.

3.1.3 Voice recognition and synthesis components

The voice recognition component is based on Julius (Kawahara et al, 2000) in combination with English and Japanese acoustic models. Our component is designed to possess the following features: (a) the ability to read grammar format in W3C-SRGS XML form and (b) the ability to output the recognized result as an extensible XML stream.

Figure 13 shows an example of voice recognition grammar specified using W3C-SRGS format. Voice recognition grammar can be visualized by combination of "srgstojulius" and "juliustographviz" tool.

The voice synthesis component is based on Festival for English and Open_JTalk for Japanese. The component accepts plain text as input and provides a data stream in the form of a synthesized voice as output.

3.2 Incremental script development

Commands given by the human to the robot are diverse. The following are the factors that cause this diversity.

The nature of language: Human language is ambiguous, and different expressions can be used to give instructions that carry the same meaning.

Tasks: Robots working in a life environment have to accept a variety of tasks. In order to cope with this, it is necessary for them to understand a variety of commands.

Ability of the robot itself: The diversity is also caused by the ability of the robot itself. A command from a human becomes effective due to the functions of the robot. For example, humans do not say "walk N steps" to a robot on wheels.

The language comprehension system of the robot must be able to deal these diversities.

In the script-based development approach, diversity has been dealt with by stacking a newly developed script onto the existing scripts. By accumulating a number of scripts, the developer can accumulate the number of commands that the system can dealt with.

Incremental development of the state-transition model has previously been conducted using the "extension-by-connection" method (described in the next section). In this section, we propose an "extension-by-unification" method that can cope with the diversities mentioned above (described in Section 3.2.4).

3.2.1 Formalization of state-transition model

A state-transition model is a modeling method in which the input and output of the system assume the following form:

$$A := < I, S, O, \gamma, \lambda, s_0 > \tag{1}$$

where I represents the input alphabet, O represents the output alphabet, S represents the internal states, γ represents the state transition function, λ represents the output function and s_0 is the initial state.

The state transition function γ is defined in association with the state to the input.

$$\gamma : S \times I \rightarrow S \tag{2}$$

The output function λ is defined in association with the state to the input.

$$\lambda : S \times I \rightarrow O \tag{3}$$

```
<?xml version="1.0" encoding="UTF-8" ?>
<grammar xmlns="http://www.w3.org/2001/06/grammar"
         xml:lang="en"
         version="1.0" mode="voice" root="main">
  <rule id="main">
    <one-of>
      <item><ruleref uri="#greet" /></item>
      <item><ruleref uri="#command" /></item>
    </one-of>
  </rule>
  <rule id="greet">
    <one-of>
      <item>hello</item>
      <item>bye</item>
    </one-of>
  </rule>
  <rule id="command">
    <one-of>
      <item>pick</item>
      <item>give me</item>
    </one-of>
    <one-of>
      <item>apple</item>
      <item>cake</item>
      <item>remote</item>
    </one-of>
  </rule>
</grammar>
```

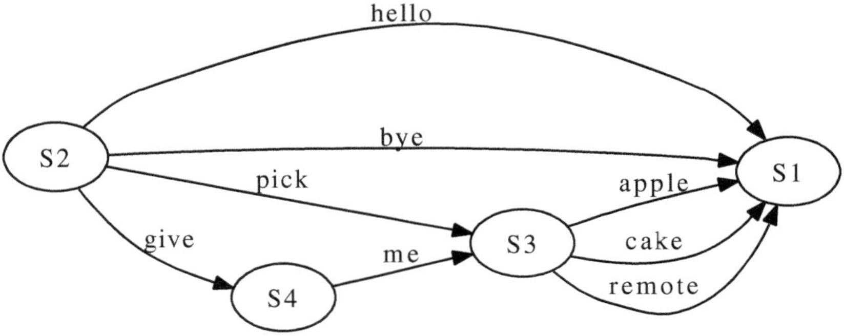

Fig. 13. Example of the voice recognition grammar and its visualization. The grammar is in W3C-SRGS form. "one-of" indicates the grammar matches either one of the child items. "ruleref" indicates reference to "rule" identified by the "id".

When the system is in state s_t and get input alphabet i_t, state transition to s_{t+1} will occour as follows:

$$s_{t+1} = \gamma_{s_t,i_t} \tag{4}$$

At the same time, we get output alphabet o_t as follows:

$$o_{t+1} = \lambda_{s_t,i_t} \tag{5}$$

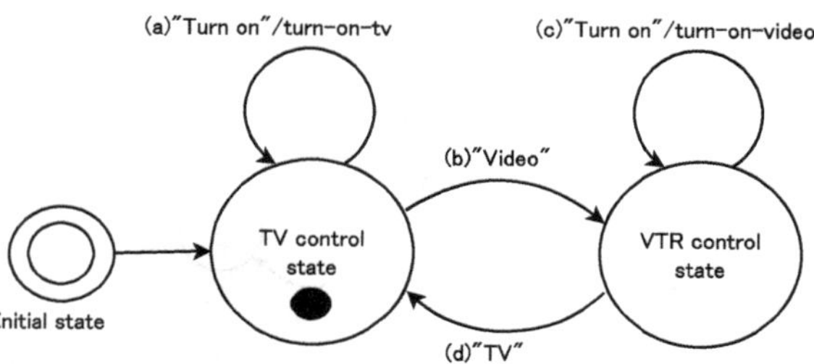

Fig. 14. Example of state-transition model.

Even the input to the system is same, the output of the sytem may be different, because the internal state s_t will be updated each time the system gets the input.
We have explained the stat-transiton model in an equation form, however, the state-transition model can be also presented in a 2-dimensional diagram called "state-transition diagram." In the diagram, each state is represented by a circle, and the transition between states is represented by arrows. In this chapter, we annotate the transition conditions and the associative actions by including text over each arrow. We use a black circle (called a "token") to represent the current state.
For example, Figure 14 represents a conversation modeled by the state-transition model.
The model presented in Figure 14, the initial state of the system is in "TV control" state. When the model gets the instruction "Turn on" as an input, it will output the command "turn-on-TV" and state transition "(a)" will occur. Then the token turns back to the same "TV control" state. When the model gets the instruction "Video" as an input, state transition "(b)" will occur and the token will move to "VTR control" state. This time when the instruction "Turn on" is given, state transition "(c)" occur and output the command "turn-on-video". In this way, we can model the context by defining an appropriate state and state transitions between the states.
Above example is expressed as follows in the equation form:

$$A := < I, S, O, \gamma, \lambda, s_0 > \tag{6}$$

$$I = (\text{"Turn on"}, \text{"Turn off"}, \text{"TV"}, \text{"Video"}) \tag{7}$$

$$S = (\text{"tv-control"}, \text{"vtr-control"}) \tag{8}$$

$$O = (\text{"turn-on-tv"}, \text{"turn-on-video"}, \text{"turn-off-tv"}, \text{"turn-off-video"}) \tag{9}$$

$$\gamma = \begin{pmatrix} s_0 & s_0 & s_0 & s_1 \\ s_1 & s_1 & s_0 & s_1 \end{pmatrix} \tag{10}$$

$$\lambda = \begin{pmatrix} o_0 & o_2 & none & none \\ o_1 & o_3 & none & none \end{pmatrix} \tag{11}$$

As we have seen here, the expression in equation form has an advantage in formalization, while the expression in diagram form has an advantage in quick understanding. In later

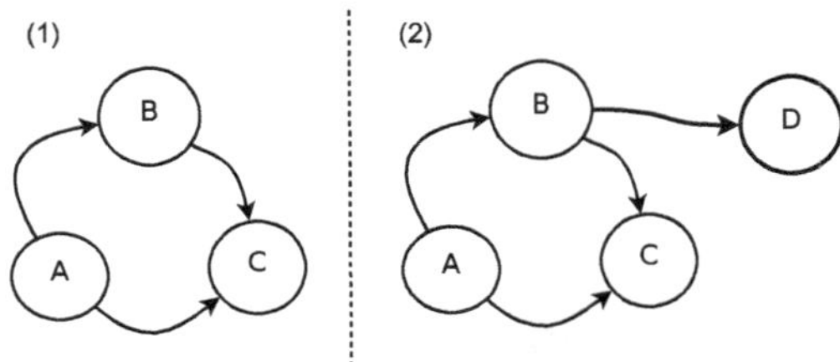

Fig. 15. Extension of a state machine using the extension-by-connection model.

discussion, we will use both the equation and the diagram forms to explain the concept quickly and formally.

State-transition model is very simple but very powerful modeling method and has been applied to very wide applications. Because the structure of state-transition model is very simple, it is frequently misunderstood that the state-transition model can only model simple behavior. However, it can model diverse behavior by applying some extensions (e.g. Huang et al (2000), Denecke (2000)).

3.2.2 Extension-by-connection method

The simplest way to extend state-transition model is as follows.

1. Add a new state to the existing state-transition model.
2. Add a new transition from the existing state to the new state.

This process is illustrated in Figure 15.

Here, we formulate the above process. Let the existing state-transition model be A and the accumulated state-transition model be A'.

As explained in Section 3.2.1, the existing state-transition model A can be represented by following form.

$$A := < I, S, O, \gamma, \lambda, s_0 > \tag{12}$$

Here, S is the set of state $s \in S$. The transition function γ can be defined in any form. In this chapter, we use the matrix of $S \times I$, in which the transition from state s_t to state s_{t+1} can occur if $\gamma_{S_t,i} = s_{t+1}$.

Similarly, we define the accumulated state-transition model A' as follows:

$$A' := < I, S', O, \gamma', \lambda', s'_0 > \tag{13}$$

Then, the new state ΔS can be calculated as follows:

$$S' = S \cup \Delta S \tag{14}$$

Here, $S' \cap \Delta S = \emptyset$.

The new state transition $\Delta\gamma$ can be calculated as follows:

$$\gamma'_{s_t,i} = \gamma_{s_t,i} \qquad (s_t \in S, i \in I) \tag{15}$$

$$\gamma'_{s'_t,i} = \Delta\gamma_{s'_t,i} \qquad (s'_t \in S', i \in I) \tag{16}$$

$$\lambda'_{s_t,i} = \lambda_{s_t,i} \qquad (s_t \in S, i \in I) \tag{17}$$
$$\lambda'_{s'_t,i} = \Delta\lambda_{s'_t,i} \qquad (s'_t \in S', i \in I) \tag{18}$$

The transition function of the accumulated part $\Delta\gamma$ needs to be defined based on the transition from the existing state S . Therefore, $\Delta\gamma$ will be a matrix of $S' \times I$. Note that the new state ΔS can be expressed only by the newly defined part, but the transition of the accumulated part $\Delta\gamma$ includes both old state S and new state ΔS in its definition.
The state-transition model is easy to understand in drawing a state-transition diagram. Extension-by-connection can also be carried out very easily by editing this diagram. There are several GUI that can add state-transition rules through the operation of mouse clicks.

3.2.3 Problems with the extension-by-connection method

Extension-by-connection is a useful method, but it has the following problems.
As we can see in Equation 14 and Equation 16, the definition of $\Delta\gamma'$ requires both S and ΔS. This causes problems in the function development of robots. For example, let us consider the following scenario:

1. Robot "A" has function A, and we have already developed a state-transition model $A^A$to realize the function.
2. For the robot "A" to accumulate function C, we have extended the state-transition model to A^{AC}.
3. We have developed another robot, "B," which has function B. And we want to add function C to this robot.

Here, the state-transition model for function C is already developed for robot A. We want to reuse the model for robot B. Here, we discuss whether such a diversion would be possible.
First, the state S^{AC} is easily separable from state S^A and state S^C:

$$S^C = S^{AC} - S^A \tag{19}$$

However, the definition of state-transition function γ^{AC} is as follows:

$$\gamma^{AC}_{s^A_t,i} = \gamma^A_{ij} \qquad (s^A_t \in S^A, i \in I) \tag{20}$$
$$\gamma^{AC}_{s^{AC}_t,i} = \gamma^C_{ij} \qquad (s^{AC}_t \in S^{AC}, i \in I) \tag{21}$$

γ^C contains state S^A in its definition.
Because states S^A and S^B are defined for different types of robots, A and B are not equal. In addition, because the transition for the function C is defined dependently on state S^A, we cannot replace variables like $S^{AC} = S^{BC}$, which means that we cannot use γ^C to extend the state-transition model A^B. The state transition of function C developed for robot A cannot be diverted for the extension of robot B.
Ideally, once a feature is developed, it would be possible to share with other robots that need the same feature. In order to achieve this, we introduce the extension-by-unification method.

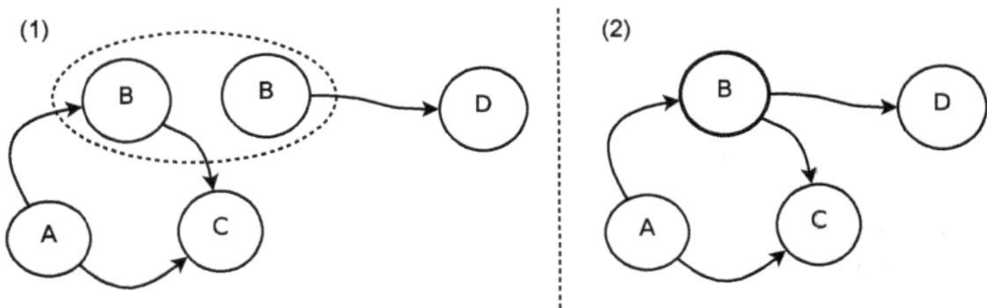

Fig. 16. Extension of the state-transition model using the extension-by-unification method.

3.2.4 Extension-by-unification method

In the extension-by-unification method, we extend the state-transition model by the following procedure:

1. Develop a state-transition model to realize a new function.
2. Unify a state with the same ID between the existing and the new state-transition models.

This process is illustrated in Figure 16.
Here, we formulate the above process.
The existing state-transition model A can be represented by state S, state transition γ, and initial state s_0:

$$A := < I, S, O, \gamma, \lambda, s_0 > \tag{22}$$

Similarly, the new state-transition model A' is represented as follows:

$$A' := < I, S', O, \gamma', \lambda', s'_0 > \tag{23}$$

We accumulate the state-transition model A'' by unifying A and A'. First, we calculate state as follows:

$$S'' = S \cup S' \tag{24}$$

Here, $S \cap S' \neq \emptyset$.
Next, the transition between the state S'' is calculated as follows:

$$\gamma''_{s_t,i} = \gamma_{s_t,i} \qquad (s_t \in S, i \in I) \tag{25}$$

$$\gamma''_{s'_t,i} = \gamma'_{s'_t,i} \qquad (s'_t \in S', i \in I) \tag{26}$$

$$\lambda''_{s_t,i} = \lambda_{s_t,i} \qquad (s_t \in S, i \in I) \tag{27}$$

$$\lambda''_{s'_t,i} = \lambda'_{s'_t,i} \qquad (s'_t \in S', i \in I) \tag{28}$$

By defining initial state s''_0 to be $s''_0 = s_0$, the extended state-transition model A'' will be as follows:

$$A'' = < I, S'', O, \gamma'', \lambda'', s''_0 > \tag{29}$$

As visible in Equation 26, the transition function γ' is an $S' \times S'$ matrix that only includes state S' in its definition. The extension-by-unification method does not require the definition of the original state in the accumulated part of the state-transition model.

As noted in Section 3.2.3, in the conventional extension-by-connection method, the definition of the accumulated part of the state-transition model depends on information on the existing state. It is limited in terms of reusing scripts for this reason. The proposed extension-by-unification method does not have this problem. Using this method, we can significantly increase the reusability of the state-transition model.

3.2.5 Visualization of unifiable states

By using the above algorithms, the possibility of unification between scripts can be identified as "Unifiable," "Unifiable (Occurrence of isolated state)," or "Conflict". Similarly, scripts can be classified as "Executable" or "Unexecutable." By comparing a script and an adaptor definition for the existing scripts, we can obtain a list of scripts annotated with 6 (3×2) classes. Our script management server displays the above list at the bottom of each wiki page. By displaying the list, the developer can easily find a script that can be included in his/her current application.

Figure 17 shows overview of the editing system and Figure 18 shows example of using the web based interface.

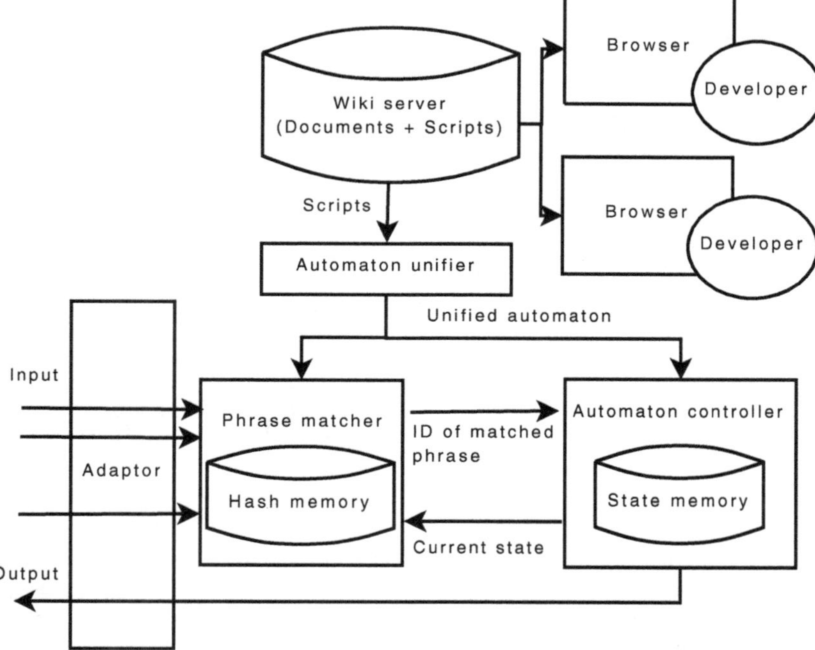

Fig. 17. Overview of the editing system.

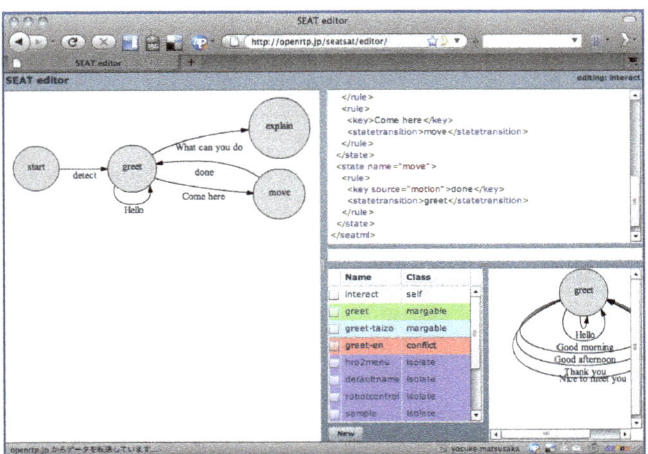

a) Overview of the development interface. Visualization of the state-transition model (left), XML based editing panel (right top), real-time annotation of existing scripts (right bottom). Editing task script. When the developer types the keyword "Hello", the existing script from the script database is annotated as "conflict" and suggest to reuse. At this step, the system only accepts 3 ("Hello", "What can you do?", "Come here") phrases.

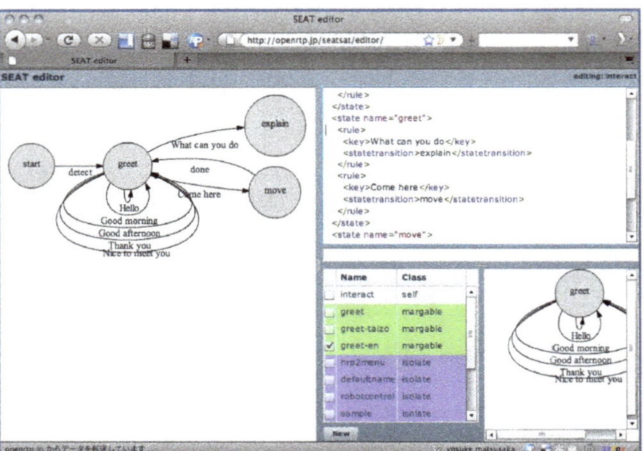

b) When the developer check the "greet" script, which already contains several vocabulary for greeting, it is unified to the task script. As a result, the developer only had to increment the application specific vocabulary to realize the whole script with many vocabularies. At this step, the system accepts 7 ("Hello", "Good morning", "Good afternoon", "Thank you", "Nice to meet you", "What can you do?", "Come here") phrases.

Fig. 18. Example of using the web based interface.

4. Summary

In this chapter, we have introduced our health exercise robot TAIZO and the development background. TAIZO has the function to accept commands by voice and keypad, and demonstrates by using its body in collaboration with human demonstrator. Through the experiment, we have measured not only the error rate of both voice and key inputs, but also, the demonstrative effect of using each method. Through real demonstrations, we have confirmed that the robot is effective for giving the incentive to engage in health exercises.
We have also discussed about some practical problems to make difficult the development of communication function for humanoid robot. By introducing the component based architecture and extension-by-unification method to develop the dialog script, scripts created in the past can easily be reused in the new application. In the conventional extension-by-connection method, the developer had to develop each function in turn, because it did not support the "merging" of scripts that had been developed simultaneously.
We believe, in future, how advanced the computer graphics are, humanoid robots still keep its strongness to affect the human to move. We hope such powerful medium would be used more in the feature, by developing practical techniques as described in this chapter.

5. References

Y.Matsusaka, H.Fujii, I.Hara: Health Exercise Demonstration Robot TAIZO and Effects of Using Voice Command in Robot-Human Collaborative Demonstration, Proc. IEEE/RSJ International International Symposium on Robot and Human Interactive Communication, [in print] (2009)

Y. Moon. (2001). Sony AIBO: The World's First Entertainment Robot," in The Harvard Case Collection, Harvard Business School Publishing

K. Wada, T. Shibata, T. Saito and K. Tanie. (2002). Robot Assisted Activity for Elderly People and Nurses at a Day Service Center," in Proceedings of the IEEE International Conference on Robotics and Automation, pp.1416-1421

Y. Matsusaka, "History and Current Researches on Building Human Interface for Humanoid Robots," in Modeling Communication with Robots and Virtual Humans, Lecture Notes in Computer Science, Springer, 2008, pp.109–124.

K. Ganeshan, "Introducing True 3-D Intelligent Ed-Media: Robotic Dance Teachers," in Proceedings of World Conference on Educational Multimedia, Hypermedia and Telecommunications 2006, pp.143–150, 2006.

Y. Katagiri, T. Takahashi, Y. Takeuchi, "Social Persuasion in Human-Agent Interaction,", in Proceedings of Second IJCAI Workshop on Knowledge and Reasoning in Practical Dialogue Systems, IJCAI-2001, pp.64–69, 2001.

K. Hayashi, T. Kanda, T. Miyashita, H. Ishiguro and N. Hagita, "ROBOT MANZAI -Robot Conversation as A Passive-Social Medium-," International Journal of Humanoid Robotics, 5(1), pp.67-86, 2008.

H. Hirukawa, F. Kanehiro, S. Kajita, "OpenHRP: Open Architecture Humanoid Robotics Platform," in Tracts in Advanced Robotics, Springer, pp.99–112, 2003.

SEAT,SAT: AIST-OpenRTP Project, http://openrtp.jp/seatsat

N.Iwahashi: Language Acquisition Through a Human-Robot Interface, In Proc. ICSLP-2000, vol.3, pp. 442–447 (2000)

D.Roy: Grounded Spoken Language Acquisition: Experiments in Word Learning, IEEE Transactions on Multimedia. vol.5, no.2, pp. 197–209 (2003)

A.Symeonidisa, I.Athanasiadisb and P.Mitkasa: A Retraining Methodology for Enhancing Agent Intelligence, Knowledge-Based Systems, vol.20, issue.4, pp. 388–396 (2007)

T.Winograd: Understanding Natural Language, Academic Press (1972)

R.Brooks: Intelligence Without Representation, Artificial Intelligence, vol.47, pp. 139–159 (1991)

L.Kaelbling: A Situated-Automata Approach to the Design of Embedded Agents, ACM SIGART Bulletin, vol.2 , issue.4, pp. 85–88 (1991)

B.Yartsev, G.Korneev, A.Shalyto, V.Kotov: Automata-Based Programming of the Reactive Multi-Agent Control Systems, Proc. IEEE International Conference on Integration of Knowledge Intensive Multiagent Systems, pp. 449–453 (2005)

T.Kanda, H.Ishiguro, T.Ono, M.Imai, R.Nakatsu: Development and Evaluation of an Interactive Humanoid Robot Robovie, Proc. IEEE International Conference on Robotics and Automation, pp. 1848–1855 (2002)

T.Kanda, M.Shiomi, Z.Miyashita, H.Ishiguro, and N.Hagita: An Affective Guide Robot in a Shopping Mall", Proc. ACM/IEEE International Conference on Human-Robot Interaction, pp. 173–180 (2009)

F.Huang, J.Yang, A.Waibel: Dialogue Management for Multimodal User Registration, In Proc. Int'l Conf. on Spoken Language Processing, Vol.3, pp. 37–40 (2000)

M.Denecke: Informational Characterization of Dialogue States, In Proc. Int'l Conf. on Spoken Language Processing, Vol.2, pp. 114–117 (2000)

Voice Extensible Markup Language (VoiceXML) Version 2.0: http://www.w3.org/TR/voicexml20/

N.Ando, T.Suehiro, K.Kitagaki, T.Kotoku, W.Yoon: RT-Middleware: Distributed Component Middleware for RT (Robot Technology), Proc. IEEE/RSJ International Conference on Intelligent Robots and Systems, pp. 3555–3560 (2005)

T. Kawahara, A. Lee, T. Kobayashi, K. Takeda, N. Minematsu, S. Sagayama, K. Itou, A. Ito, M. Yamamoto, A. Yamada, T. Utsuro and K. Shikano: Free Software Toolkit for Japanese Large Vocabulary Continuous Speech Recognition, In Proc. Int'l Conf. on Spoken Language Processing, Vol. 4, pp. 476–479, (2000)

I.Hara, F.Asano, H.Asoh, J.Ogata, N.Ichimura, Y.Kawai, F.Kanehiro, H.Hirukawa, K.Yamamoto: Robust Speech Interface Based on Audio and Video Information Fusion for Humanoid HRP-2, Proc. IEEE/RSJ International Conference on Intelligent Robots and Systems, pp. 2404–2410 (2004)

F.Asano, K.Yamamoto, I.Hara, J.Ogata, T.Yoshimura, Y.Motomura, N.Ichimura, H.Asoh: Detection and Separation of Speech Event Using Audio and Video Information Fusion and Its Application to Robust Speech Interface, EURASIP Journal on Applied Signal Processing, vol.2004, issue.11, pp. 1727-1738 (2004)

9

Exoskeleton and Humanoid Robotic Technology in Construction and Built Environment

T. Bock, T. Linner and W. Ikeda
Technische Universität München,
Germany

1. Introduction

The human being is the only living organism which steadily uses "tools". We have used tools to cultivate our land, grow our food, build up cities and communication infrastructures - tools are the basis for phenomena as culture and globalization. Some even argue that tools (and especially the wealth they are able to create for a huge amount of people) are the basis for today's global spread of freedom and democracy [1].

Especially tools which enhance our power in the field of mobility have played an important role in human history. The bicycle, an archetype of the assistance in physical ability and mobility, is based on the combination of human power and an artificial, technical system and was introduced by C. Drais in 1817. Later on, the car pressed ahead with this approach and supplemented human force by motor technology, a kind of actuator. Ergonomics and the research on efficient man-machine cooperation developed during First and Second World War in order to maximize the efficiency of man controlled artifacts as motor cycles, cars, airplanes, ships and other war equipment. After the Second World War, systematic science in improving man-machine systems led to airplanes and cars which more and more reduced the physical and cognitive workload of the human users. Today's cars take over driving maneuvers in critical situations and electric cars equipped with sensor-actuator systems provide a multitude of possibilities to assist the driver and driving efficiency. Within the scope of research on the next generation fighter jet control an autopilot is used which is able to set its degree of autonomy in real-time based on the measured cognitive workload of the pilot [2]. An even closer relation between man and machine is represented by so called mobile suits envisaged by Japanese technology visionaries (e.g. in Japanese Mangas) since the 60 's. In 1963, the Rancho Arm was developed by Rancho Los Amigos Hospital (California) as an artificial limb for handicapped and later on integrated with computer technology by Stanford University. Experiments with whole mobile suits and power assistance devices were conducted by Japanese robotic scientists since the 70 's. Today's version of HAL (Hybrid Assistive Limb) is controlled by bio-electric signals thus blurring the borders between man and machine. Further, modern power suits allow a stepwise regulation of the suits' assistive power according to user's individual needs. Finally, Toyota calls its next generation of downsized, personal, and electrical mobility devices like iReal and iSwing explicitly "Mobility Robots" and closely cooperates with top robotic researches to make them as intuitively operated as possible.

Meanwhile, the ICT (Information and Communication Technology) and robotic technology no longer only focus on upgrading devices for mobility on middle and long distance (e.g. mobility from city to city, within a city) [3] but enhance more and more devices for mobility on a short distance and on the level of centimeters (mobility in the neighborhood, within the building, and individual motions). Especially in ageing societies, aforementioned robotic power assisting "tools" might transform our way of thinking about how to utilize robot technology. A multitude of robotic devices able to restore, support, augment, and supplement human abilities has been developed up to now. In order to support a systematic development of future concepts, new application scenarios and technologies, we have mapped the state-of-the-art of robotic power assisted "tools" supporting and augmenting human abilities. Particularly, we will show in this article, that advancing robot technology has a growing potential to gain great influence in the construction and building sector and as assistants in our built environment.

Most major industries have already extensively made use of robotic technology. Robotics has transformed production system technology in automotive industry, aircraft industry and in the electrical appliances' sector. Rapid advancements are currently made in ICT (Information and Communication Technology) and robotics in the medical field. Furthermore, in the US companies, e.g. John Deer, make advancements in applying field robotics to partly and fully autonomous farming machines. In the future, we see a huge potential for robotics - wearable cooperative systems as well as fully autonomous systems to permeate the field of construction and building technology. As construction technology we define tools and processes needed to erect a building. Whereas building technology refers to the buildings' or environment's performance and stands for tools and processes that assist people within the built environment from the scale of individual buildings up to neighborhoods or cities.

1.1 Construction technology

Up to now, automation and robotic technology has been applied in construction mainly for processing raw materials and production of building parts and building modules. Parts and modules had to be prefabricated in a structured and standardized environment for a safe and robust operation of the robots. In unstructured and not-standardized environments as on the construction site or in service environments, autonomous humanoids or service robots were difficult to operate. However, robot technology advances. Scientists as e.g. T. Hasegawa find ways to structure environments for robots [4] and also cognition and control technology become more advanced. Shimizu Corporation, a big Japanese construction company, cooperates with Yasukawa Electric Corporation, Kawada Industries and the national research institute AIST for introducing Humanoid robots to construction work for more than eight years already [5]. It has already been shown that humanoid robots as HRP-2 can carry a joinery bench together with a construction worker, fit an interior wall, and drive forklifts or diggers. Groups of HRP-2s can cooperate, move over a gradient of around five degrees and compensate for up to two centimeters on uneven surfaces [6]. They can straighten up themselves when they fall over. When carrying a component with a human, they use an adaptive and flexible arm system. An image processing system with a mobile portable control system has been developed to allow location detection. When the robots move over uneven surface, a force sensor in the sole of the foot and a balance sensor in the body register the difference and so, the sole of the foot can adapt to the surface.

Yokoyama, K., Maeda, J., Isozumi, T., Kaneko, K. (2006) Application of Humanoid Robots for Cooperative Tasks in the Outdoors

Fig. 1. Humanoid Robot HRP-2 assisting in construction environment in carrying and installing building parts and building modules. [5]

1.2 Building technology and service tasks

Experts and masterminds, as for example Bill Gates, announce the era of service robotics and estimate that service robotics as part of assisted environments will undergo a similarly fast and rigid development as the spread of personal computers in private and economic areas since the nineties. In 1961, Joe Engelberger already wondered, whether using robotic technologies only as industrial applications makes any sense. "The biggest market will be service robots," [7] asserted Engelberger, who started the industrial robotics era, when his firm Unimation delivered GM's first robot. Today, the application of robotics and distributed robotic sub-systems finally starts to extend into our home, office and town surroundings. This transformation, which has to be understood as a natural part of the evolution of robotics, will become visible especially when robots enter the field of service, assistance and care [8]. We think that modern robotics assisting and serving human beings will permeate into the "surroundings" of daily life and thus become an integral part of our built environment. Although building's interior environments and service environments tend to be less structured and standardized, increasingly autonomous robot systems can be applied to those environments. However, from a short term perspective, it will be easier to deploy not fully autonomous robotic systems as e.g. Suits for Power Assistance because they exploit human receptiveness and flexibility for robotic service.

2. Concepts and technologies

Exoskeleton and humanoid robot technology applied in construction and building technology demands for key concepts and technologies. At fists the degree of autonomy of the designed system has to be considered. Further, the fusion of speed, power and accuracy of robotic systems with human intelligence and flexibility within one system and the operation of humans and robots in dynamic environments can be supported by recent advancements in sensing and interface technology, actuator and control system technology and system design strategies. Further, a slow but continuous break up of strict borders between professions helps to create interdisciplinary cooperation and consortia which are able handling the complex challenges of man-robot cooperation. At the end of this chapter we present a categorization of exoskeleton and humanoid robot technology applied in construction and building technology based on the system complexity.

2.1 Exoskeletons, humanoids and autonomy

Robotic systems can have varying degrees of autonomy. Robots with a low degree of autonomy require detailed pre-programming or detailed real-time operation of a human person. Robots with a medium degree of autonomy only require supervision and an operator only has to assign tasks for which the robot autonomously finds sufficient solutions. Robots with a high degree of autonomy are capable of performing tasks and making decisions without major human interference. Especially in the area of construction and building technology the degree of autonomy of a system plays an important role as e.g. construction sites and service environments within buildings often provide dynamic environments and unstructured, complex work tasks. One can address this problem by modifying or structuring the environment or work task on the one hand or by advancing robot control technology or the application of artificial intelligence on the other.

2.2 Interface technology for human-robot cooperation

In task oriented systems where humans and robots closely cooperate a close link between the man's sensing and motion system and the robot's sensors and actuators is created ideally. With every advance in sensing technology and signal interpretation methods, these cooperative approaches become more practical. Following control strategies based on sensing human motions, feelings and intentions can be distinguished:

Conventional Control

- Steering Wheel
- Joystick
- Buttons
- Touch Screens

Intuitive Control

- Motion
- Gesture
- Eye Movement
- Force

Control by Bio Signals

- Bio-electric Signals
- Vital Data (EKG, Blood Pressure, Respiration Frequency)
- Brainwaves
- Electrons transmitted from Nervous System

2.3 Tele-Existence & Tele-Control

Concepts of Tele-Existence and Tele-Control to be used in the field of construction and building technology were advocated by Prof. Susumu Tachi at the University of Tokyo, already in 1980s. Tele-existence can be seen concept of advanced Tele-operation. Real world applications for tele-operated construction machinery as e.g. excavators and trucks had been developed in Japan since the Mount Unzen incident in 1991. A Vulcan eruption covered a large area with dust which would be health threatening for humans removing it. Thus a number of construction machines with the ability to be remote controlled from a save place had to be developed. Mt. Fugen is the main peak of Unzen Volcano, which is the collective name of a group of volcanic cones constituting the main part of the Shimabara Peninsula. Its phreatic eruption on 17 November 1990 caused a number of pyroclastic flows, which killed 44 people and destroyed 820 houses. The area around Mt. Fugen was deadly damaged by debris flow and pyroclastic flow. The restoration works to remove much stone and sand and the bank protection works were done by unmanned construction machines in order to avoid the risk of further catastrophes. Tele-operators manipulated machines from the operation room, which was more than 2km apart. Wearing special goggles, operators were watching 3D-images of the site sent by cameras equipped with machines. The efficiency of these remote-controlled works was estimated to be 70 percent of usual works [9]. Due to this incident Japanese researchers and construction companies realized the importance of tele-operation technology.

Today intelligent excavators with the ability for tele-operation and even partly autonomous operation capability are under development in the R&D sections of all major Japanese and Korean contractors. Further Japanese researchers and construction companies have tried to control construction machinery by teleported humanoids (Figure 02). This approach has the advantage that standard construction machinery can be used without modification.

Tele-existence and Tele-control can not only be used for 1:1 real time control of a single robot or intelligent construction machine by one assigned operator. With rising degree of autonomy of the robot systems used the tele-operator becomes a sort of supervisor able to control multiple construction machines at once. Already in the 80s the vision of multiple cooperating construction robots are operated by a single human supervisor from a central existed (Figure 03). Today indeed more and more researchers succeed in developing fully functioning and highly autonomous construction machines that can be tele-supervised (Figure 03).

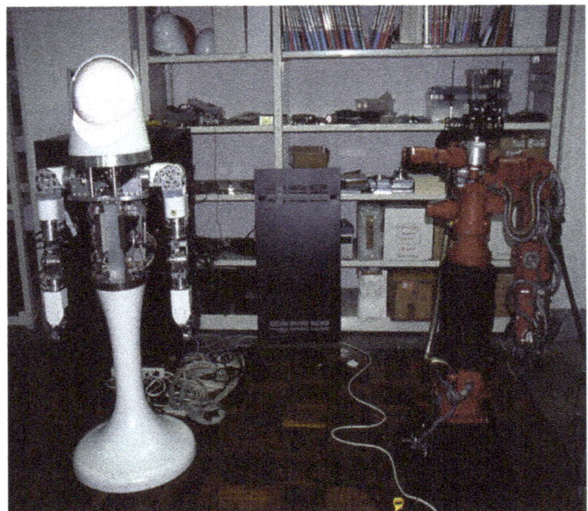

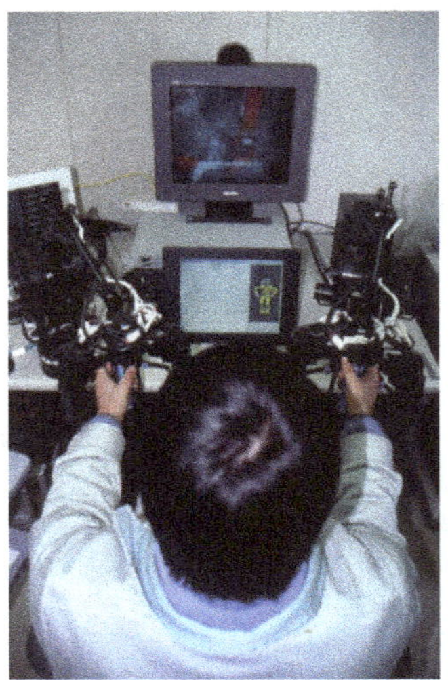

Fig. 2. Left: Prof. S. Tachi, Tele-existence Mechanical Engineering Laboratory (MEL) and MITI, 1986; Middle and Right: Contol of Honda ASIMO Humanoid; Tokyu Construction, Kawasaki Heavy Industries,and AIST

Society of Civil Engineers, Construction Robotics Commission, Prof. Shigeyuki Obayashi, 1985

Fig. 3. Multiple cooperating construction robots are operated by a single human supervisor from a central box, Vision Sketch Japanese Research Institute, 1980

Kajima, Pictures taken form website:
http://www.kajima.co.jp/gallery/civil_kajima/bousai/bousai01.html, last visited 24/07/2011.

Fig. 4. Real world applications for tele-operated construction machinery as e.g. excavators and trucks had been developed in Japan since the Mount Unzen incident in 1991. A Vulcan eruption covered a large area with dust which would be health threatening for humans removing it. Thus a number of construction machines with the ability to be remote controlled from a save place had to be developed. Kajima Corporation, Japan, 1991

Fig. 5. Fully functioning system for tele-operation of robotic excavators, the excavators can operate on a high level of autonomy; the excavation process is monitored by separate laser module (picture right side) providing information to the robotic excavator. Hanyang University, Korea, 2011.

2.4 Actuator and control system technology

Complex systems of actuators, joints and links are controlled based on information sensed and interpreted by internal and external sensor systems. Actuators create the activity and movement within robotic systems. Today following actuation systems are used in a robotic power, motion/sensing and cognition augmentation:

- Electric Motors
- Series Elastic Actuators
- Air Pressure
- Muscle Wire (e.g. Shape Memory Alloy)
- Electroactive Polymers
- Piezoelectric Actuators

Besides the increasing ability to downsize motors it is by now possible to improve precision and speed. Advances in robot kinematics and robot dynamics are important for developing robust and save control system technology for more complex man-robot systems in construction and building technology.

2.5 Energy supply

Energy Supply is a crucial issue in developing exoskeleton and humanoid robotic applications for construction and building technology. Unlike to robotic applications in other industries, many tools and assistive devices need to be independent from connecting cables. However, battery packs necessary to supply energy for the actuators represent heavy load. Thus, on the one hand the battery systems need to be developed so that they support mobility and wear-ability of robotic systems but on the other hand robotic applications and systems have to be designed to be highly energy efficient.

2.6 Development complexity

Only interdisciplinary cooperation can handle the complexity associated with advanced man-robot cooperation systems. Besides knowledge from fields related to robotics (electrical engineering, mechanical engineering, and informatics), knowledge from various anthropological sciences as psychology, ergonomics, neuroscience and psychology is needed to design such systems [10]. Moreover, the blurring of borders between man and machine within a single system gives rise to philosophical and ethical questions. Finally, in order to receive subsidies from investing enterprises and to manage complex system developments, entrepreneurs with the ability to lead highly interdisciplinary teams and complex innovations have to be educated.

2.7 Categorization according to system complexity

In order to be able to design work tasks and application scenarios for exoskeletons and humanoids in construction and building technology we classify robotic systems according to system complexity. With complex systems we mean systems that consist of a number of sub-systems and sub-elements. Accordingly, element technologies are basic technologies. They can be applied as standalone systems or combined as sub-elements to more complex subsystems. Subsystems denote e.g. partial exoskeletons (exoskeleton for lower body part/feet, Exoskeleton for upper body part). A total system consists of several sub-systems; here we mean e.g. total exoskeletons or mobility robots. Autonomous robot systems (humanoid robots, service robots) and distributed robot systems can operate highly autonomous and are able to support robot service on city scale. They stand for highly complex robot systems built up by multitude of element technologies, subsystems and autonomous robot systems.

1. **Element Technology**
   - Power Augmentation
   - Sensing and Motion Augmentation
   - Cognition Augmentation
2. **Subsystems**
   - Assistive Devices and Partial Exoskeletons
3. **Total Systems**
   - Exoskeletons
   - Mobility Robots
4. **Autonomous Robot Systems**
   - Android/Humanoid Robots
   - Service Robots (Service in Buildings)
5. **Distributed Robot Systems**
   - Town Robotics & Space Robotics

3. Examples according to system complexity

In this section we outline several examples of each of the categories introduced above. All examples contain information about the developing institution and about the systems' performance. We also go into the target groups and the development stage of each system. Each category is introduced by a short description of the status quo in the field. Further, we outline applications in construction and building technology for each category.

3.1 Element technology

Element technologies are basic technologies that can be applied as standalone systems or combined as sub-elements to more complex subsystems. We denote technologies for power augmentation, sensing and motion augmentation and cognition augmentation as element technologies.

3.1.1 Power augmentation

"Power Effector" developed by MMSE Project Team is a robot which augments the strength of a part of human body, but its concept is different from others. Most wearable robots must be compact and light in order to be comfortable for the users and be suitable for the surroundings which are designed for the dimensions of the human body. On the other hand, another approach is to be bigger and heavier so that operations can be carried out which a person itself could never accomplish. Mr. Katsuya Kanaoka, Ritsumeikan Univ. has proposed the concept "Man-Machine Synergy Effector" (MMSE), which combines flexible human skills with precision and high power of machines [11]. "Power Effector" can amplify human power 1 to several thousand times. This Technology is expected to be introduced to heavy physical work that is not programmable and requires not only powerfulness but also intelligence, facility, and experience.

Power Effector	
Developer	MMSE Project Team
Leading Researcher	Katsuya Kanaoka, Ritsumeikan University
Purpose	Augmentation of the strength in upper limbs
Output	Arm: 50 kgf, Grip: 500 kgf
Height	1550 mm
Width	1200 mm
Length	3360 mm
Weight	120 kg
Driving System	AC Servo Motor, Ball Screw
Power Supply	AC Power Supply
Sensor	6-Axis Force Sensor

Power Effector: Scanned from Takashi, Y. (2005) Collected Data on Partner Robot Technologies, NTS. INC.

Power Pedal	
Developer	MMSE Project Team
Leading Researcher	Katsuya Kanaoka, Ritsumeikan University
Purpose	Augmentation of the strength in upper legs
Output	7 times of human power
Commercial Launch	2015
Price	20 million yen
Degree of Freedom	Leg: 3 DOF x 2 Sole: 3 DOF x 2
Sensor	6-Axis Force Sensor

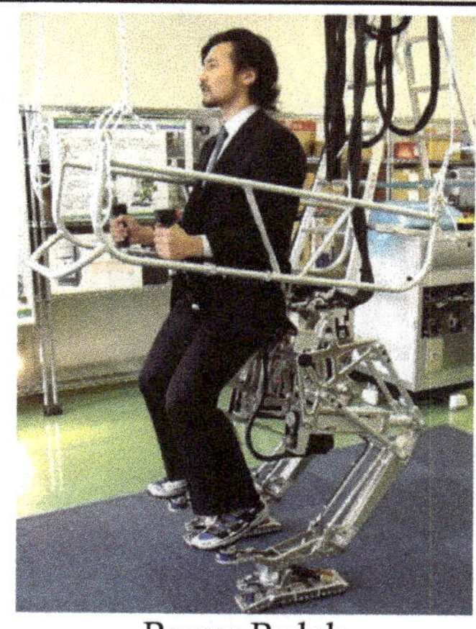

Power Pedal:
http://robonable.typepad.jp/trendwatch/2008/07/post-483b.html

Application in Construction: Pre-fabrication, handling and assembly of heavy building components in factory and on-site installation of heavy panels to walls and facades.

Left: Copyright T. Bock, Right: Copyright T. Bock Komatsu Construction Machinery Division, applied at Kajima Construction

Fig. 6. Left: handling robot used in building prefabrication, Germany. Right: Power Effector used in high-rise construction for façade element installation, Japan.

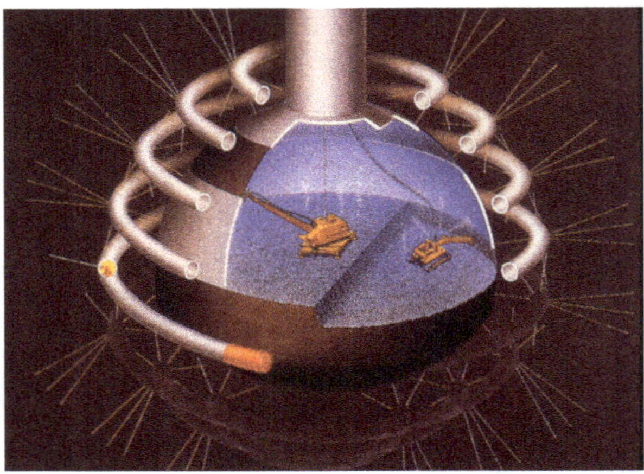

Left: Copyright T. Bock, Telerobotic Caisson Construction Project, Right: Copyright T. Bock, MITI Chikakukan Project, 1985

Fig. 7. Tele-operated Power Effectors used in mining, Japan

3.1.2 Sensing and motion augmentation

This category represents robotic devices which are equipped with a part of human body and support its movements. These systems should be designed accurately not to interfering complex movements on joints. An exoskeleton developed by University of Tsukuba works only when the wearer needs its help so that it doesn't disturb wearer's delicate works [12]. Researchers in Okayama University developed some wearable robots called "Power Assist Wear" [13]. Their actuator is a pneumatic rubber artificial muscle which is light, soft and fitted for users. "Power Assist Glove" is made from a curved type of artificial muscle which is a combination of materials with different stretch, e.g. rubbers and cloths. Although they are mainly used as rehabilitation tools at the moment because of their limited effectiveness, some products aim at being adapted to construction works and enabling elderly or female workers to work with less physical efforts.

Exoskeleton hand and wrist support system	
Developer	University of Tsukuba
Leading Researcher	Yasuhisa Hasegawa
Technology Readiness	Prototype, Research and Development Project
Target user	People with weakened holding force
Purpose	Assist for motions of hand and wrist without decreasing DOF
Weight	1850 g
Sensor	Bio-electric potential measurement
Actuator	DC Motor x 12

Exoskeleton Hand and Wrist Support System: http://www.edu.esys.tsukuba.ac.jp/~hase/ForearmSupport.html

Power Assist Glove	
Developer	Okayama University
Leading Researcher	Noritsugu Toshiro
Technology Readiness	Prototype, Research and Development Project
Target User	Elderly and female workers, Heavy workers
Purpose	Assist for bending motion, Augment of the grasping force
Weight	120 g
Actuator	Two-Joint Curved Pneumatic Rubber Artificial Muscle

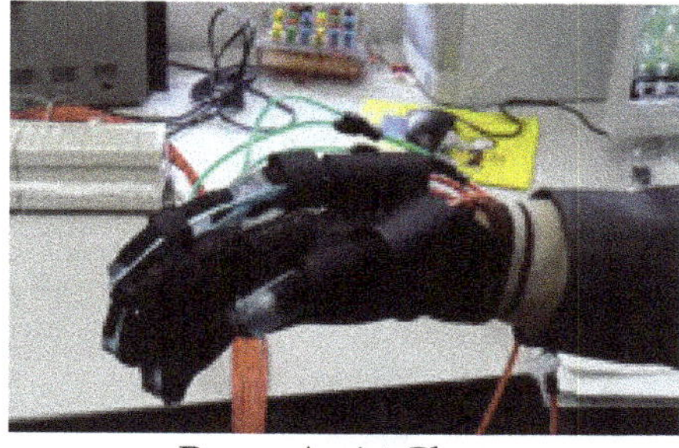

Power Assist Glove: http://www.smrj.go.jp/incubation/od-plus/labolist/055057.html

Application in Construction: Support of workers simple and continuous movements such as grasping control sticks or lifting heavy building materials up. Enabling weakened workers because of aging or injuries continue to work.

3.1.3 Cognition augmentation

Wearable computing systems are systems which are attached to a person's body during use. A main goal of researchers and developers is that this systems work seamlessly in the background. They shall assist a person in various situations but not distracting him or the environment - at the best they are invisible. Wearable Computing technologies have initially been developed for monitoring astronauts: Life Guard [14] by NASA and Stanford University, USA, Health Gear [15] by Microsoft Research, E-Watch [16] by Technical University of Munich, Germany and Carnegie Melon University, USA, V-Mote [17] by Virginia Commonwealth University. Today, wearable computing systems are increasingly applied in the industry and service scenarios. A multitude of applications are envisioned in the military, too. This category "Wearable Computing" mainly represents technologies that support or augment human sight, hearing and cognition but not human's physical motion power. Compared to mobile robots and humanoids, these wearable computing devices generally have a lower degree of autonomy as they are directly connected to the human activity.

Application in Construction: Augmented and Mixed Reality applications can support workers off-site and on-site to perform assembly operations. Wearable sensors devices attached to workers can be used to monitor their construction acidity as well as their health. Various AR and MR application have been developed at the laboratory of the authors in a project called MARY [18].

Liteye LE-700	
Developer	Liteye Systems, Inc.
Product Type	Head Mounted Display
Size	80mm x 24mm x 31mm
Weight	80 g
Display Technology	OLED
Resolution	800 x 600 and 640 x 480
Luminance	Color:>70cd/m2 White:>270cd/m2 Yellow:>650cd/m2 Green:>600cd/m2
Power	5 - 6 v DC input 400 mW Typical

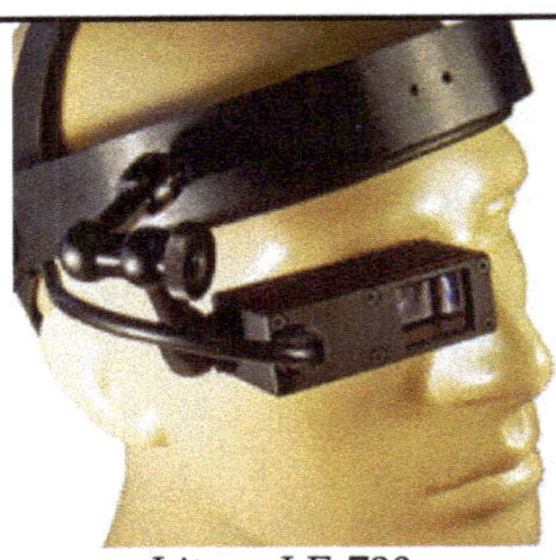

Liteye LE-700: http://www.inition.co.uk/inition/dispatcher.php?URL_=product_hmd_liteye_700&SubCatID_=15&model=products&action=get&tab=summary

Anti-RSI Garment	
Project	the Con Text project: Contactless sensors for body monitoring incorporated in textiles
Developer	Philips Research Technische Universität Berlin Katholieke Universiteit Leuven Textile Research Institute Thüringia-Vogtlandia Netherlands Organization for Applied Scientific Research Clothing Plus Oy
Product Type	Wearable Computing
Purpose	Prevention against repetitive strain injuries
Sensor	Contactless EMG Sensors

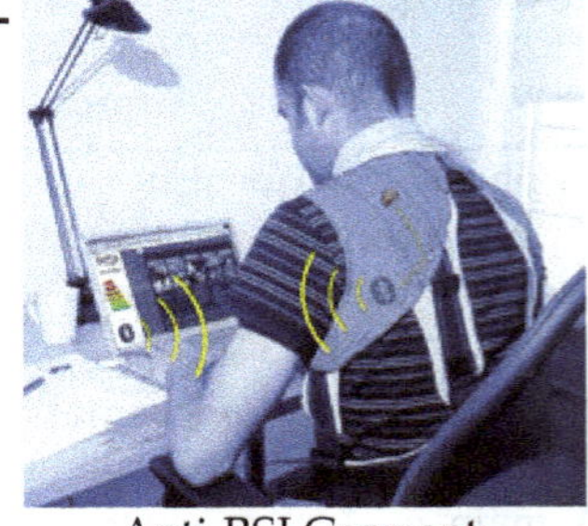

Anti-RSI Garment: http://www.gizmag.com/smart-fabrics-medical-applications/10242/picture/56113/

3.2 Subsystems (assistive devices and partial exoskeletons)

This category represents wearable robots which assist wearers during laborious and continuous work. Their output is not very strong, but these devices are effective in preventing workers from getting injuries such as backaches. Honda has developed several walking assist devices based on the technology utilized in ASIMO, their famous Humanoid robot. "Walking Assist Device with Bodyweight Support System" supports a part of the wearer's weight while walking, going upstairs and downstairs and keeping in a hard position. The supposed users are not disabled but need support for certain works. Smart Support is a business company from Hokkaido University, which is aimed to popularize their product called "Smart Suit". It's a light and comfortably wearable power assist system motivated "Semi-Active Assist Mechanism". This product has been already used for restoration works after the big earthquake in Tohoku Japan.

Walking Assist Device with Bodyweight Support System	
Developer	Honda
Target User	Walker, Factory Workers
Technology Readiness	Prototype, Tested in own Factories
Weight	6.5 kg
Drive System	Motor x 2
Power Supply	Rechargeable Lithium-ion Battery
Operating time	2 hours
Support Motions	Walking, going up and down stairs, in a semi-crouching position
Sensor	Shoes: Foot force sensors
Based Technology	Honda Humanoid Robot ASIMO

Walking Assist Device Honda: http://world.honda.com/news/2008/c081107Walking-Assist-Device/

KAS: Knee-assistive System	
Developer	Hanyang Univ. Korea
Leading Researcher	Chang Soo Han
Target User	Construction Workers
Purpose	Prevention against impairment on knees
Support Motions	Level walking and Step walking while carrying heavy materials
Technology Readiness	Prototype, Research and Development Project
Strength of Assistance	45 kg
Sensor	Muscle Stiffness Sensor
Actuator	Flat motor, Harmonic drive

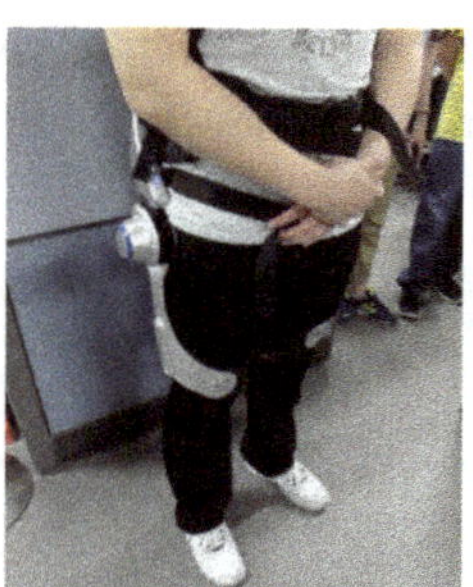

KAS: Prof. Thomas Bock

Smart Suit	
Developer	Smart Support
Leading Researcher	Takayuki Tanaka, Hokkaido University
Target User	Agricultural workers, Care workers
Technology Readiness	Prototype, Used for Restoration Works
Model	Smart Suit / Smart Suit Light
Weight	1 kg (goal) / 400g
Power Supply	Dry battery
Reduction of Fatigue	14%
Sensor	Back: Bending sensor
Actuator	Elastic Material, small motor / Elastic Material

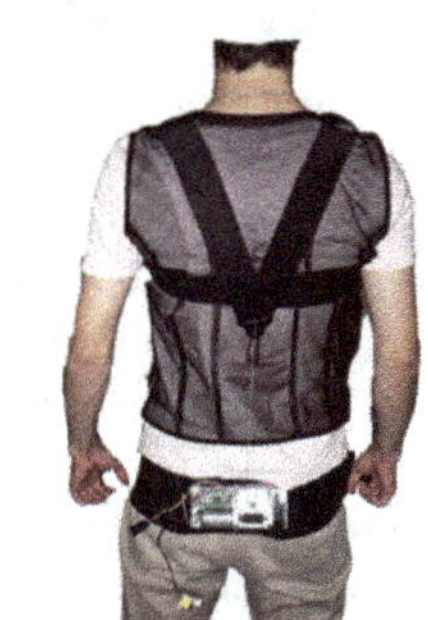

Smart Suit: http://smartsuit.org/

Application in Construction: Support of workers physical abilities in light construction and restoration tasks. Support of workers in prefabrication factories for industrialized building construction (Sekisui House, Sekisui Heim and Toyota Home)

Smart Suit, Figure taken form Website: http://smartsuit.org/, last visited 24/07/2011

Fig. 8. Smart Suit developed by Hokkaido University and Smart Support Company is used for restoration works after the big earthquake in Tohoku, Japan.

Left: Walking Assist Device Honda, Figure taken form Website: http://world.honda.com/news/2008/c081107Walking-Assist-Device/, last visited 24/07/2011, Right: Copyright T. Bock, T. Linner

Fig. 9. Left: Honda is now testing the usability of its Body Weight Assist Device in its own factories. Right: Devices like the Body Weight Assist Device can support existing industrialized and production line based prefabrication of buildings, Sekisui Heim, Japan.

3.3 Total systems

Element technologies as described above can be combined with sub-elements and subsystems (e.g. partial exoskeletons, exoskeleton for lower body part/feet, and exoskeleton for upper body part) to more complex total systems as full body exoskeletons or mobility robots.

3.3.1 Exoskeletons

"Robot Suit HAL" is a well-known Japanese Exoskeleton which is specialized on detecting very weak corporal signals on the surface of the skin which are generated when a person attempts to move. In 2008, Daiwa House Industry started the renting of "HAL for Welfare-being". The product is now used in several nursing homes and welfare facilities in Japan to assist elderly or disabled people in walking. There are also some other prototypes of exoskeleton in Japan, and each of them uses different actuators, e.g. ultrasonic motors, pneumatic rubber artificial muscles, and air bag actuators[19][20][21]. They are tackling some common challenges such as down-sizing, long-time operations, and low-cost manufacturing in order to bring their product to market. These exoskeletons will get further usability when they are combined with some other element technologies. Prof. Shigeki Toyama, who made "Wearable Agri Robot", plans to develop Augmented Reality goggles which show information of vegetables and fruits, the health condition of workers, and the working hours and inform workers when to have a break. Although each project team expects to introduce own products into a specific working area, it's relatively easy to apply one them to other fields, especially construction works, because they support mainly same movements such as bending down or lifting heavy things up and have a common purpose; preventing workers from repetitive strain injuries.

HAL: Hybrid Assistive Limb	
Developer	CYBERDYNE
Leading Researcher	Yoshiyuki Sankai
Type	Full Body / Lower body
Target User	Physically weakened people, Disabled people
Technology Readiness	Lease Rental in nursing home and welfare facility
Price	4 - 5 million yen
Height	1600mm
Weight	23 kg/ 15 kg
Power Supply	AC100V Charged battery
Operating time	2 hours 40 minutes
Sensor	Corporal Signal Sensors Angle Sensor of joints Floor Reaction Force Sensor
Drive System	Power Units

HAL: Prof. Thomas Bock

Wearable Agri Robot	
Developer	Tokyo Agriculture and Technology University
Leading Researcher	Shigeki Toyama
Target User	Agricultural Workers
Technology Readiness	Prototype, Tested in Farmland
Commercial Launch	2012
Price	1 million yen
Type	Heavy/ Light
Support Motions	ex. Harvesting vegetables / Picking fruits
Weight	23 kg/ 30 kg
Strength of Assistance	62 % (average)
Interface	Voice Recognition
Sensor	4 types of sensors (Angle, Pressure)
Actuator	Ultrasonic Motor x 8

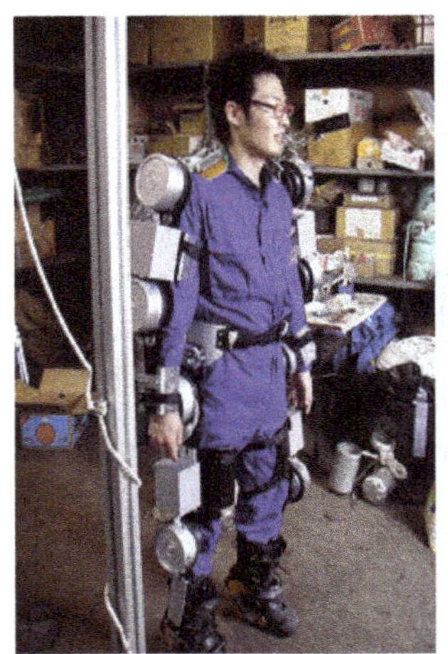

Wearable Agri Robot: http://www.tuat.ac.jp/~toyama/research_assistancesuit.html

Muscle Suits	
Developer	Tokyo University of Science, Hitachi Medical Corporation
Leading Researcher	Hiroshi Kobayashi
Target User	Heavy Workers
Technology Readiness	in the phase of Commercialization
Type	Arm & Back/ Back
Weight	7.5 kg/ 3.5 kg
Total DOF	6 DOF/ 1 DOF
Support Torque	Elbow: 45 Nm/ - Shoulder: 45 Nm/ - Back: 90Nm/ 90 - 360 Nm
Interface	Motion Playback by Switch / Switch
Actuator	Pneumatic Rubber Artificial Muscle

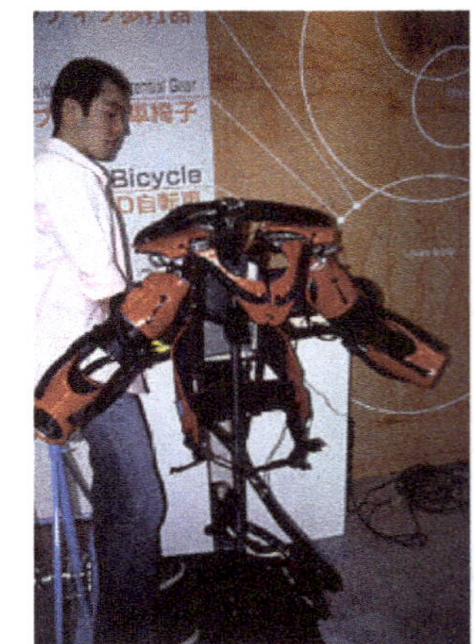

Muscle Suit: Prof. Thomas Bock

Power Assist Suit for nursing care	
Developer	Kanagawa Institute of Technology
Leading Researcher	Keijirou Yamamoto
Target User	Nurses, Care-workers
Support Motion	Lifting up a care-gaver
Technology Readiness	Prototype
Weight	30 kg
Power Supply	Ni-MH batteries
Operating time	20 minutes
Strength of Assistance	50 % (for safety measure)
Sensor	Muscle Hardness Sensor
Actuator	Air Bag Actuators driven by micro air pump

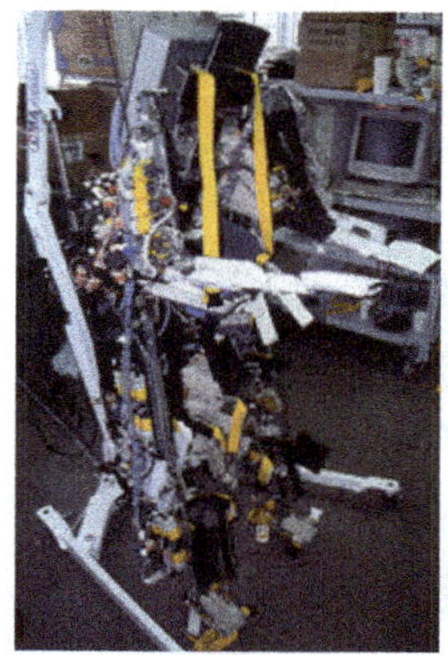
Power Assist Suit: Prof. Thomas Bock

Wearable Robot Suit Version 2	
Developer	Univ. Korea
Leading Researcher	
Target User	
Technology Readiness	
Weight	
Power Supply	
Operating Time	
Strength of Assistance	
Sensor	
Actuator	

Wearable Robot Suit: Prof. Thomas Bock

Application in Construction: Support of workers physical abilities in prefabrication factories or on the construction site. Support in lifting and assembly of heavy and bulky construction components [22].

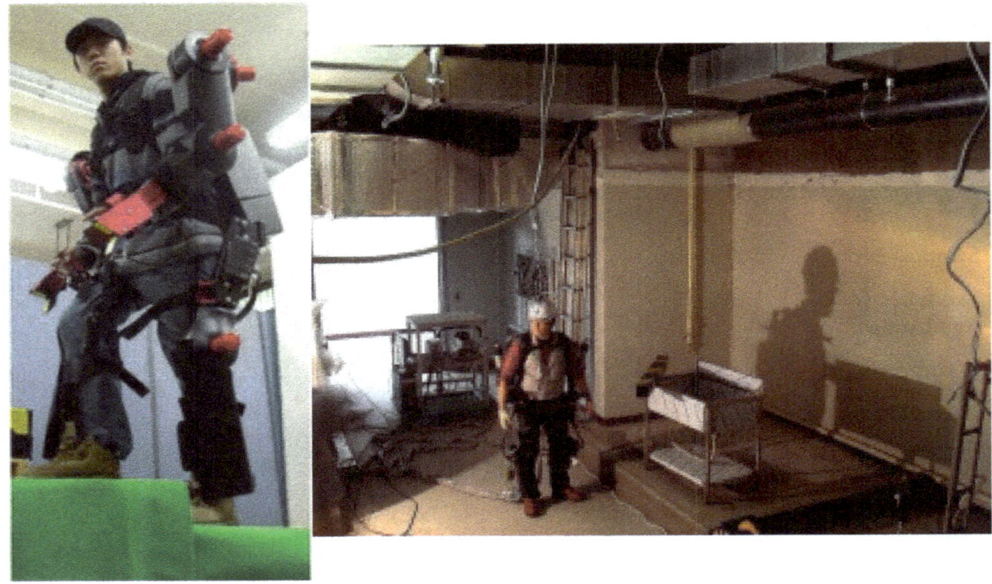

Fig. 10. Wearable robotic exoskeleton system for construction workers. The system can e.g. support workers to carry and assemble heavy steel bars. Hanyang University, Korea

3.3.2 Mobility robots

Robots for lifting people are applied at the homely environment to support people with immobility (elderly, patients or disabled) and their caregivers. Lifting is a basic activity of daily life, meaning it is an event that is indispensable for bathing, dressing, going onto the toilet and feeding. Patient transfer robots were in the focus of researchers and commercial developers since the beginnings of the research upon nursing in the 70s. Several types of transfer can be identified and various types of robots have been developed. Robots for lifting people from the bed, robotic wheelchairs and robotic walking frames are just a few basic examples to be named among a series of robotic patient transfer systems, which have been developed up to now. However, recently robotic technology is also applied to personal mobility following a "design for all" strategy. Toyota calls its next generation of downsized, personal, and electrical mobility devices like iREAL and i-Swing explicitly "Mobility Robots" and for that closely cooperates with top robotic researchers making these devices as intuitively controllable as possible. Further, also mobile suits as Toyota's i-foot and KAIST's HUBO-FX1 [23] belong to the category of mobility robots. Mobility Robots can be considered as a special type of mobile suits. They not only augment or multiply human power but they equip human beings with a completely new capability. Mobility robots can communicate with each other and the environment (car-to-x communication) and have a high potential for autonomous or autopilot control. Therefore, in our categorization we place mobility robots between Exoskelettons and fully autonomous Humanoids.

i-REAL	
Developer	Toyota
Driving Mode	Low/ High
Height	1430 mm/ 1125 mm
Width	700 mm
Length	995 mm/ 1510 mm
Maximum cruising speed	6 km/ 60 km
Power Supply	Lithium-ion Rechargeable Battery
Charging Time	2 hours
Cruising range	30 km
Interface	Drive Controller
Other Technology	Communication Display

i-REAL: Prof. Thomas Bock

Personal Mobility for Indoor Use	
Developer	The University of Tokyo IRT, Toyota
Height	1300 mm
Width	600 mm
Length	640 mm
Weight	45 kg
Sensor	Seat: 6-Axis F/T Sensor Seat and Footrest: Pressure Sensor
Other Technology	Perception of pattern on the floor containing information about position

Mobility for Indoor Use: Prof. Thomas Bock

CHRIS: Cybernetic Human-Robot Interface System	
Developer	Hiroshima University
Height	1400 mm
Width	1000 mm
Length	750 mm
Weight	70 kg
Maximum Moving Speed	Forward: 2.5 km / h Backward: 1.8 km / h
Power Supply	Lead Storage Battery x 3
Driving System	DC Brushless Motors x 2
Interface	Cybanetic Interface

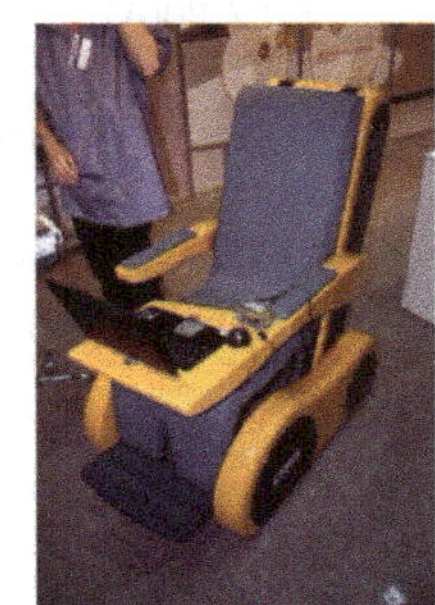
CHRIS: Prof. Thomas Bock

Walking Assist Device	
Developer	HITACHI
Height	
Width	
Length	
Weight	
Power Supply	
Driving System	
Interface	

HITACHI: Prof. Thomas Bock

i-foot	
Developer	Toyota
Height	2360 mm
Weight	200 kg
Total DOF	12 DOF
Load Capacity	60 kg
Cruising speed	1.35 km/h
Interface	Joystick Controller
Other Ability	Navigating Staircase

i-foot: Prof. Thomas Bock

HUBO FX-1	
Developer	KAIST
Height	1750 mm
Weight	150 kg
Total DOF	12 DOF
Load Capacity	100 kg
Driving System	400 / 800W AC Servo Motor with Driver
Sensor	3-Axis F/T Sensor at feet Inertial Sensor at Torso 2-Axis Accelerometers on Soles
Interface	Joystick

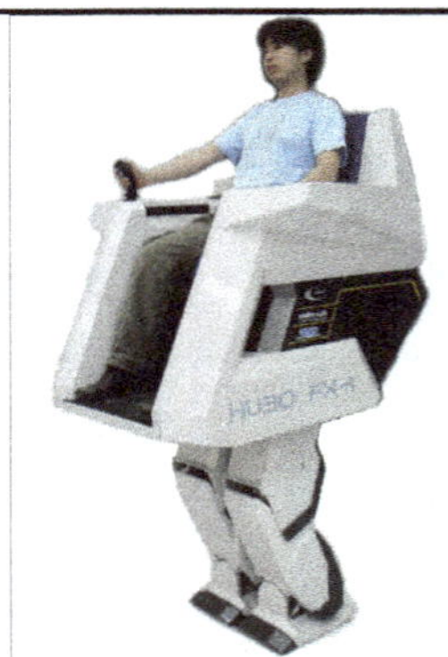

HUBO FX-1: KAIST Humanoid Robot Series, Public Demonstration

Application in Construction: Support of material and element delivery and installation. Support of factory logistics and construction site logistics. Adaptability of technologies like a recognition system of floor surface which some personal mobility robots already have into logistics on construction site or prefabrication factories.

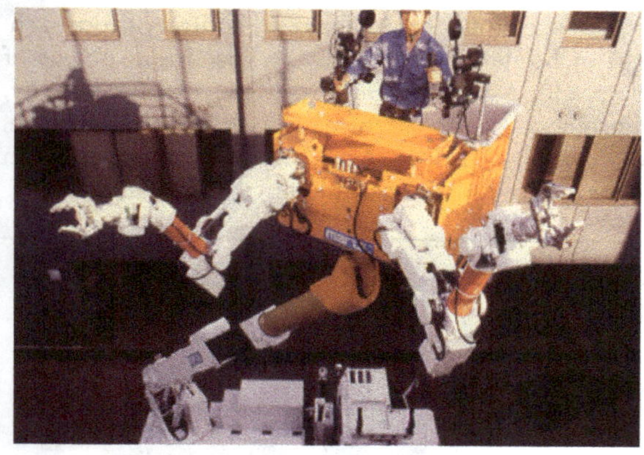

Copyright T. Bock

Fig. 11. Left: Mobile Construction and Maintenance Robot, TEPCO, Japan; Mobile and Remote Controlled Transportation Robot for Construction Sites, Obayashi, Japan

Copyright Dr. S. Lee, Prof. Han, Hanyang University

Fig. 12. Mobile Robotic System for Human-Robot cooperative work tasks (Ceiling Panel Installation), Samsung Construction and Hanyang University, Korea.

3.4 Autonomous robots

Autonomous robots stand for highly complex and autonomous robot systems built up by multitude of element technologies and subsystems. In our categorization we consider android/humanoid robot system and service robots as autonomous robot systems.

3.4.1 Android/Humanoid robots

Humanoid robots are complex autonomous systems that can adapt to changes in the environment. Their appearance, function and motion capability are entirely depending on the equivalent in the human body. Androids not only interpret the human body's function but are designed to imitate human appearance and behavior. For both humanoids and androids service scenarios can easily be envisioned, yet, due to their technical complexity, real world applications are still rare. Exoskeletons come from a contrary approach, combining the flexibility and intelligence of human beings with the speed and power of robotic systems. Today, Wearable Robots and Assistance Suits provide more flexibility, however, in the future, considering advancements in robot control and artificial intelligence, autonomous humanoids, androids and inhuman service robots are likely to increase in flexibility and the ability to adapt to various unstructured tasks and environments.

HRP-2 Promet	
Developer	Kawada Industries
Height	1540 mm
Width	620 mm
Weight	58 kg (including batteries)
Walking speed	0~2 km/h
Holding Force	2kgf (one hand)
Total DOF	30 DOF
Drive System	48V 20A(Imax), 2axes/driver x 16
Power Supply	NiMH Battery DC 48V, 18Ah
Sensor	Joint: Incremental Encoder Visual Input: Trinocular Stereo Camera Body: 3-axis Vibrating Structure Gyro, 3DOF Acceleration Sensor Arm: 6-axis F/T Sensor Leg: 6-axis F/T Sensor

HRP-2: Prof. Thomas Bock

HRP-1S	
Developer	Honda
Height	1600 mm
Width	600 mm
Weight	99 kg (excluding batteries)
Walking speed	0 ~ 6 km/h
Total DOF	30 DOF
Drive System	Brushless DC servo motor
Power Supply	Ni-Zn Battery
Sensor	Body: Inclination Sensor (Gyro-scopes and G-force sensors) Foot and wrist: F/T Sensor Head: 2 Video Cameras

HRP-1S: Prof. Thomas Bock

KHR-3 (HUBO)	
Developer	KAIST
Height	1250 mm
Width	417mm
Depth	210mm
Weight	56 kg
Walking Speed	0 ~ 1.25 km/h
Grasping Force	0.5 kg / finger
Total DOF	41 DOF
Power Supply	300W NiMH Battery
Operation Time	90 min.
Sensor	3-axis F/T Sensor Tilt Sensor CCD Camera Pressure Sensor

KHR-3: http://hubolab.kaist.ac.kr/KHR-3.php

Application in Construction: Control of existing standard construction machinery by tele-operated humanoids (Figure 13). Humanoid Robots (as e.g. HRP-2) can assist workers in construction environments by carrying and installing building parts and building modules (Figure 01).

Copyright T. Bock

Fig. 13. Left: HARP Humanoid Robot driving forklift delivering construction material. Right: Honda's Asimo controlling an excavator.

3.4.2 Service robots (service in buildings)

Especially in Japan, a multitude of so called Entertainment Robots and Service Robots are developed and sold. Entertainment robots are designed to amuse, communicate, and perform simple tasks in the household. Mitsubishi's Wakamaru and Sony's Aibo for example had primarily been designed to communicate with household members and play music, not for providing care or household services. Yet, as the upkeep of social interaction increasingly becomes an integral part of care strategies, the taking over of entertainment and communication tasks by robots is envisioned by researchers and developers. Furthermore, homemaking robots are robots which take over simple tasks as cleaning, transport of objects or informing about intruders or the pet's well-being. Often, the robot's performing tasks in the household contain elements of both entertainment and homemaking.

Wakamaru

Developer	MITSUBISHI HEAVY INDUSTRIES
Business Model	Home Service Robot
Height	1000 mm
Width	450 mm
Length	470 mm
Weight	30 kg
Maximum Moving speed	1 km/h
Total DOF	13 DOF
Drive System	DC Servo Motor
Power Supply	Lithium-ion Battery
Operation Time	2 hours
Communication	Human detection, Individual recognition, Voice Recognition, Speech synthesis

Wakamaru: Prof. Thomas Bock

RIDC-01

Developer	Tmsuk
Business Model	Guidance & Cleaning Robot
Height	1300 mm
Width	700 mm
Length	960 mm
Weight	100 kg
Maximum Moving speed	3.0 km/h
Total DOF	10 DOF
Power Supply	DC-24V Lithium-ion Battery
Operation time	2 hours

RIDC-01: Prof. Thomas Bock

PBDR: Partner Ballroom Dance Robot	
Developer	Tohoku University NOMURA UNISON TroisO
Business Model	Dance Partner Robot
Height	1650 mm
Width	1000 mm
Length	1000 mm
Weight	100 kg
Degree of Freedom	15 DOF
Drive System	Servo Motor
Power Supply	Battery

PBDR: Prof. Thomas Bock

Application in Construction/Building: Service Robots can assist to carry out or can (partly) autonomously carry out household tasks and care tasks in an ageing society. Transfer of technologies (which some entertainment robots already have) towards humanoid robots thus gaining communication and cooperation ability. Further service robots can be used to maintain buildings, inspect nuclear power plants [24] and assist in homes for the elderly.

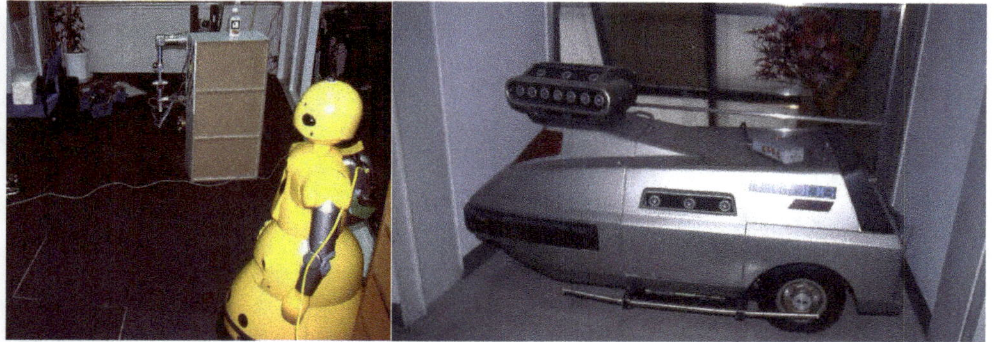

Copyright T. Bock

Fig. 14. *Left:* Wakamaru acting as edutainment and communication robot in a home environment. *Right:* Robot for guiding and helping blinded and disabled people at home

3.5 Distributed robot systems

Urban robotics is a research field situated between smart/sensible city research and robotics research. Its goal is to develop cutting-edge technologies as well as application scenarios for urban life supported by robotic devices. The research field is pioneered by T. Hasegawa and his Town Management System enabling robots to outsource complexity to sensors and vision systems distributed in the city environment [4]. Other interesting impulses in this research field are coming from research on smart cars and e-government. Furthermore, NASA accounts controlled traffic systems and smart grid energy systems as so called "Immobile Robots" [25].

Application in Construction/Building: Distributed robot systems enable robots to execute various tasks for ordinary human life on building and city scale. Further, they can be used to operate highly automated construction sites (this application of robotics we describe in detail in [26]). Tele-operated robot and construction system consisting of multiple subsystems can be used for automated construction on moon, mars or deep sea and underwater mining operations.

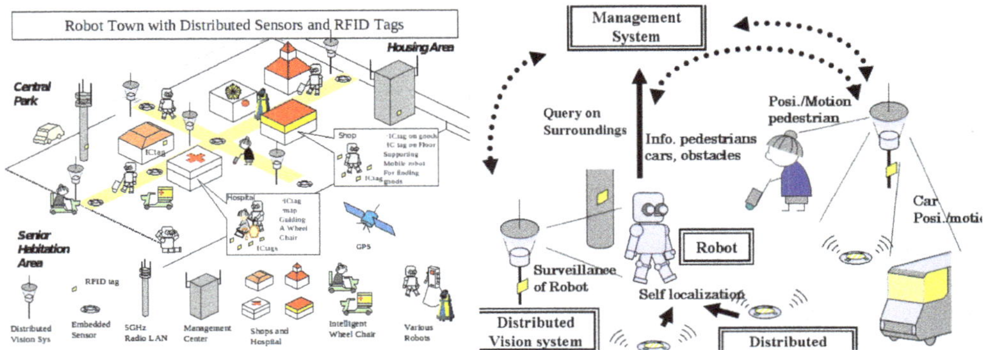

Copyright T. Hasegawa

Fig. 15. The Robot Town enables robots to execute various tasks for ordinary human life by creating an urban environment well structured in informative way for robots and service systems. T. Hasegawa, Kyushu University [4].

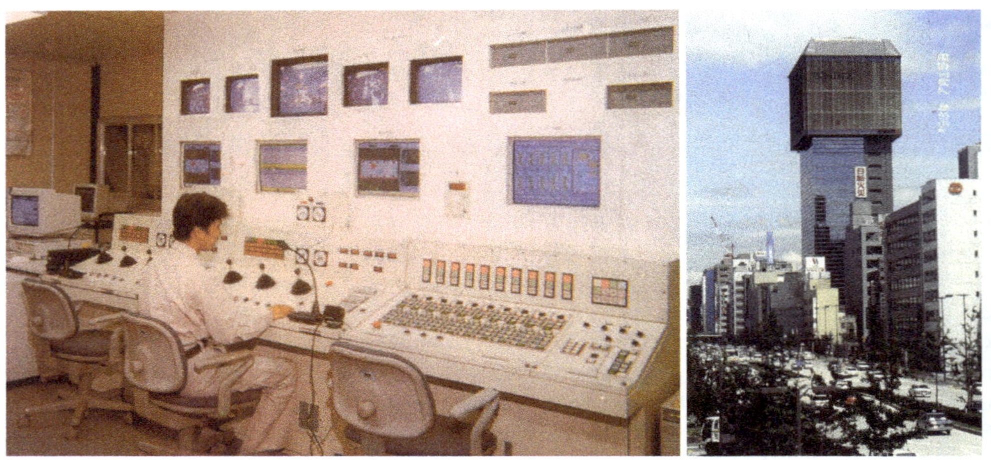

Copyright T. Bock

Fig. 16. Control Center of Shimizu's automated construction system for highly automatic erection of high-rise buildings, Japan, Shimizu Corporation.

Fig. 17. Tele-operated robot and construction system consisting of multiple subsystems for automated construction on moon or mars, Japan, Shimizu Corporation.

4. Relation of system complexity and work task complexity

By implementing robotic technology in construction and building technology, the degree of autonomy of the robotic system has to be considered. In general, the degree of autonomy of a robotic system is closely correlated to its work tasks it can perform. Work tasks can be classified into work tasks which are structured and standardized on the one hand and unstructured and not standardized work tasks on the other hand. For example, on the lowest level, resources and materials are processed using robots in standardized conditions. However, the assembly of building kits is done in a less structured environment and thus needs robotic systems which are more flexible. Up to today, it was difficult to apply humanoids to other autonomous complex robot technology in work tasks as building kit assembly and service. Yet, advancements in structuring environments and information about the environment for robotic systems on the one hand, and robot control technology and artificial intelligence on the other hand, lead to the fact that all highly autonomous systems can increasingly be applied in service environments.

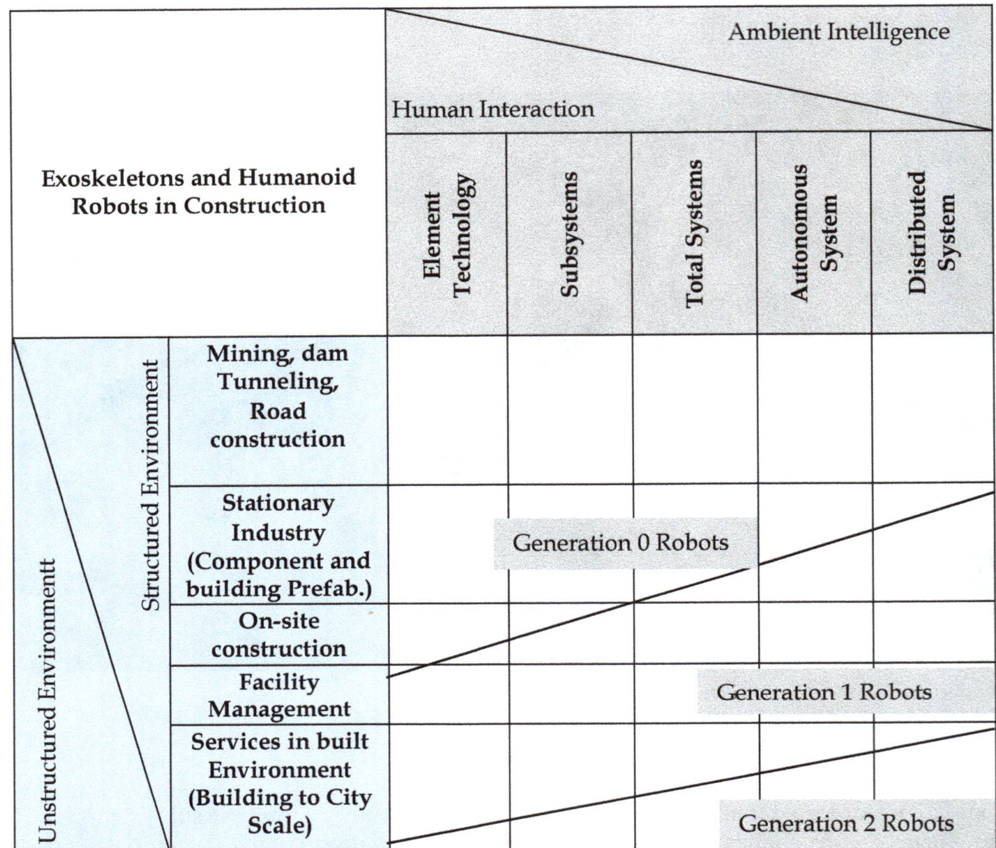

Table 1. Up to today, it was difficult to apply humanoids to other autonomous service robots in work tasks as building kit assembly and service. Yet, advancements in structuring environments and information about the environment for robotic systems on the one hand, and robot control technology and artificial intelligence on the other hand, lead to the fact that all highly autonomous systems can increasingly be applied in well planned service environments.

The notion of Generation Robots was introduced by Professor H. Moravec, Carnegie Mellon University, in order to describe the evolution of robot technology in near future. First Generation Robots refer to robot systems have an autonomy and intellectual capacity that is compare able to that of a lizard (available: 2010). Second Generation Robots are capable of learning and their intelligence is comparable to that of a mouse (available: 2020). Further, intellectual abilities of Third Generation Robots shall be comparable to that of a monkey (available 2030) and that of Fourth Generation Robot's intelligence finally shall be comparable to that of human beings (available: 2040). In order to be able to describe earlier developments in robot technology we introduce generation zero in our graphic.

5. Modularity and compatibility of element technology

The authors are currently working on applying and seamlessly integrating distributed robotic technology and mechatronic systems into home, care and city environments [27] [28] [29]. When people are assisted in close correlation by a robotic system, it is necessary to acquire as much data as possible about the person in real-time (e.g. activity, movement, vital signs) in order to understand and be able to predict mental and physical stat at any time. The authors currently develop a chair which is in real-time monitoring and interpreting vital data and is beyond that able to serve as a control station for games and home automation. The chair is developed within GEWOS, a University-Industry collaborative project financed by the German ministry (Runtime: 2010-2013) [30]. Its objective is to upgrade furniture components with sensors and other mechatronic components in order to support a healthy, save and active life at home. Among the partners are the Fraunhofer Institute for integrated circuits (section medical sensors) and EnOcean GmbH, a forerunner in energy harvesting and sensor applications. The first target of the consortium is to develop a "Fitness Chair" which is measuring people's vital signs, then makes those vital signs transparent to the user and finally try to activate the user to become more active (Figure 15), do sports and meet friends.

Fig. 18. Sensor Chair developed within the authors' R&D (Research & Development) Project GEWOS. The "Fitness Chair" is measuring people's vital signs, makes those vital signs then transparent to the user and finally try's to activate the user to become more active, do sports and meet friends.

Fig. 19. Similarity and interchangeability of underlying basic technologies between robots of different categories. From left to right: Kaist's Humaniod Robot HUBO, Kaist's Mobile HUBO FX-1 suit built upon the HUBO platform, TUM's GEWOS sensor chair serving as control interface, IRT'S and Toyota's r intuitively controllable robotic wheelchair.

Above, the chair provides an open server platform which allows doctors, physical therapists and other health professionals to develop service applications for customers. Beyond that, the chair with its variety of integrated sensors serves as a controller for virtual reality games and home automation. Companies as well as researchers are interested in bringing this solution to the market. In March 2011 it has even been covered by the German issue of Technology Review. The chair contains following systems:

EKG-Module: Measuring heart rate variability

SPO2-Module: Measuring blood pressure and oxygen saturation of the blood by infrared and special signal processing algorithms

Activity-Module: Sensor system for analyzing the user's activity in the proximity of the chair

Weight-Module: Measuring weight and weight distribution on the chair

Data Platform with GUI: Allows third parties (doctors, physical therapists and other health professionals) to develop service applications for customers

Gaming Aspect: Chair itself can be used as controller and training application to enhance the user's activity at home.

The technology applied to the GEWOS Sensor chair has the potential to be applied to Mobility Robots (e.g. IRT'S and Toyota's r intuitively controllable robotic wheelchair) and mobile suits (e.g. Toyota's i-foot, Kaist's Mobile HUBO FX-1 suit built upon the HUBO platform) for more users being accumulated and indirectly controlled. Further, HUBO FX-1 is good example that it is possible to apply technological platforms to robots of various categories. The HUBO leg platform has been applied to the Humanoid robot HUBO as well as to the Mobility Robot HUBO FX-1. It can be assumed that in the future this interchangeability of technologies will increase. So that, for example wearable computers (e.g. head up displays) and Single Joint Assistance Devices can support users to control Mobility Robots and Humanoids.

6. Conclusion

We have argued that human beings are steadily using and advancing tools. Exoskeletons and especially humanoid robotic technology in ill defined construction and built service environment as a whole or its subsystems/elements can be seen as a highly advanced tool or cooperating set of tools. Exoskeletons and humanoid robotic technology not only allows augmenting human abilities but creates tools that are capable of autonomous decision-making and performance in order to achieve certain goals as agent of a human being especially in dangerous, dirty and tedious construction activities. Most major industries have already extensively made use of robotic technology, which transforms production system technology in automotive industry, aircraft industry, the electrical appliance's sector, the medical field, farming and even recently construction. For the near future, we see a huge potential for robotics - wearable cooperative systems as well as fully autonomous systems- to permeate the field of construction and building technology. We have presented a categorization distinguishing between mechatronic, robotic, microsystemic element technology (power augmentation, sensing and motion augmentation, and cognition augmentation), subsystems (assistive devices and partial exoskeletons), total systems (exoskeletons, mobility robots), autonomous robots (humanoids, service robots) and highly complex distributed robot systems. Further, we have shown that with each new generation of robots, the applicability of robots in rather unstructured environments as on the construction sites or in building service environment advances. Finally, new sensing and interface technologies allow that robotic systems can be fully integrated in complex human-machine interaction systems and tasks. Based on the findings presented in this article, we assume that more and more flexible and autonomous exoskeletons and humanoid robotic technology will continue to permeate our in terms of complexity and work tasks rather unstructured domain of construction and building environment. Ultimately those exoskeletons and humanoid robotic technologies even will open up completely new possibilities for mankind in extreme and highly unstructured environments such as deep sea under water mining/habitat and construction and mining in space.

7. References

[1] Lenk, H. (2010) Humans- flexible multi-ability beings. An introduction modern philosophical Antropology situated between bio, techno and culture related science, Vielbrück Wissenschaft, Weilerswist

[2] Dobson, D.L., Hollnagel, E (2004) Handbook of Cognitive Task Design (Human Factors and Ergonomics), Lawrence Erlbaum Associates, New Jersey.

[3] The European Ambient Assisted Living Program distinguishes between assistive devices for hectometric, metric and centimetric mobility or motion ability.

[4] Murakami, K., Hasegawa T., Karazume R., Kimuro, Y. (2008). A Structured Environment with Sensor Networks for Intelligent Robots. IEEE Sensors 2008 Conference

[5] Yokoyama, K., Maeda, J., Isozumi, T., Kaneko, K. (2006) Application of Humanoid Robots for Cooperative Tasks in the Outdoors

[6] Bock, t. (2004) Humanoid Construction Robots instead of Low Wage Labor. In: Concrete Plant and Precast Technology, Bauverlag, Gütersloh.

[7] Englberger, J.F. (1989). Robotics in Service. Massachussets: MIT Press.

[8] T. Linner, M. Kranz, L. Roalter, T. Bock (2011) *Robotic and Ubiquitous Technologies for Welfare Habitat. In: Journal of Habitat* Engineering, Vol. 03, Number 1, pp. 101-110

[9] Department of Earth and Planetary Sciences, Faculty of Science, Kyushu University, Website: http://133.5.170.64/Museum/Museum-e/Museum-e.html, last visited 24/07/2011.

[10] Prof. Sankai, Cyberdyne, "Cyerbernics" Science

[11] Katsuya Kanaoka, "A Study on Man-Machine Synergy Effect in Non-Programmable Heavy Physical Work", Proceedings of the 11th Symposium on Construction Robotics in Japan, September 2, 2008, pp.119-124

[12] Yasuhisa Hasegawa, Kosuke Watanabe, Yasuyuki Mikami, Yoshiyuki Sankai, "Exoskeleton hand and wrist support system" , Proceedings of the 11th Symposium on Construction Robotics in Japan, September 2, 2008, pp.81-92

[13] Toshiro Noritsugu, Masahiro Takaiwa, Daisuke Sasaki, "Power assist wear driven with pneumatic rubber artificial muscles", Proceedings of the 11th Symposium on Construction Robotics in Japan, September 2, 2008, pp.109-118
[14] K. Montgomery, C. Mundt, G. Thonier, A. Tellier, U. Udoh, V. Barker, R. Ricks, L. Giovangrandi, P. Davies, Y. Cagle, J. Swain, J. Hines, G. Kovacs, "Lifeguard- A Personal Physiological Monitor For Extreme Environments".
[15] Nuria Oliver & Fernando Flores-Mangas, "HealthGear: A Real-time Wearable System for Monitoring and Analyzing Physiological Signals".
[16] Uwe Maurer, Anthony Rowe, Asim Smailagic, Daniel P. Siewiorek, "eWatch: A Wearable Sensor and Notification Platform".
[17] Hughes, E.; Masilela, M.; Eddings, P.; Raflq, A.; Boanca, C.; Merrell, R "VMote: A Wearable Wireless Health Monitoring System".
[18] Timmermanns, J. (2006) MARY - Konzeption und Bewertung eines Augmented-Reality-gestützten Systems zur Optimierung des Bauprojektzyklus. Editor: Prof. Dr.-Ing./Univ. Tokio Thomas Bock, Fraunhofer IRB Verlag, Stuttgart.
[19] Shigeki Toyama, Junichiro Yonetake, UltraSonicMotor Powered Assisted Suit System, Society of Biomechanism Japan, Vol. 30, No.4, 2006, pp189-193
[20] Hiroshi Kobayashi, Sho Hasegawa, Hirokazu Nozaki, "Development and Application of a Muscle Force Enhancement Wear: Muscle Suit", Proceedings of the 11th Symposium on Construction Robotics in Japan, September 2, 2008, pp.93-100
[21] Mineo Ishii, Keijiro Yamamoto, Kazuhito Hyodo, "Stand- Alone Wearable Power Assist Suit -Development and Availability-", Journal of Robotics and Mechatronics Vol.17 No.5, 2005
[22] KAIST Humanoid Robot Series, Public Demonstration
[22] Han, C. (2011) "Human Robot Cooperation Technology" - An ideal midway Solution heading toward the Future of Robotics and Automation in Construction, International Symposium on Automation and Robotics in Construction, Korea
[23] KAIST Humanoid Robot Series, Public Demonstration
[24] Arai, T. (2011) Advanced Robotics and Mechatronics and their applications in Construction Automation, International Symposium on Automation and Robotics in Construction, Korea
[25] McCandless, J.W., McCann, R.S., Marshi, I. Kaiser, M.K., Andre,A.D.(2006) Human Factors Technologies for Space Exploration. In proceedings of Space 2006. San Jose, 2006
[26] Bock, T., (2007) Construction Robotics, in Autonomous Robots, Springer Science
[27] Bock, T., Linner, T., Lee, S. (2010) Ambient Integrated Robotics: new Approach for supporting Elderly People with integrated Technology in Living Environments. ISR 2010 International Symposion on Robotics. Munich , June 2010
[28] Kranz, M., Linner, T., Ellmann, B., Bittner, A. (2010) Robotic Service Core for Ambient Assisted Living. 4th International Conference on Pervasive Computing Technologies for Healthcare 2010, Munich, March.
[29] T. Linner, B. Ellmann, T. Bock (2011) *Ubiquitous Life Support Systems for an Ageing Society in Japan.* In: Ambient Assisted Living: Advanced Technologies and Societal Change.

Edited by R. Wichert, B. Eberhardt, Springer Science + Business Media, ISBN 978-3-642-18167-2, pp. 31- 48

[30] GEWOS (2011) BMBF/ VDI/VDE funded Project "GEWOS, Gesund Wohnen mit Stil". Project Partners: EnOcean GmbH, Fraunhofer IIS, ISA Informationssysteme GmbH, Sportkreativwerkstatt, SOPHIA GmbH, Technical University Munich, Runtime: 2010-2013, Further Information: www.br2.ar.tum.de and www.gewos.org

10

Implementation of a Framework for Imitation Learning on a Humanoid Robot Using a Cognitive Architecture

Huan Tan
Vanderbilt University
USA

1. Introduction

Robots are designed to assist human beings to complete tasks. As described in many research journals and scientific fictions, researchers and the general public expect that one day robots can complete certain tasks independently and autonomously in our world.

A typical implementation of Artificial Intelligence (AI) in robotics research is expected to incorporate Knowledge Representation, Automated Reasoning, Machine Learning, and Computer Vision [Russell and Norvig, 2010].

From the beginning of this century, the importance of cognitive sciences is highlighted by the researchers in both robotics and artificial intelligences[Russell and Norvig, 2010]. Real Cognitive Science is based on the testing and experiments on animals and humans to obtain the understanding and knowledge of the cognition, and AI research is based on designing and testing algorithms on a computer based artificial system. In the development, research in both Cognitive Sciences and AI benefit each other. Cognitive Science provides theoretical foundation and solutions to the problems in AI, and AI enhances the research in Cognitive Science and provides possible research directions for Cognitive Science.

Recently, an emerging field, named Cognitive Robotics, is developed for robots to generate human-like behaviors and intelligences, which integrates perception, action, learning, decision-making, and communication[Kawamura and Browne, 2009]. Cognitive Robotics largely incorporates the concepts in the research of cognitive sciences and tries to simulate the architecture of the information processing of human cognition. Currently, some long term requirements are proposed for Cognitive Robotics, which put concentration on the embodiment of cognitive concepts [Anderson, 2003; Sloman et al., 2006]. Increasing achievements in Cognitive Robotics began to grab the attention of the robotics research community. However, currently, it is still difficult to design truly cognition, especially human-like cognition, on a robotic platform due to the limitations in mechanism, computation, architecture, etc. [Brooks, 1991]. Therefore, on one side, researchers try to place robots in a human-existing dynamic environment to complete tasks[Brooks, 1991]; on the other side, researchers still do not obtain a general architecture to generate complex behaviors in such area for robots[Sloman, 2001]. In recent research, researchers try to place robots in robotic aid area, where robots can assist humans to complete tasks[Tan et al., 2005].

An important feature of robotics research is to design human-like mechanism for robots [Brooks et al., 1999]. Humanoid robots, which are designed human-like, have been gradually selected as the platform for experimentally implementing the conceptual design of Cognitive Robotics. An increasing number of humanoid robots have been designed to fulfill the dream of the researchers (such as NASA Robonaut, Aldebaran's NAO and ISAC)[Tan and Kawamura, 2011]. The human-like mechanism provides a possible platform for researchers to design human-like behaviors [Tan and Liao, 2007b]and it is expected that one day such human-like artificial creatures can live in our real world. However, there is no defying the fact that the essential part of designing a truly human-like robot is to provide robots abilities and skills to generate human-like behaviors.

Even a robot with six degrees-of-freedom (DOF) is not easy to plan the motion in human-existing environment [Wang et al., 2006; Yang et al., 2006]. Futher more, the mechanism of a humanoid robot is much more complicated than an industrial robot, because the degree-of-freedom is much higher. Therefore, it is difficult for researchers to use traditional methods to plan the motions for humanoid robots. When humanoid robots are asked to carry out some human-like behaviors, they are much more difficult to be programmed or controlled because such behaviors are more complex. Additionally, as stated in former paragraphs, it is expected that humanoid robots should be placed in a dynamic and human-existing environment to complete tasks. Therefore, researchers began to find solutions from cognitive sciences.

Recently, imitation learning has been considered as a powerful tool for rapidly transferring skills between humans and robots [Uchiyama, 1978; Atkeson and Schaal, 1997; Schaal, 1999]. Human beings often try to imitate the behaviors of other people in the environment. This kind of ability is especially important for children. Concepts of imitation learning can be found from the investigation of the psychology research and biology research [Billard, 2001]. Therefore, it is reasonable to implement imitation learning for robots since we are struggling to develop cognitive skills for robots.

In a typical imitation learning, first, human teachers demonstrate a behavior sequence to complete certain task in a specified situation. Second, this behavior sequence is learned by recording the behavior either by using the encoders on the robot (if the demonstration is given by manually moving arms of robots) or by recording the trajectory in the task space (if the demonstration is given by showing the robot the task in the task space). Third, the robot will generate similar behavior in similar but different situations to complete tasks.

Imitation learning aims to train robot the behaviors and skills from the demonstrations and let the robots adapt them to similar situations, using which robots reproduce similar movements to complete similar tasks [Billard et al., 2007]. Imitation learning algorithms can be divided into two categories [Calinon et al., 2007]: one is trying to train robots to extract and learn the motion dynamics [Schaal and Atkeson, 2002; Ijspeert et al., 2003; Calinon and Billard, 2007; Tan and Kawamura, 2011], and the other is trying to let robots to learn higher-level behaviors and action primitives [Dillmann et al., 2000; Bentivegna and Atkeson, 2001] by imitation. Both methods require a set of predefined basis behaviors to ensure convergence.

For most tasks, a human teacher shows robots several behaviors in a behavior sequence to complete a task. That means the behavior sequence is composed of several behaviors and each behavior has its own parameters. Recently, imitation learning methods mostly put concentration on reproducing new movements with dynamics similar to one single behavior in a behavior sequence[Ijspeert et al., 2002] or generating a behavior which has dynamics similar to the whole behavior sequence[Dillmann et al., 2000]. Therefore, the problem is if robots are required to generate new behaviors in a similar situation but the task constraints have been changed a lot, sometime robots may fail to complete this task using current methods.

In a Chess-Moving-Task, a humanoid robot, named ISAC, is required to grasp and move the Knight two steps forward and subsequently one step left.

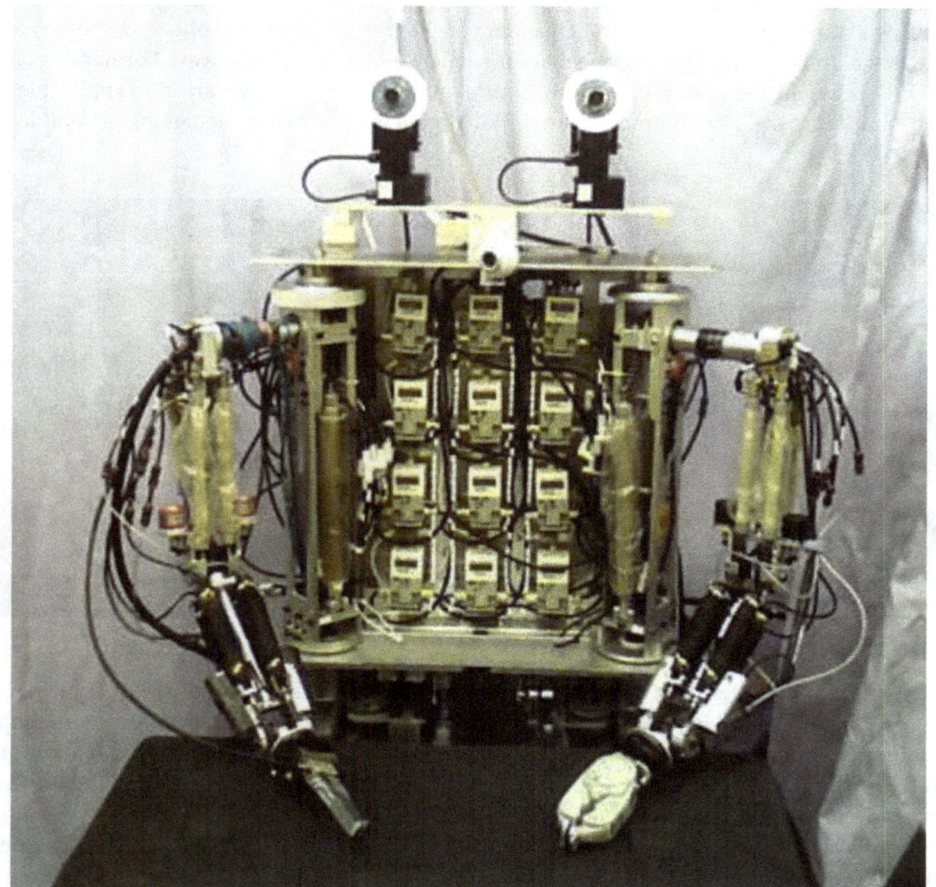

Fig. 1. ISAC Robot (peneumatic driven and stationery).

The demonstration by a human teacher is shown in the left picture of Fig. 2. In the middle picture of Fig.1, if the Knight is moved not too far away from the place in the demonstration, ISAC can use current imitation methods to complete the task. But if the Knight is moved far away from the place in the demonstration, ISAC fails to grasp and to move this piece in an expected style as shown in the right picture of Fig.2. That is because current learning methods can guarantee the convergence of achieving the global goal, but the local goals have not been taken into consideration.

Therefore, in a biologically inspired way, robots should understand the task as a behavior sequence. The reason for choosing the segmentation of a behavior sequence is: first, behavior based cognitive control method provides a robust method for manipulating the object and complete a task through learning, because behavior based methods can train robots to understand the situation and the task related information; second, the segmentation provides a more robust approach for robots to handle completed tasks. As in Fig.2, if ISAC knows that this task is composed of several parts, the grasping sub-task can be guaranteed by adapting the first reaching behavior. However, robots do not have knowledge on how to segment the behavior sequence in a reasonable way. Former research methods segment behaviors based on its dynamics. However, they are not applicable to some complex tasks. For complex tasks, a global goal is achieved by achieving several local goals.

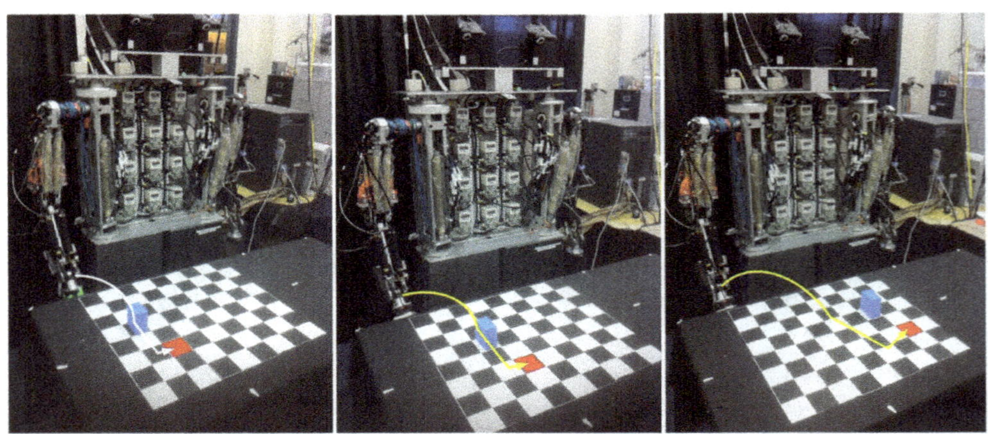

Fig. 2. Chess-moving task

Fuzzy methods[Dillmann et al., 1995] and the Hidden Markov Model (HMM) [Yang et al., 1997] are normally used for segmentation. Another type of segmentation methods is to detect change-points on the trajectory[Konidaris et al., 2010]. Kulic and Nakamura [Kulic et al., 2008]proposed a segmentation method based on the optical flow in the environment, which is a cluster based method. These methods are not robust, because they rely on the analysis of the dynamics of the demonstration and pre-defined behavior primitives, which are not suitable for extension. Readers can simply image a new situation where the pre-defined behavior primitives cannot be used, and a noises existing environment, where the sensory information of the demonstration can be largely affected.

In this paper, we propose a cognitive segmentation method.

Imitation learning provides a possible solution for behavior generation for humanoid robots. However, researchers gradually found that it is possible to design algorithms and architectures for a specific task in a specific situation, but it seems difficult to design a

general imitation learning method to generate behaviors in a large number of different situations. Researchers starts to find the solution from the cognitive sciences, because the research of cognitive sciences investigate the stable learning processes in human or animal brains and possibly it can provide solution to current behavior based robotics research, especially for the research on humanoid robots[Tan and Liang, 2011].
Recently, the cognitive architecture received broad attention from the robotics research community, because it provides a kind of methods of using cognitive processes[Tan and Liang, 2011].
Current cognitive architectures can be divided into four categories: Symbolic, Connectionist, Reactive (Behavior Based Connectionist), and Hybrid. Some well-known Symbolic architectures include: ACT-R [Anderson et al., 1997], SOAR [Laird et al., 1987], EPIC [Keiras and Meyer, 1997], Chrest[Gobet et al., 2001], and Clarion [Sun, 2003], in which symbolic knowledge are stored for automated reasoning. For Connectionist, BICS[Haikonen, 2007], Darwinism[Krichmar and Reeke, 2005], and CAP2[Schneider, 1999] are developed, in which connectionism method are implemented to generate behaviors or decisions. A typical Reactive architecture is Subsumption[Brooks, 1986; Brooks, 1991], which directly couples the sensory information and the behavior primitives. Some researchers have begun to recognize the need of both deliberative interaction and reactive interaction for cognitive robots, which motivates the research on hybrid architectures[Kawamura et al., 2004]. Such integration offers the promise of robots which are both fluent in routine operations and capable of adjusting their behavior in the face of unexpected situations or demands. Typical Hybrid architectures include: RCS [Albus and Barbera, 2005], and JACK[Winikoff, 2005]. In our lab, we developed a hybrid cognitive architecture, named ISAC Cognitive Architecture [Kawamura et al., 2004]. In this paper, we propose to use ISAC cognitive architecture to implement the imitation learning for a humanoid robot.
The rest of the papers are organized as follows: Section II introduces the design of the proposed system, Section III describes the experimental setup, procedure and the results, Section IV discuss the proposed system and the future work, and Section V summarizes the work in this paper.

2. System design

For ISAC, demonstrations are segmented into behaviors in sequences. The segmented behaviors are recognized based on the pre-defined behavior categorizes. The recognized behaviors are modeled and stored in a behavior sequence. When new task constraints are given to the robot, ISAC generates same behavior sequences with new parameters on each behavior, the dynamics of which are similar to the behaviors in demonstration. These behaviors are assembled into a behavior sequence and sent to the low-level robotic control system to move the arm and control the end-effectors to complete a task.
Fig.1 is the system diagram of ISAC Cognitive Architecture, which is a multi-agents hybrid architecture. This cognitive architecture provides three control loops for cognitive control of robots: *Reactive, Routine* and *Deliberative.* Behaviors can be generated through this cognitive architecture. Imitation learning basically should be involved in the Deliberative control loop. Three memory components are implemented in this architecture, including: Working Memory System (WMS), Short Term Sensory Memory (STM), Long Term Memory (LTM).

The knowledge learned from the demonstrations is stored in the Long Term Memory (LTM). When given a new task in a new situation, ISAC retrieves the knowledge from the LTM and generates a new behavior according to the sensed task information in the STM and WMS[Kawamura et al., 2008].

- Perceptual Agents (PA)

The PA obtains the sensory information from environment. Normally, encoders on the joints of the robot, cameras on the head of the robot, and the force feedback sensor on the wrist of the robot are implemented in this agent.

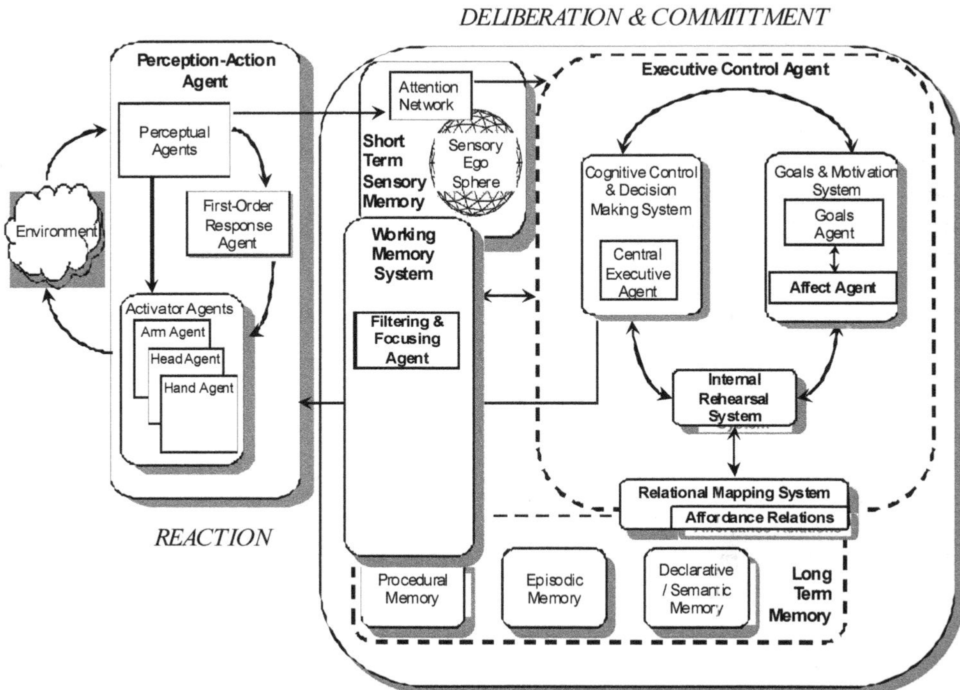

Fig. 3. ISAC cognitive architecture.

- Short Term Memory (STM)

The obtained information is sent to and stored in the STM. The Sensory Ego Sphere (SES) is implemented in the STM, which performs spatio-termporal coincidence detection, mediates the salience of each percept, and facilitates perceptual binding.

- Working Memeory System (WMS)

The WMS stores the task-related information in chunks. This component is especially important in the generation stage.

- Central Executive Agent (CEA)

The CEA provides central processing, decision making, and control policy generating for different task goals which is stored in the Goals Agent (GA). In hierarchy architecture, this component accesses all of the sensed information and makes decision for tasks.

- Goal Agent (GA)

Correspondingly, the GA stores the motivations or goals of tasks in situations.

- Long Term Memory (LTM)

The LTM stores the memory especially the knowledge for long term use. Procedural, semantic, and episodic knowledge are stored in this component. In imitation learning, the learned skill or knowledge is stored as procedural and episodic knowledge using a mathematical model.

- Internal Rehearsal System (IRS)

The IRS evaluates the results of the decisions made from the CEA through internal rehearsal.

The activity of ISAC can be divided into two stages: Learning from Demonstration, Generation by Imitating. Figure 2 displays the control loop for the first stage: Learning from Demonstration.

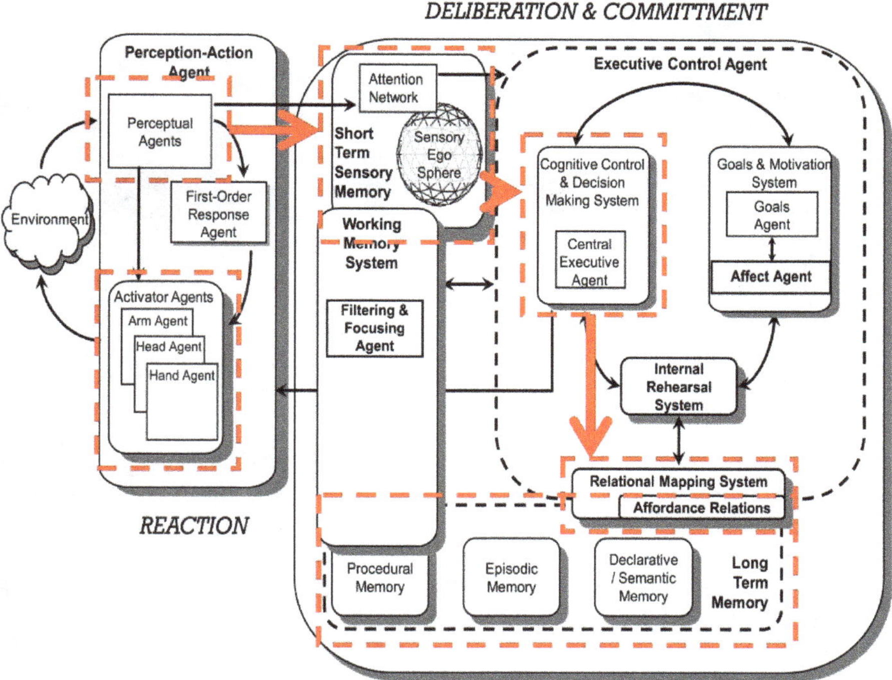

Fig. 4. Control loop of the learning from demonstration.

2.1 Learning from demonstration

This stage is divided into three steps: demonstration, segmentation, and recognition.

Demonstration

Assumption-We assume 1) human teachers are well-trained, 2) behaviors of the same part in different demonstrations should have similar dynamics, and 3) the behavior sequences of the demonstrations should be composed of the same number of behaviors.

For the Chess-Moving-Task, the Knight (We used blue cubes for easy grasping because the control of ISAC's arm is not very precise) was placed on several different places and several demonstrations have been given to the robots by manually moving robot's arm to complete the task. For each demonstration, joint angles were sampled using the encoders on the joints, and the position values were sampled using a camera on the robot's head.
A human teacher shows the demonstration by manually moving the right arm of ISAC. The PA senses the movement of right arm using the encoders on the joints of the right arm of ISAC, and records the movement of the Knight using the camera mounted on the head of ISAC.

Fig. 5. Demonstration.

Sensed information is stored in the STM as data matrices. Three data structures are used for the recorded demonstration: θ_{rs} and P_{os}. $\theta_{rs} = \{\theta_r, t\}$, which is a $N \times 7$ matrix, records the joint angles of the robot's right arm and the temporal information on sampling points, and is used to calculate the position of the end-effector in the Cartesian space. $P_{os} = \{P_o, t\}$ which is a $M \times 4$ matrix records the positions of the Knight on the chess board in the Cartesian space in the demonstration and the temporal information related to the sampling points.
$P_{rs} = \{P_r, t\}$, which is a $N \times 4$ matrix, records the position values of the robot's effector and the temporal information related to the sampling points.

$$P_{rs} = forward\ kinematics(\theta_{rs}) \tag{1}$$

These data arrays are stored in the STM for the CEA to process.

2.2 Segmentation

Segmentation is important for the system proposed in this paper. We propose that the segmentation is based on the change of world (environmental) states. The change of world states can be considered as the static object in the environment begins to move, the signal in the environment varies from one area to another area, some objects comes into the environment, etc.

The CEA obtains the segmentation method from the LTM and segments the sensed information in the STM.

In the Chess-Moving-Task, we define the change or the world states as the status of the Knight changes from static to moving. Upon that, the behavior sequence can be segmented. (The position value of the Knight was obtained using cameras)

Starting from the task, the position values of the Knight were recorded. At time step t, the array of the position values is $P = \{p_{o0}, p_{o1}, p_{o2}, p_{o3}, \dots p_{o(t-1)}, p_{ot}\}$.

Therefore, the position of the Knight can be considered as a Gaussian distribution. The mean of this Gaussian distribution is p_o, which can be considered as the estimated position value of the Knight:

$$p_o(x,y) = \frac{\sum_{i=0}^{t} p_{oi}}{t+1} \tag{2}$$

And the variance of this probabilistic distribution is:

$$v_o(x,y) = \left\{ \sqrt{\frac{\sum_{i=0}^{t}\left(p_{oi}(x)-p_o(x)\right)^2}{t}}, \sqrt{\frac{\sum_{i=0}^{t}\left(p_{oi}(y)-p_o(y)\right)^2}{t}} \right\} \tag{3}$$

At time step t+1, $d(p_{o(t+1)}|p_o,v_o)$, the distance between $p_{o(t+1)}$ and the mean of this the distribution, can be calculated through simple calculation:

$$P(p_{o(t+1)}|p_{o,}v_o) = \sqrt{\left(p_{o(t+1)} - p_o(x,y)\right)^T \left(p_{o(t+1)} - p_o(x,y)\right)} \tag{4}$$

Heuristically, if

$$P(p_{o(t+1)}|p_{o,}v_o) > 3|v_o| \tag{5}$$

the Knight can be considered as having been moved from its initial position.

Initial knowledge of the segmentation method is given to the robot. In the experiment, ISAC observed the environment and set an array T to record the temporal information, the elements t_{t+1} in this array are the temporal information of that point when the state of the knights has been changed. The recorded array T is used as segmentation parameters.

Assumption-In order to implement the behavior recognition, the 'grasping' behavior is defined in advance because it is almost impossible to observe the grasping using cameras. When the position of the Knight began to move along the trajectory of the end-effector on the robot's arm, we can assume that a grasping happened just before this movement. In real implementation, a grasping behavior is added into the 'static' status of the Knight and the 'begins to move' of the Knight.

A behavior sequence is obtained as {*Behavior 1, Grasping, Behavior 2*}.

2.3 Recognition

The CEA obtains the criterion of recognition from the STM and recognize the behaviors in the segmented behavior sequences.

For each behavior, the sampled points should be pre-processed before processing. Prior to that, it is necessary to determine what types of behavior they belong to. Therefore the data of sampled points are aligned and dimensionally reduced into low-dimensional space to extract the features. PCA [Wold et al., 1987], LLE [Roweis and Saul, 2000], Isomap [Tenenbaum et al., 2000], and etc. can be used to realize this step. For this special Chess-Moving-Task, the sampled trajectory is simply projected into X-Y plane, because the variations on Z-direction are very small.

Behaviors are divided into two types: one is the 'Common Behavior', which means that parameter can be modified according to different task constraints, e.g., different position values of the Knight on the chess board. The other is the 'Special Behavior', which means parameters remain the same in different task constraints.

The criterion for judgment is based on the scaling. In the latent space, if the axes of the trajectories of the behaviors satisfy the following equation, we consider it as a 'Special Behavior'; otherwise, it is a 'Common Behavior'.

$p(c)$, the probability of a common behavior is calculated as follow:

$$m(c) = \frac{\sum_{i=1}^{n} |max(x_i) - min(x_i)|}{n} \tag{6}$$

$$v(c) = \sqrt{\frac{\sum_{i=0}^{n} \left(|max(x_i) - min(x_i)| - m(c)\right)^2}{n-1}} \tag{7}$$

$$p(c) = v(c)/m(c) \tag{8}$$

$p(s)$, the probability of a special behavior is calculated as follow:

$$p(s) = 1 - p(c) \tag{9}$$

Heuristically, if $p(c) > p(s)$, this behavior can be considered as common behavior; if $p(c) < p(s)$, this behavior can be considered as special behavior. However, this criterion is not applicable in all situations. It should be modified through an iterative machine learning process.

Intuitively, the 'Common Behavior' means we can modify the parameters of behaviors to adapt them to different situations, and the 'Special Behavior' means that in specific tasks, such behaviors should not be changed, and robots should follow the demonstration strictly.

After that, points are interpolated to align the signals in order to extract the common feature of the behaviors. Based on the assumption in section 2.1, the position values on the obtained behavior trajectory are obtained by calculating the average of the sampled position values on the common timing points as:

$$P_n(t) = average\left(P_{demo1}(t) + P_{demo2}(t) + P_{demo3}(t)\right) \tag{10}$$

The learned knowledge is stored in the LTM for the stage of generation.

2.4 Generation by Imitating

Robots obtains the environmental information from the PA and sends it to the STM. The CEA analyze the sensed information from the STM and sent to the GA to generate the goal of the task. The CEA gets the generation method from the LTM and generates same behavior sequences. In the behavior sequences, behaviors have similar dynamics to the behaviors in the demonstrations. The generated behavior sequences are sent to the WMS.

Subsequently they are sent to the actuators to complete tasks in similar but slightly different situations.

In behavior modeling, classical Locally Weighted Projection Regression (LWPR) method is used to model the trajectory of a 'common Behavior'. Heuristically, 10 local models are chosen to model the trajectory with specific parameters. The general method of LWPR can be referred to[Atkeson et al., 1997; Vijayakumar and Schaal, 2000].

Given new task constrains, specifically the new position value of the Knight on the chess board, ISAC uses the same behavior sequence to reach, grasp and move the Knight. The behavior sequence is obtained in section 2.2 and DMP is chosen to generate a new trajectory for behavior 1.DMP method originally proposed by Ijspeert [Ijspeert et al., 2003] , records the dynamics of demonstrations and generates new trajectories with similar dynamics, is an effective tool for robots to learn the demonstrations from humans .

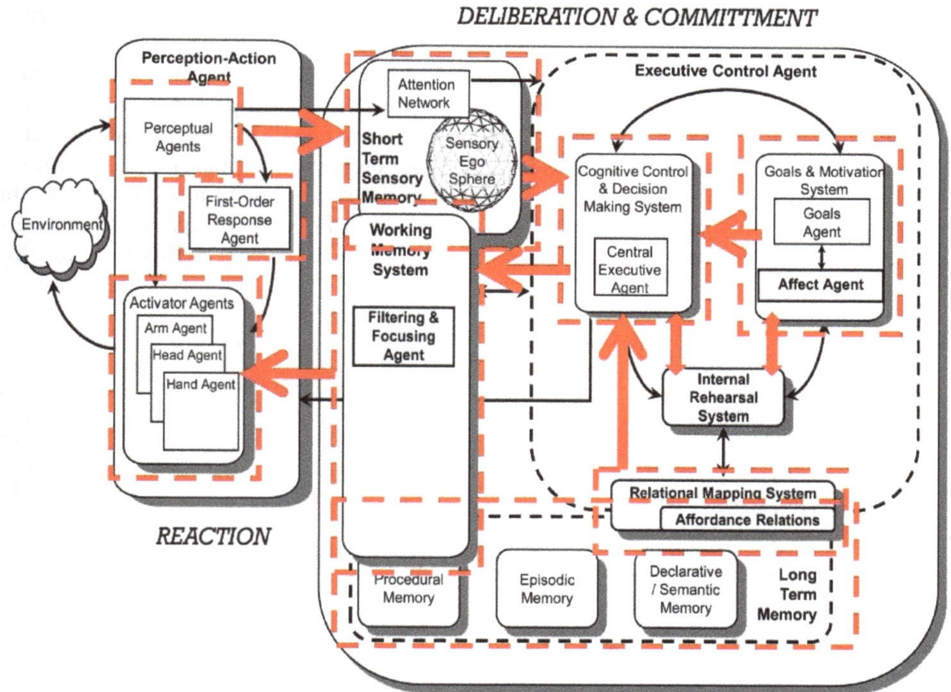

Fig. 6. Control Loop of the Generation by Imitating.

The formulation of the original DMP algorithm is shown as differential equations:

$$\tau\dot{z} = \alpha_z(\beta_z(g - y) - z) \tag{11}$$

$$\tau\dot{y} = z + f \tag{12}$$

where g is the goal state, z is the internal state, f (a LWPR model) is calculated to record the dynamics of the demonstration, y is the position generated by the differential equations, and

$\dot{y}$ is the generated velocity correspondingly. α_z, β_z, and τ are constants in the equation. From the original paper of the DMP, α_z, β_z, and τ are chosen as 1, ¼ and 1 heurisitically to achieve the convergence.
When the position value of the Knight is given, the CEA generates a new trajectory which has similar dynamics to the demonstration of behavior 1.
After the grasping, behavior 2 is generated strictly following the obtained behavior in section.

3. Experimental results

Proposed experimental scenario is: 1) a human teacher demonstrates a behavior by manually moving the right arm of ISAC to reach, grasp and move the Knight on the piece; 2) ISAC records the demonstrations using encoders on its right arm and the cameras on the head; 3) ISAC generates several behaviors to complete tasks in a similar but different situations;
Using the cognitive segmentation method proposed in this paper, each of the three demonstrations are segmented into 2 parts as shown in Fig.6. Each part is considered as a behavior. The black circle demonstrates the point of the change of the world states because the Knight began to move with the end-effector. Based on the assumption in section 2.1, the first behavior of each demonstration is considered as the same type of a behavior, and the same consideration for the second behavior. As shown in Fig.6, three demonstrations are segmented into 2 behaviors with a changing point which is considered as the grasping behavior.

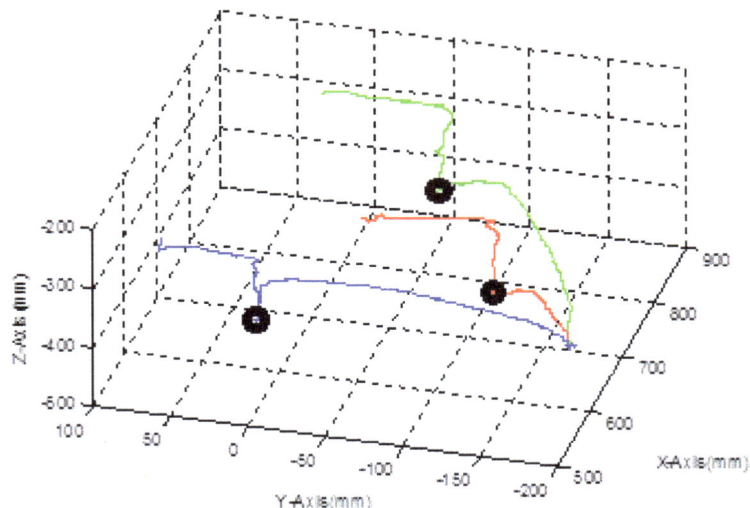

Fig. 7. Segmented behavior sequences.

In Fig. 7 and Fig. 8, Behaviors are categorized according to the formula in Section 2.4. Behavior 1is a Common Behavior and Behavior 2 is a Special Behavior.

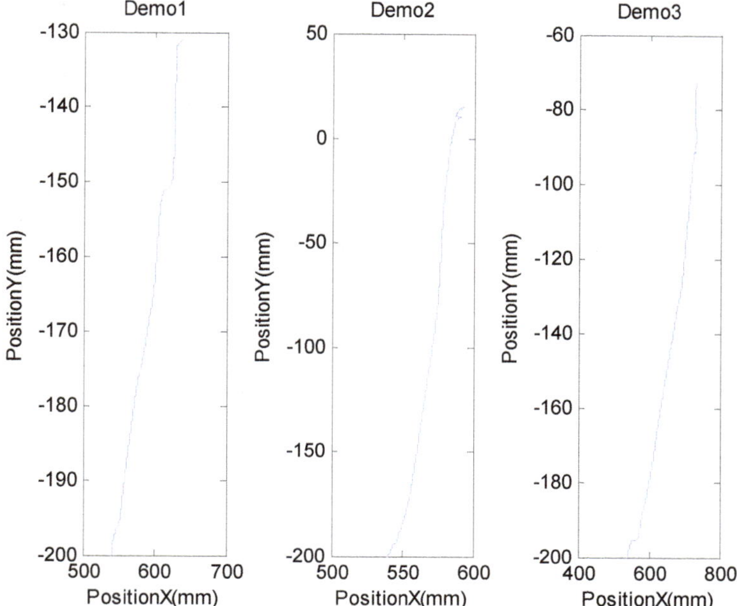

Fig. 8. Recorded behavior 1.

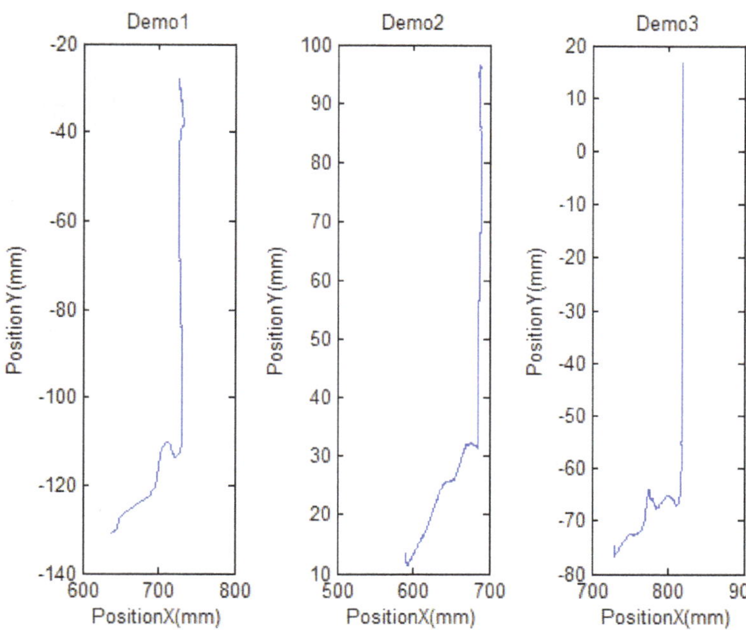

Fig. 9. Recorded Behavior 2.

In Fig. 9, on each dimension, the sampled trajectories of different behaviors in each demonstration are normalized in time and magnitude. The magnitude was normalized in the range of (0, 1) for each dimension of the trajectory. Because DMP is chosen in the trajectory generation part and DMP only requires the dynamics of the trajectory. In practical, DMP algorithm automatically normalizes the sampled trajectory, therefore different time periods and magnitudes will not affect the results of the experiments.
In Fig.7, the first row are the trajectories of Behavior 1 in X-direction, the second row are the trajectories of Behavior 1 in Y-direction, and the third row are the trajectories of Behavior 1 in Z-direction. The first column is the trajectory of Behavior 1 in demonstration 1, the second column is the trajectory of Behavior 1 in demonstration 1, the third column is the trajectory of Behavior 1 in demonstration 1, and the fourth column is the processed trajectory of Behavior 1 which will be used for future behavior generation.

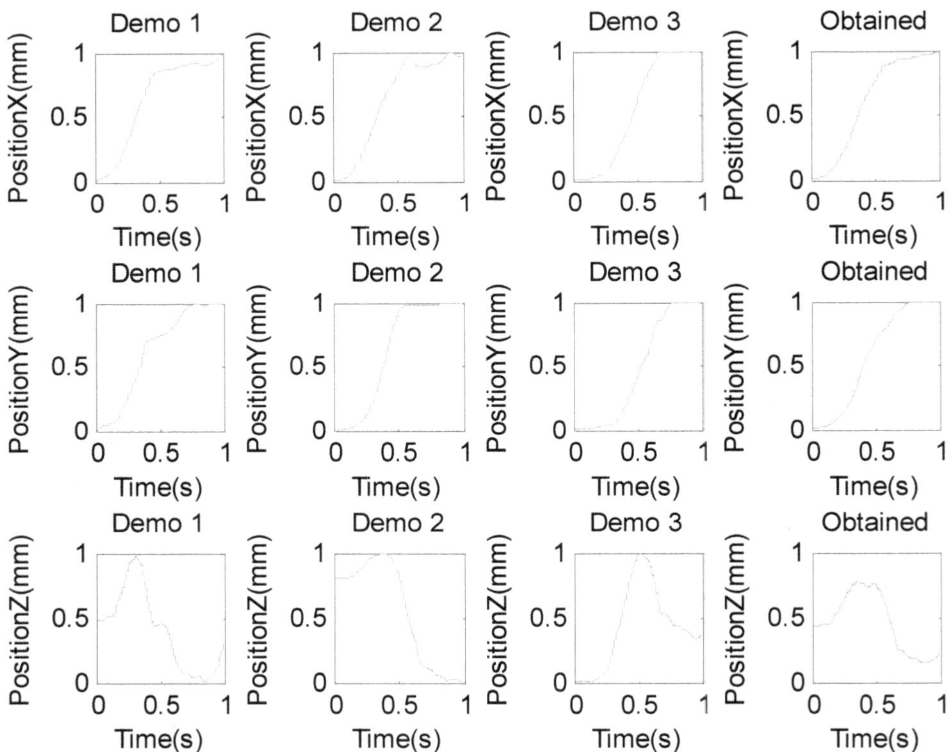

Fig. 10. Stored knowledge of behavior 1.

Behavior 2 is calculated by getting the average value of the sampled position values on the common timing points.
In Fig. 9, the left figure is the trajectory of Behavior 2 in demonstration 2 on X-Y plane, the left middle figure is the trajectory of Behavior 2 in demonstration 2 on X-Y plane, the right middle figure is the trajectory Behavior 2 on X-Y plane, and the right figure is the processed trajectory of Behavior 2 which will be used for future behavior generation.

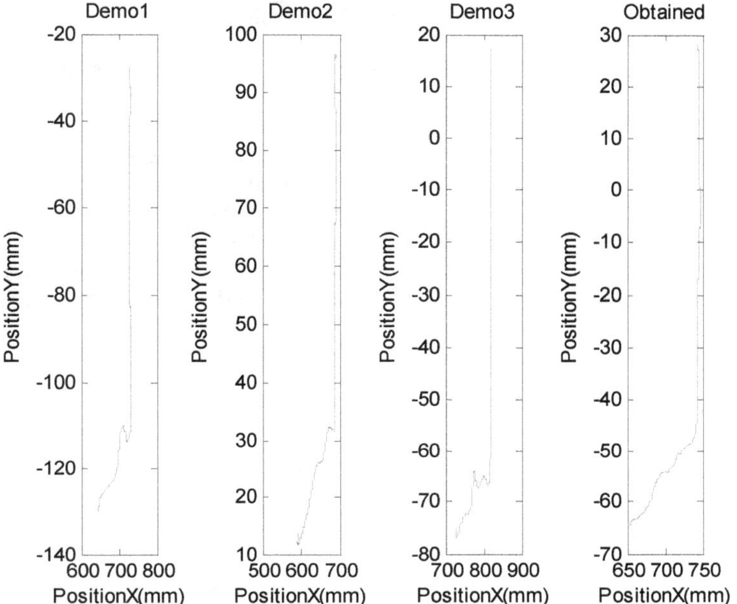

Fig. 11. Stored knowledge of behavior 1.

Fig. 11 displays the generated behaivor 1 when ISAC is asked to reach, grasp, and move the Knight on the board, the coordinates of which is (450, 215, -530).

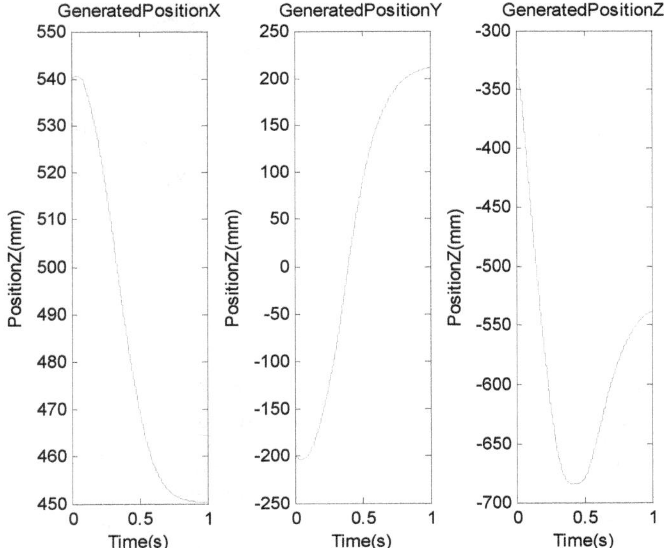

Fig. 12. Generated New Behavior 1 in X, Y, and Z direction.

By comparing Fig. 11 with the fourth column of Fig. 9, similar dynamics of the two trajectories in X, Y, Z-direction can be found.
Grasping is added between Behavior 1 and Behavior 2 based on the assumption in section 2.2.
Fig.12 shows the generated behavior sequences for the new task constraints. The red line is the behavior 1, the blue line is the behavior 2, and the intersection point of the two behaviors is the 'grasping' behavior. The simulation results shows that the end-effector moves to reach the Knight at (450, 215, -530) with similar dynamics to the behavior 1 in the demonstration, then it move the Knight in the same way as behavior 2. Although the position of the Knight has been changed, the movement of the Knight is the same as the demonstration. From Fig. 11, it is concluded that this algorithm successfully trained the robot to reach, grasp and move the chess in the expected way.

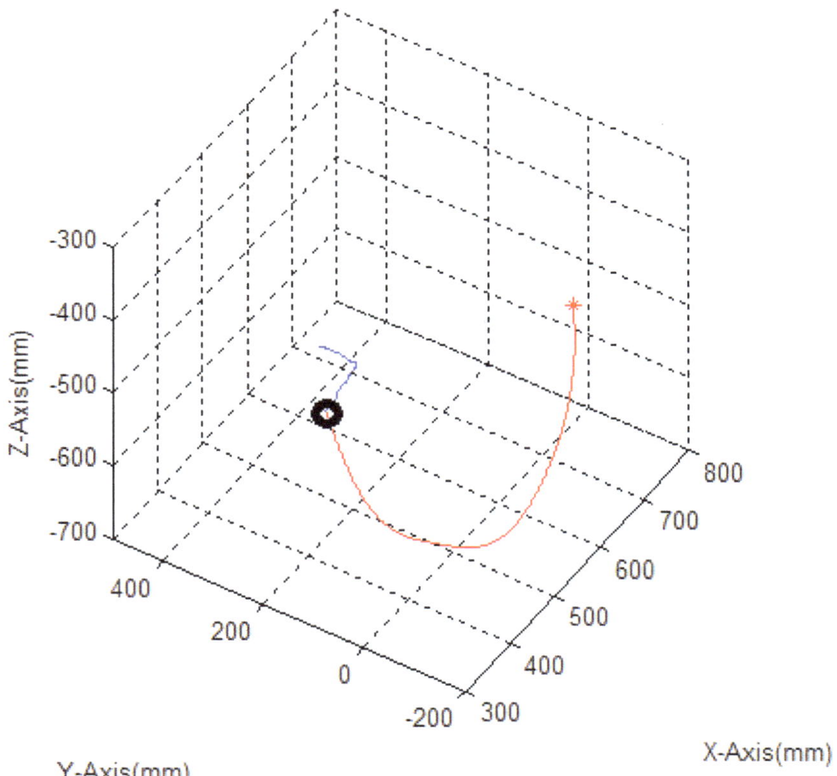

Fig. 13. A generated new behavior sequence in cartesian space

Fig. 13 shows the generated behavior sequences for the new task constraints. The red line is the behavior 1, the blue line is the behavior 2, and the intersection point of the two behaviors is the 'grasping' behavior. The simulation results shows that the end-effector moves to reach the Knight at (450, 215, -530) with similar dynamics to the behavior 1 in the demonstration, then it move the Knight in the same way as behavior 2. Although the position of the Knight

has been changed, the movement of the Knight is the same as the demonstration. From Fig. 13, it is concluded that this algorithm successfully trained the robot to reach, grasp and move the chess in the expected way.
Practical experiments were carried out on ISAC robot as shown in Fig.13. The Knight (blue piece) was placed at (450, 215, -530) which is far away from the place in the demonstrations and ISAC was asked to reach, grasp and move it to the red grid in an expected way which is similar to the demonstrations.

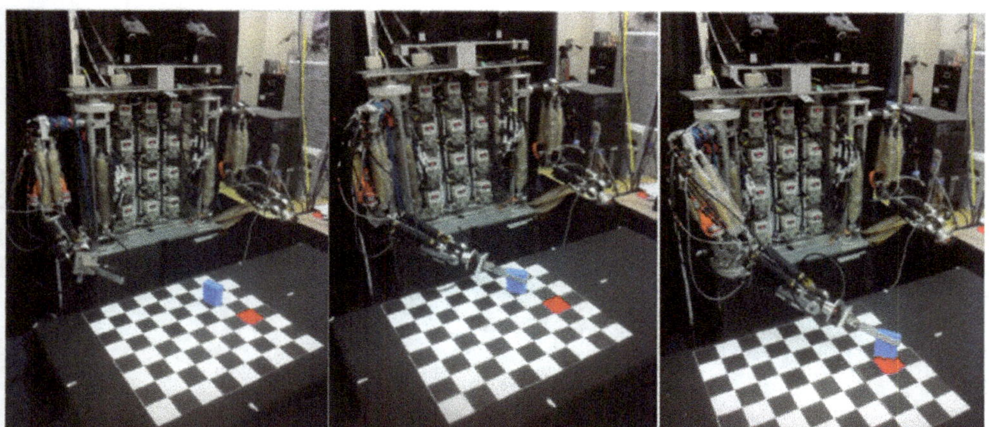

Fig. 14. ISAC reach, grasp, and move the knight on the board

The experimental results demonstrates that ISAC successfully learns this behavior sequence, and it can reach, grasp, and move the piece in an expected way.

4. Discussion and future work

The main idea of this paper is to demonstrate the effectiveness of cognitive segmentation of the behavior sequence. From the simulation and the practical experiments, we can conclude that this method is successful and it is useful for future study on cognitive imitation learning. Because ISAC is pneumatically driven and it is not very precise, although it can move the end-effector according to the behavior sequence and complete the movement successfully, it cannot grasp the piece every time. Fig.12 only shows one successful experiment.
We have finished current work on applying the proposed method on simulation platform and ISAC robot. Our long term goal is to design a robust behavior generation method which enables robots to work in a dynamic human-robot interaction environment safely. This requires that system have the ability to learn knowledge from demonstration, develop new skills through learning and adapt the skills to new situations. Therefore, robots should learn the demonstration, store the knowledge and retrieve the knowledge for future tasks. Therefore, in cognitive robotics, the functions of robotic control, perception, planning, and interaction are always incorporated into cognitive architectures. Upon that, researchers can use general cognitive processes to enhance imitation learning processes and stores the knowledge for other cognitive functions.

Some researchers are inrested in transferring skills and behaviors between robots. The skill transfer between robots does not implemented in this paper. However, it is still can be incorporated in this cognitive architecture.

Skill Transfer is divided into two parts: demonstration and observation.

Assume that ISAC is asked to transfer the skills to another robot, named Motoman. ISAC demonstrates the behavior sequences to Motoman strictly follows the demonstrations from the human teacher. Therefore, there are three demonstrations to reach, grasp and move the Knight.

Fig. 15. ISAC demonstrates the behavior sequences with the knight at different locations

The only difference between the observation by ISAC and Motoman is that Motoman should observe the demonstrations using the camera and convert the recorded data in the tasks space, which is the movements of the right hand of ISAC, to the joint space.

$O_{rs} = \{O_r, t\}$, which is a $N \times 4$ matrix, records the position values of the end-effector on the right arm of ISAC and the temporal information related to the sampling points.

$$\theta_{ors} = \text{inverse kinematics}(O_{rs}) \tag{13}$$

$P_{os} = \{P_o, t\}$ is still a $M \times 4$ matrix recording the positions of the Knight on the chess board in the Cartesian space in the demonstration and the temporal information related to the sampling points

After the observation, Motoman uses the same process as described in former paragraphs to segment, recognize, and generates new behaviors in simialr but slightly different situations.

In the future, our lab intend to implemente the skill stransfer using this framework and cognitive architecture described in this paper.

The existing problem is this cognitive framework largely relies on the vision system. In this paper, we proposed a cognitive segmentation method using the visual information from the cameras. As known, vision system is not very stable, and it is easy to be affected by the environmental issues, e.g., light, perspective, and noises. Therefore, how to design a stable vision system is crucial for robotics research, especially for humanoid robots and cognitive robots[Tan and Liao, 2007a].

Another possible future work is to design a probabilistic imitation learning method especially for the generation stage. It is known that the sensed information, and the results of the actions are uncertain. Therefore, robots should make a probabilistic decision in the imitation learning to complete tasks. Generally, the imitation learning is considered to be a learning of control policies[Argall, 2009]. Therefore, how can robots choose a policy from many candidate policies can be design in a probabilistic way.
Machine learning normally is a iterative process, in which, computers or robots obtain the experience from the exercises, either good or devil. The decision making is improved by utilizing the results fromthe iterative learnings in the practice. In Atkeson's famous experiment, inverted pendulum experiment, the robot tried to improve the performance through a feedback process. Initially, the robot may fail serveral times to hold pendulum in a balanced position. However, it can obtain the experiences from the failure, and finally got success. Currently, there are two types of teaching methods can be used for robots to improve and fasten the learning process: one is to provide the demonstration at the begining and robots try to complete a similar but different task from reproducing the demonstration; the other one is to train robots to learn strarting from scratch, and correct the behavior of the robots in the learning process.Both methods are effective for robotic imitation learning. And we expect that we can combine the two kinds of methods by simulating the way in which human beings learn from the demonstrations from childhood.
Currently, Cognitive Sciences receives broad attention from the robotics research community. It is expected that robots can behave like a real human and live with us in the future, and it is reasonable that researchers may seek the solution or motivation from the interdiciplinary research.

5. Conclusion

This paper proposes a cognitive framework to incorporate cognitive segmentation and the DMP algorithm in a cognitive architecture to deal with generating new behaviors in similar but different situations using imitation learning method. The simulation and experimental results demonstrates that this method is effective to solve basic problems in imitation learning when the task constraints were changed a lot.
The main contribution of this paper is that it provides a framework and architecture for robots to complete some complicated tasks, especially in the situation where several task constraints have been changed. A cognitive segmentation method is proposed in this paper. And the experimental resutls demonstrates that the integration of robotic cognitive architectures with the imitation learning technologies is successful.
Basically, the current research in imitation learning for robots is still a control problem in which the sensory information increases largely. Cognitive robots should understand the target of the task, incorporate the perceptual information, and use cognitive methods to generate suitable behaviors in a dynamic environment. This paper provides a possible solution, which can be used in different cognitive architectures, for the future cognitive behavior generation.

6. References

Albus, J. and Barbera, A. (2005). RCS: A Cognitive Architecture for Intelligent Multi-Agent Systems. *Annual Reviews in Control,* Vol.29, No.1, pp. 87-99, 2005

Anderson, J. R., Matessa, M. and Lebiere, C. (1997). ACT-R: A Theory of Higher Level Cognition and Its Relation to Visual Attention. *Human-Computer Interaction,* Vol.12, No.4, pp. 439-462, 1997

Anderson, M. (2003). Embodied Cognition: A Field Guide. *Artificial Intelligence,* Vol.149, No.1, pp. 91-130, 2003

Argall, B., Chernova, S, Veloso, M, and Browning, B (2009). A Survey of Robot Learning from Demonstration. *Robotics and Autonomous Systems,* Vol.57, No.5, pp. 469-483, 2009

Atkeson, C., Moore, A. and Schaal, S. (1997). Locally Weighted Learning. *Artificial intelligence review,* Vol.11, No.1, pp. 11-73, 1997

Atkeson, C. and Schaal, S. (1997). Robot Learning from Demonstration. *Proceedings of the Fourteenth International Conference on Machine Learning,* pp.11-73, Morgan Kaufmann, 1997

Bentivegna, D. and Atkeson, C. (2001). Learning from Observation Using Primitives. *Proceedings of the 2011 International Conference on Robotics & Automation,* pp.1988-1993, Seoul, Korea, 2001

Billard, A. (2001). Learning Motor Skills by Imitation: A Biologically Inspired Robotic Model. *Cybernetics and Systems,* Vol.32, No.1, pp. 155-193, 2001

Billard, A., Calinon, S., Dillmann, R. and Schaal, S. (2007). Robot Programming by Demonstration. In: *Handbook of Robotics*. B. Siciliano and O. Khatib (Ed.), Springer. New York, NY, USA

Brooks, R. (1986). A Robust Layered Control System for a Mobile Robot. *IEEE Journal of Robotics and Automation,* Vol.2, No.1, pp. 14-23, 1986

Brooks, R. (1991). How to Build Complete Creatures Rather Than Isolated Cognitive Simulators. In: *Architectures for Intelligence*. K. VanLehn (Ed.), 225–239, Erlbaum. Hillsdale, NJ

Brooks, R., Breazeal, C., Marjanovi , M., Scassellati, B. and Williamson, M. (1999). The Cog Project: Building a Humanoid Robot. In: *Computation for Metaphors, Analogy, and Agents*. C. Nehaniv (Ed.), 52-87, Springer-Verlag. Heidelberg, Berlin, Germany

Brooks, R. A. (1991). Intelligence without Representation. *Artificial Intelligence,* Vol.47, No.1-3, pp. 139-159, 1991

Calinon, S. and Billard, A. (2007). Incremental Learning of Gestures by Imitation in a Humanoid Robot. *Proceedings of the 2007 ACM/IEEE International Conference on Human-robot interaction,* pp.255-262, New York, NY, USA, 2007

Calinon, S., Guenter, F. and Billard, A. (2007). On Learning, Representing, and Generalizing a Task in a Humanoid Robot. *IEEE Transactions on Systems, Man, and Cybernetics, Part B,* Vol.37, No.2, pp. 286-298, 2007

Dillmann, R., Kaiser, M. and Ude, A. (1995). Acquisition of Elementary Robot Skills from Human Demonstration. *Proceedings of the 1995 International Symposium on Intelligent Robotic System,* pp.1-38, Pisa, Italy, 1995

Dillmann, R., Rogalla, O., Ehrenmann, M., Zollner, R. and Bordegoni, M. (2000). Learning Robot Behaviour and Skills Based on Human Demonstration and Advice: The Machine Learning Paradigm. *Proceedings of the Ninth International Symposium of Robotics Research (ISRR-1999),*, pp.229-238, Snowbird, UT, USA, 2000

Gobet, F., Lane, P., Croker, S., Cheng, P., Jones, G., Oliver, I. and Pine, J. (2001). Chunking Mechanisms in Human Learning. *Trends in cognitive sciences,* Vol.5, No.6, pp. 236-243, 2001

Haikonen, P. (2007). Essential Issues of Conscious Machines. *Journal of Consciousness Studies,* Vol.14, No.7, pp. 72-84, 2007

Ijspeert, A., Nakanishi, J. and Schaal, S. (2002). Movement Imitation with Nonlinear Dynamical Systems in Humanoid Robots. *Proceedings of the 2002 IEEE International Conference on Robotics and Automation*, pp.1398-1403, Washington, DC, USA, 2002

Ijspeert, A., Nakanishi, J. and Schaal, S. (2003). Learning Attractor Landscapes for Learning Motor Primitives. In: *Advances in Neural Information Processing Systems*. S. Becker, S. Thrun and K. Obermayer (Ed.), 1547-1554, MIT Press. 15

Kawamura, K. and Browne, W. N. (2009). Cognitive Robotics. In: *Encyclopedia of Complexity and System Science*. R. A. Meyers (Ed.), 1109-1126, Springer Science. Heidelberg, Germany

Kawamura, K., Gordon, S., Ratanaswasd, P., Erdemir, E. and Hall, J. (2008). Implementation of Cognitive Control for a Humanoid Robot. *International Journal of Humanoid Robotics,* Vol.5, No.4, pp. 547-586, 2008

Kawamura, K., Peters II, R., Bodenheimer, R., Sarkar, N., Park, J., Spratley, A. and Hambuchen, K. (2004). Multiagent-Based Cognitive Robot Architecture and Its Realization. *International Journal of Humanoid Robotics,* Vol.1, No.1, pp. 65-93, 2004

Keiras, D. E. and Meyer, D. E. (1997). An Overview of the Epic Architecture for Cognition and Performance with Application to Human-Computer Interaction. *Human-Computer Interaction,* Vol.12, pp. 391-438, 1997

Konidaris, G., Kuindersma, S., Barto, A. and Grupen, R. (2010). Constructing Skill Trees for Reinforcement Learning Agents from Demonstration Trajectories. *Proceedings of Advances in Neural Information Processing Systems (NIPS 2010)*, pp.1-11, Vancouver, BC, 2010

Krichmar, J. and Reeke, G. (2005). The Darwin Brain-Based Automata: Synthetic Neural Models and Real-World Devices. In: *Modelling in the Neurosciences: From Biological Systems to Neuromimetic Robotics*. G. N. Reeke, R. R. Poznanski, K. A. Lindsay, J. R.Rosenberg and O. Sporns (Ed.), 613–638, Taylor and Francis. Boca, Raton, FL, USA

Kulic, D., Takano, W. and Nakamura, Y. (2008). Combining Automated on-Line Segmentation and Incremental Clustering for Whole Body Motions. *Proceedings of IEEE International Conference on Robotics and Automation*, pp.2591-2598, Pasadena, CA, USA, 2008

Laird, J., Newell, A. and Rosenbloom, P. (1987). Soar: An Architecture for General Intelligence. *Artificial Intelligence,* Vol.33, pp. 1-64, 1987

Roweis, S. and Saul, L. (2000). Nonlinear Dimensionality Reduction by Locally Linear Embedding. *Science,* Vol.290, No.5500, pp. 2323-2326, 2000

Russell, S. J. and Norvig, P. (2010). Artificial Intelligence : A Modern Approach. Upper Saddle River, N.J., Prentice Hall/Pearson Education.

Schaal, S. (1999). Is Imitation Learning the Route to Humanoid Robots. *Trends in Cognitive Sciences,* Vol.3, No.6, pp. 233-242, 1999

Schaal, S. and Atkeson, C. (2002). Robot Juggling: Implementation of Memory-Based Learning. *IEEE Control Systems Magazine,* Vol.14, No.1, pp. 57-71, 2002

Schneider, W. (1999). Working Memory in a Multilevel Hybrid Connectionist Control Architecture (Cap2). In: *Models of Working Memory: Mechanisms of Active Maintenance and Executive Control*. A.Miyake and P. Shah (Ed.), 340–374, Cambridge University Press. New York, NY, USA

Sloman, A. (2001). Varieties of Affect and the Cogaff Architecture Schema. *Proceedings of Symposium on Emotion, Cognition, and AffectiveComputing at the AISB 2001 Convention*, pp.39–48, York, UK, 2001

Sloman, A., Wyatt, J., Hawes, N., Chappell, J. and Kruijff, G. (2006). Long Term Requirements for Cognitive Robotics. *Proceedings of Cognitive Robotics 06 Workshop, AAAI' 06*, pp.143–150, 2006

Sun, R. (2003). A Tutorial on Clarion. Technical Report. Cognitive Science Department, Rensselaer Polytechnic Institute. 15: 2003.

Tan, H. and Kawamura, K. (2011). A Framework for Integrating Robotic Exploration and Human Demonstration for Cognitive Robots in Imitation Learning. *Proceedings of the 2011 IEEE International Conference on System, Man and Cybernetics*, Anchorage, AK, USA, 2011

Tan, H. and Liang, C. (2011). A Conceptual Cognitive Architecture for Robots to Learn Behaviors from Demonstrations in Robotic Aid Area. *Proceedings of 33rd Annual International Conference of the IEEE Engineering in Medicine and Biology Society* Boston, MA, USA, 2011

Tan, H. and Liao, Q. (2007a). Design of Double Video Signal Based Location and Stereo Video Signal Generation System for Humanoid Robot. *Proceedings of the 2007 IEEE-RAS International Conference on Humanoid Robots*, pp.466-470, Pittsburgh, USA, 2007

Tan, H. and Liao, Q. (2007b). Improved Task-Oriented Force Feedback Control Method for Humanoid Robot. *Proceedings of the 2007 IEEE-RAS International Conference on Humanoid Robots*, pp.461-465, Pittsburgh, USA, 2007

Tan, H., Zhang, J. and Luo, W. (2005). Design of under-Layer Control Sub-System for Robot Assisted Microsurgery System. *Proceedings of the 2005 International Conference on Control and Automation*, pp.1169-1174, Budapest, Hungary, 2005

Tenenbaum, J., Silva, V. and Langford, J. (2000). A Global Geometric Framework for Nonlinear Dimensionality Reduction. *Science*, Vol.290, No.5500, pp. 2319-2323, 2000

Uchiyama, M. (1978). Formation of High Speed Motion Pattern of Mechanical Arm by Trial. *Transactions, Society of Instrument and Control Engineers*, Vol.19, No.5, pp. 706-712, 1978

Vijayakumar, S. and Schaal, S. (2000). Locally Weighted Projection Regression: An O (N) Algorithm for Incremental Real Time Learning in High Dimensional Space. *Proceedings of the Seventh International Conference on Machine learning*, pp.288–293, 2000

Wang, S., Zhang, J., Yang, T., Tan, H. and Luo, W. (2006). Workspace Analysis on the Robot Assisted Micro-Surgery System (in Chinese). *Journal of Machine Design*, Vol.3, No.3, pp. 25-27, 2006

Winikoff, M. (2005). Jack™ Intelligent Agents: An Industrial Strength Platform. In: *Multi-Agent Programming*. Bordini (Ed.), 175-193, Kluwer

Wold, F., Esbensen, K. and Geladi, P. (1987). Principal Component Analysis. *Chemometrics and Intelligent Laboratory Systems*, Vol.2, No.1-3, pp. 37–52, 1987

Yang, J., Xu, Y. and Chen, C. (1997). Human Action Learning Via Hidden Markov Model. *IEEE Transactions on Systems, Man and Cybernetics, Part A: Systems and Humans*, Vol.27, No.1, pp. 34-44, 1997

Yang, T., Zhang, J., Wang, S., Tan, H. and Luo, W. (2006). Motion Planning and Error Analysis in Robot Assistant Micro-Surgery System. *Proceedings of the 6th World Congress on Intelligent Control and Automation*, pp.8819-8823, Dalian, Liaoning, China, 2006

11

User, Gesture and Robot Behaviour Adaptation for Human-Robot Interaction

Md. Hasanuzzaman[1] and Haruki Ueno[2]
[1]*Department of Computer Science & Engineering, University of Dhaka,Dhaka*
[2]*National Institute of Informatics (NII), Tokyo*
[1]*Bangladesh*
[2]*Japan*

1. Introduction

Human-robot interaction has been an emerging research topic in recent year because robots are playing important roles in today's society, from factory automation to service applications to medical care and entertainment. The goal of human-robot interaction (HRI) research is to define a general human model that could lead to principles and algorithms allowing more natural and effective interaction between humans and robots. Ueno [Ueno, 2002] proposed a concept of Symbiotic Information Systems (SIS) as well as a symbiotic robotics system as one application of SIS, where humans and robots can communicate with each other in human friendly ways using speech and gesture. A Symbiotic Information System is an information system that includes human beings as an element, blends into human daily life, and is designed on the concept of symbiosis [Ueno, 2001]. Research on SIS covers a broad area, including intelligent human-machine interaction with gesture, gaze, speech, text command, etc. The objective of SIS is to allow non-expert users, who might not even be able to operate a computer keyboard, to control robots. It is therefore necessary that these robots be equipped with natural interfaces using speech and gesture.

There are several researches on human robot interaction in recent years especially focussing assistance to human. Severinson-Eklundh et. al. have developed a fetch-and-carry-robot (Cero) for motion-impaired users in the office environment [Severinson-Eklundh, 2003]. King et. al. [King, 1990] developed a 'Helpmate robot', which has already been deployed at numerous hospitals as a caregiver. Endres et. al. [Endres, 1998] developed a cleaning robot that has successfully been served in a supermarket during opening hours. Siegwart et. al. described the 'Robox' robot that worked as a tour guide during the Swiss national Exposition in 2002 [Siegwart, 2003]. Pineau et. al. described a mobile robot 'Pearl' that assists elderly people in daily living [Pineau, 2003]. Fong and Nourbakhsh [Fong, 2003] have summarized some applications of socially interactive robots. The use of intelligent robots encourages the view of the machine as a partner in communication rather than as a tool. In the near future, robots will interact closely with a group of humans in their everyday environment in the field of entertainment, recreation, health-care, nursing, etc.

Although there is no doubt that the fusion of gesture and speech allows more natural human-robot interaction, for single modality gesture recognition can be considered more reliable than speech recognition. Human voice varies from person to person, and the system needs to take care of large number of data to recognize speech. Human speech contains three types of information: who the speaker is, what the speaker said, and how the speaker said it [Fong, 2003]. Depending on what information the robot requires, it may need to perform speaker tracking, dialogue management or even emotion analysis. Most systems are also sensitive to mis-recognition due to the environmental noise. On the other hand, gestures are expressive, meaningful body motions such as physical movements of head, face, fingers, hands or body with the intention to convey information or interact with the environment. Hand and face poses are more rigid, though its also varies little from person to person. However, humans will feel more comfortable in pointing at an object than in verbally describing its exact location. Gestures are an easy way to give geometrical information to the robot. However, gestures are varying among individuals or varying from instance to instance for a given individual. The hand shape and human skin-color are different for different persons. The gesture meanings are also different in different cultures. In human-human communications, human can adapt or learn new gestures or new users using own intelligence and contextual information. Human can also change each other behaviours based on conversation or situation. Achieving natural gesture-based interaction between human and robots, the system should be adaptable to new users, gestures and robot behaviors. This chapter includes the issues regarding new users, poses, gestures and behaviours recognition and adaptation for implementing human-robot interaction in real-time.

Adaptivity is the biological property in all creatures to survive in the biological world. It is the capability of self-modification that some agents have, which allows them to maintain a level of performance in front of environmental changes, or to improve it when confronted repeatedly with the same situation [Torras, 1995]. Gesture-based human-robot natural interaction system could be designed so that it can understand different users, their gestures, meaning of the gestures and the robot behaviours. Torras proposed robot adaptivity technique using neural learning algorithm. This method is computationally inexpensive and there is no way to encode prior knowledge about the environment to gain the efficiency. It is essential for the system to cope with the different users. A new user should be included using on-line registration process. When a user is included the user may wants to perform new gesture that is ever been used by other persons or himself/herself. In that case, the system should include the new hand poses or gestures with minimum user interaction.

In the proposed method, a frame-based knowledge model is defined for gesture interpretation and human-robot interaction. In this knowledge model, necessary frames are defined for the known users, robots, poses, gestures and robot behaviours. The system first detects a human face using a combination of template-based and feature-invariant pattern matching approaches and identifies the user using the eigenface method [Hasanuzzaman 2007]. Then, using the skin-color information of the identified user three larger skin-like regions are segmented from the YIQ color spaces, after that face and hand poses are classified by the PCA method. The system is capable of recognizing static gestures comprised of face and hand poses. It is implemented using the frame-based Software Platform for Agent and Knowledge Management (SPAK) [Ampornaramveth, 2001]. Known gestures are defined as frames in SPAK knowledge base using the combination of face and hand pose frames. If the required combination of the pose components is found then corresponding gesture frame will be activated. The system learns new users, new poses using multi-clustering approach and combines computer vision and knowledge-based approaches in order to adapt to different users, different gestures and robot behaviours.

New robot behaviour can be learned according to the generalization of multiple occurrence of same gesture with minimum user interaction. The algorithm is tested by implementing a human-robot interaction scenario with a humanoid robot "Robovie" [Kanda, 2002].

2. Related research

In this chapter we have described a vision and knowledge-based user and gesture recognition as well as adaptation system for human–robot interaction. Following subsections summarize the related works.

2.1 Overview of human-machine interaction systems

Both machines and human measure their environment through sense or input interfaces and modify their environment through expression or output interfaces. Most popular mode of human-computer or human-intelligence machine interaction is simply based on keyboards and mice. These devices are familiar but lack of naturalness and do not support remote control or telerobotics interface. Thus in recent years researchers are giving tedious pressure to find attractive and natural user interface devices. The term natural user interface is not an exact expression, but usually means an interface that is simple, easy to use and seamless as possible. Multimodal user interfaces are a strong candidate for building natural user interfaces. In multimodal approaches user can include simple keyboard and mouse with advance perception techniques like speech recognition and computer vision (gestures, gaze, etc.) as user machine interface tools.

Weimer et. al. [Weimer, 1989] described a multimodal environment that uses gesture and speech input to control a CAD system. They used 'Dataglove' to track the hand gestures and presented the objects in three-dimension onto the polarizing glasses. Yang et. al. [Yang, 1998] have implemented a camera-based face and facial features (eyes, lips and nostrils) tracker system. The system can also estimate user gaze direction and head poses. They have implemented two multimodal applications: a lip-reading system and a panoramic image viewer. The lip-reading systems improve speech recognition accuracy by using visual input to disambiguate among acoustically confusing speech elements. The panoramic image viewer uses gaze to control panning and tilting, and speech to control zooming. Perzanowski et. al. [Perzanowski, 2001] proposed multimodal human-robot interface for mobile robot. They have incorporated both natural language understanding and gesture recognition as a communication mode. They have implemented their method on a team of 'Nomad 200' and 'RWI ATRV-Jr' robots. These robots understand speech, hand gestures and input from a handheld Palm pilot to other Personal Digital Assistant (PDA).

To use the gestures in the HCI or HRI it is necessary to interpret the gestures by computer or robot. The interpretation of human gestures requires that static or dynamic modelling of the human hand, arm, face and other parts of the body that is measurable by the computers or intelligent machines. First attempt is to measure the gesture features (hand pose and/or arm joint angles and spatial positions) are by the so called glove-based devices [Sturman, 1994]. The problems regarding gloves and other interface devices can be solved using vision-based non-contract and nonverbal communication techniques. Numbers of approaches have been applied for the visual interpretation of gestures to implement human-machine interaction [Pavlovic, 1997]. Torrance [Torrance, 1994] proposed a natural language-based interface for teaching mobile robots about the names of places in an indoor environment. But due to the lack of speech recognition his interface system still uses keyboard-based text input. Kortenkamp et. al. [Kortenkamp, 1996] have developed gesture-based human-mobile robot interface. They have used static arm poses as gestures. Stefan Waldherr et. al. proposed gesture-based interface for human and service robot interaction [Waldherr, 2000]. They combined template-based approach and Neural Network based approach for tracking a person and recognizing gestures involving arm motion. In their work they proposed illumination adaptation methods but did not consider user or hand pose adaptation. Bhuiyan et. a.l detected and tracked face and eye for human robot interaction [Bhuiyan, 2004]. But only the largest skin-like region for the probable face has been considered, which may not be true when two hands are present in the image. However, all of the above papers focus primarily on visual processing and do not maintain knowledge of different users nor consider how to deal with them.

2.2 Face detection and recognition

A first step of any face recognition or visually person identification system is to locate the face in the images. After locating the probable face, researchers use facial features (eyes, nose, nostrils, eyebrows, mouths, leaps, etc.) detection method to detect face accurately [Yang, 2000]. Face recognition or person identification compares an input face image or image features against a known face database or features databases and report match, if any. Following two subsections summarize promising past research works in the field of face detection and recognition.

2.2.1 Face detection

Face detection from a single image or an image sequences is a difficult task due to variability in pose, size, orientation, color, expression, occlusion and lighting condition. To build a fully automated system that extracts information from images of human faces, it is essential to develop and efficient algorithms to detect human faces. Visual detecting of face has been studied extensively over the last decade. Face detection researchers summarized the face detection work into four categories: template matching approaches [Sakai, 1996] [Miao, 1999] [Augusteijn, 1993] [Yuille, 1992] [Tsukamoto, 1994]; feature invariant approaches [Sirohey, 1993]; appearance-based approaches [Turk, 1991] and knowledge-based approaches [Yang, 1994] [Yang, 2002] [Kotropoulous, 1997]. Many researches also used skin color as a feature and leading remarkably face tracking as long as the lighting conditions do not varies too much [Dai, 1996], [Crowley, 1997] [Bhuiyan, 2003], [Hasanuzzaman 2004b]. Recently, several researchers combined multiple features for face localization and detection and those are more robust than single feature based approaches. Yang and Ahuja [Yang, 1998] proposed a face detection method based on color, structure and geometry. Saber and Tekalp [Saber, 1998] presented a frontal view-face localization method based on color, shape and symmetry.

2.2.2 Face recognition

During the last few years face recognition has received significant attention from the researchers [Zhao, 2003] [Chellappa, 1995]. Zhao [Zhao, 2003] et. al. have summarized the past recent research on face recognition methods with three categories: Holistic matching methods, Feature-based matching methods and Hybrid methods. One of the most widely used representations of the face recognition is eigenfaces, which are based on principal component analysis (PCA). The eigenface algorithm uses the principal component analysis (PCA) for dimensionality reduction and to find the vectors those are best account for the distribution of face images within the entire face image spaces. Turk and Pentland [Turk, 1991] first successfully used eigenfaces for face detection and person identification or face recognition. In this method from the known face images training image dataset are prepared. The face space is defined by the "eigenfaces" which are eigenvectors generated from the training face images. Face images are projected onto the feature space (or eigenfaces) that best encodes the variation among known face images. Recognition is performed by projecting a test image onto the "facespace" (spanned by the m number of eigenfaces) and then classified the face by comparing its position (Euclidian distance) in face space with the positions of known individuals.

Independent component analysis (ICA) is similar to PCA except that the distributions of the components are designed to be non-Gaussian. The ICA separates the high-order moments of the input in addition to the second order moments utilized in PCA [Bartlett, 1998]. Face recognition system using Linear Discriminant Analysis (LDA) or Fisher Linear Discriminant Analysis (FDA) has also been very successful [Belhumeur, 1997]. In feature-based matching methods, facial features such as the eyes, lips, nose and mouth are extracted first and their locations and local statistics (geometric shape or appearance) are fed into a structural classifier [Kanade, 1977]. One of the most successful of these methods is the Elastic Bunch Graph Matching (EBGM) system [Wiskott, 1997]. Other well-known methods in these systems are Hidden Markov Model (HMM) and convolution neural network [Rowley, 1997].

2.3 Gesture recognition and gesture based interface

A gesture is a motion of the body parts or the whole body that contains information [Billinghurst, 2002]. The first step in considering gesture-based interaction with computers or robots is to understand the role of gestures in human-human communication. Gestures are varying among individuals or vary from instance to instance for a given individual. The Gesture meanings also follow one-to-many mapping or many-to-one mapping. Two approaches are commonly used to recognize gestures. One is a glove-based approach that requires wearing of cumbersome contact devices and generally carrying a load of cables that connect the devices to computers [Sturman, 1994]. Another approach is a vision-based technique that does not require wearing any contact devices, but uses a set of video cameras and computer vision techniques to interpret gestures [Pavlovic, 1997]. Although glove-based approaches provide more accurate results, they are expensive and encumbering. Computer vision techniques overcome these limitations. In general, vision-based systems are more natural than glove-based systems and are capable of hand, face and body tracking but do not provide the same accuracy in pose determination. However, for general purposes, achieving a higher-level accuracy may be less important than a real-time and inexpensive method. In addition, many gestures involve two hands, but most of the research efforts in glove-based gesture recognition use only one glove for data acquisition. In vision-based systems, we can use two hands and facial gestures at the same time.

Vision-based gesture recognition systems have three major components: image processing or extracting important clues (hand or face pose and position), tracking the gesture features (related position or motion of face and hand poses), and gesture interpretation. Vision-based gesture recognition system varies along a number of dimensions: number of cameras, speed and latency (real-time or not), structural environment (restriction on lighting conditions and background), primary features (color, edge, regions, moments, etc.), user requirements etc. Multiple cameras can be used to overcome occlusion problems for image acquisition but this adds correspondences and integration problems. The first phase of vision-based gesture recognition task is to select a model of the gesture. The modelling of gesture depends on the intent-dent applications by the gesture. There are two different approaches for vision-based modelling of gesture: Model based approach and Appearance based approach. The Model based techniques are tried to create a 3D model of the user hand (parameters: Joint angles and palm position) [Rehg, 1994] or contour model of the hand [Shimada, 1996] [Lin, 2002] and use these for gesture recognition. The 3D models can be classified in two large groups: volumetric model and skeletal models. Volumetric models are meant to describe the 3D visual appearance of the human hands and arms. Skeletal models are related to the human hand skeleton.

Once the model is selected, an image analysis stage is used to compute the model parameters from the image features that are extracted from single or multiple video input streams. Image analysis phase includes hand localization, hand tracking, and selection of suitable image features for computing the model parameters. Two types of cues are often used for gesture or hand localization: color cues and motion cues. Color cue is useful because human skin color footprint is more distinctive from the color of the background and human cloths [Kjeldsen, 1996], [Hasanuzzaman, 2004d]. Color-based techniques are used to track objects defined by a set of colored pixels whose saturation and values (or chrominance values) are satisfied a range of thresholds. The major drawback of color-based localization methods is that skin color footprint is varied in different lighting conditions and also the human body colors. Infrared cameras are used to overcome the limitations of skin-color based segmentation method [Oka, 2002].

The motion-based segmentation is done just subtracting the images from background [Freeman, 1996]. The limitation of this method is considered the background or camera is static. Moving objects in the video stream can be detected by inter frame differences and optical flow [Cutler, 1998]. However such a system cannot detect a stationary hand or face. To overcome the individual shortcomings some researchers use fusion of color and motion cues [Azoz, 1998]. The computation of model parameters is the last step of the gesture analysis phase and it is followed by gesture recognition phase. The type of computation depends on both the model parameters and the features that were selected. In the recognition phase, parameters are classified and interpreted in the light of the accepted model or the rules specified for the gesture interpretation. Two tasks are commonly associated with the recognition process: optimal partitioning of the parameter space and implementation of the recognition procedure. The task of optimal partitioning is usually addresses through different learning-from-examples training procedures. The key concern in the implementation of the recognition procedure is computation efficiency. A recognition method usually determines confidence scores or probabilities that define how closely the image data fits each model. Gesture recognition methods are divided into two categories: static gesture or hand poster and dynamic gesture or motion gesture.

Eigenspace or PCA is also used for hand pose classification similarly its used for face detection and recognition. Moghaddam and Pentland used eigenspaces (eigenhands) and principal component analysis not only to extract features, but also as a method to estimate complete density functions for localization [Moghaddam, 1995]. In our previous research, we have used PCA for hand pose classification from three larger skin-like components that

are segmented from the real-time capture images [Hasanuzzaman, 2004d]. Triesch et. al. [Triesch, 2002] employed the elastic graph matching techniques to classify hand posters against complex backgrounds. They represented hand posters by label graphs with an underlying two-dimensional topology. Attached to the nodes are jets, which are a sort of local image description based on Gabor filters. This approach can achieve scale-invariant and user invariant recognition and does not need hand segmentation. This approach is not view-independent, because use one graph for one hand posture. The major disadvantage of this algorithm is the high computational cost.
Appearance based approaches use template images or features from the training images (images, image geometry parameters, image motion parameters, fingertip position, etc.) which use for gesture recognition [Birk, 1997]. The gestures are modeled by relating the appearance of any gesture to the appearance of the set of predefined template gestures. A different group of appearance-based model uses 2D hand image sequences as gesture templates. For each gestures number of images are used with little orientation variations [Hasanuzzaman, 2004a]. Appearance based approaches are generally computationally less expensive than model based approaches because its does not require translation time from 2D information to 3D model. Dynamic gestures recognition is accomplished using Hidden Markov Models (HMMs), Dynamic Time Warping, Bayesian networks or other patterns recognition methods that can recognize sequences over time steps. Nam et. al. [Nam, 1996] used HMM methods for recognition of space-time hand-gestures. Darrel et. al. [Darrel, 1993] used Dynamic Time Warping method, a simplification of Hidden Markov Models (HMMs) to compare the sequences of images against previously trained sequences by adjusting the length of sequences appropriately. Cutler et. al. [Cutler, 1998] used a ruled-based system for gesture recognition in which image features are extracted by optical flow. Yang [Yang, 2000] recognizes hand gestures using motion trajectories. First they extract the two-dimensional motion in an image, and motion patterns are learned from the extracted trajectories using a time delay network.

3. Frame-based knowledge-representation system for gesture-based HRI

The 'frame-based approach' is a knowledge-based problem solving approach based on the so called, 'Frame theory', first proposed by Marvin Minsky [Minsky, 1974]. A frame is a data-structure for representing a stereotyped unit of human memory including definitive and procedural knowledge. Attached to each frame there are several kinds of information about the particular object or concept it describes such as name and a set of attributes called slots. Collections of related frames are linked together into frame systems. Framed-based approach has been used successfully in many robotic applications [Ueno, 2002]. Ueno presented the concepts and methodology of knowledge modeling based on Cognitive Science for realizing the autonomous humanoid service robotics arm and hand system HARIS [Ueno, 2000]. A knowledge-based software platform called SPAK (Software Platform for Agent and Knowledge management) has developed for intelligent service robots under the internet-based distributed environment [Ampornaramveth, 2004]. SPAK has been developed to be a platform on which various software components for different robotic tasks can be integrated over a networked environment. SPAK works as a knowledge and data management system, communication channel, intelligent recognizer, intelligent scheduler, and so on. Zhang et. al. [Zhang, 2004b] have developed an Industrial Robot Arm control system using SPAK. In that system SPAK works as a communication channel and intelligent robot actions scheduler. Kiatisevi et. al. [Kiatisevi, 2004] has proposed a distributed architecture for knowledge-based interactive robots and through SPAK they have implemented dialogue-based human-robot ('Robovie') interaction for greeting scenarios. This system employs SPAK, a frame-based knowledge engine, connecting to a group of network software agents such as 'Face recognizers', 'Gesture recognizers', 'Voice recognizers', 'Robot Controller', etc. Using information received from these agents, and based on the predefined frame knowledge hierarchy, SPAK inference engine determines the actions to be taken and submit corresponding commands to the target robot control agents.

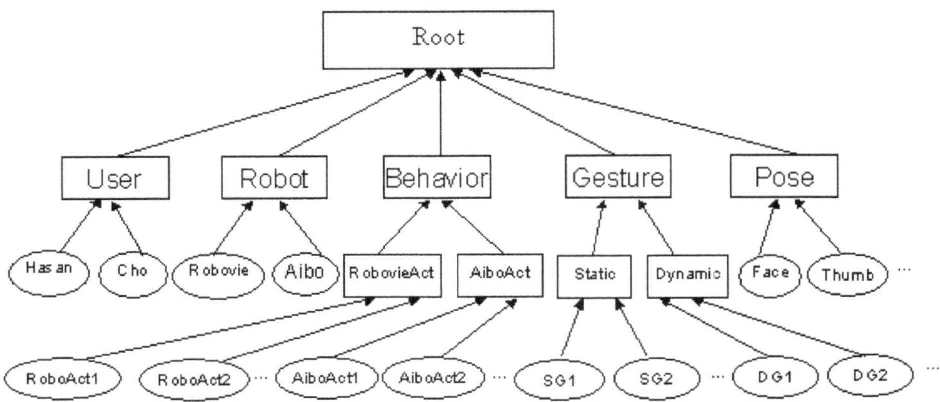

Fig. 1. Frame hierarchy for gesture-based human-robot interaction system (SG=Static Gesture, DG=Dynamic Gesture)

Figure 1 shows the frame hierarchy of the knowledge model for the gesture-based human-robot interaction system, organized by the IS_A relationship indicated by arrows connecting upper and lower frame boxes. Necessary frames are defined for the users, robots, poses, gestures and robot behaviors (actions). The user frame includes instances of all known users (instance "Hasan", "Cho", ...); robot frame includes instances of all the robots ("Aibo", "Robovie",....) used in the system. The behavior frame can be sub-classed further into "AiboAct" and "RobovieAct", where "AiboAct" frame includes instances of all the predefined 'Aibo' actions and "RobovieAct" frame includes instances of all the predefined 'Robovie' actions. The gesture fame is sub-classed into "Static" and "Dynamic" for static and dynamic gestures respectively. Examples of static frame instances are, "TwoHand", "One", etc. Examples of dynamic frame instances are "YES", "NO", etc. The pose frame includes all recognizable poses such as "LEFTHAND", "RIGHTHAND", "FACE", "ONE", etc. Gesture and user frames are activated when SPAK receives information from a network agent indicating a gesture and a face has been recognized. Behavior frames are activated when the predefined conditions are met. In this model each recognizable pose is treated as an instance-frame under the class-frame "Pose". All the known poses are defined as frames. If a predefined pose is classified by vision-based pose classification module, then corresponding pose-frame will be activated. The static gestures are defined using the combination of face and hand poses. These gesture frames has three slots for the gesture components.

Each robot behaviour, includes a command or series of commands for a particular task. Each robot behavior is mapped with user and gesturer (user-gesture-action) as we proposed person-centric interpretation of gesture. In this model same gesture can be used to activate different actions of a robot for different persons even the robot is same.

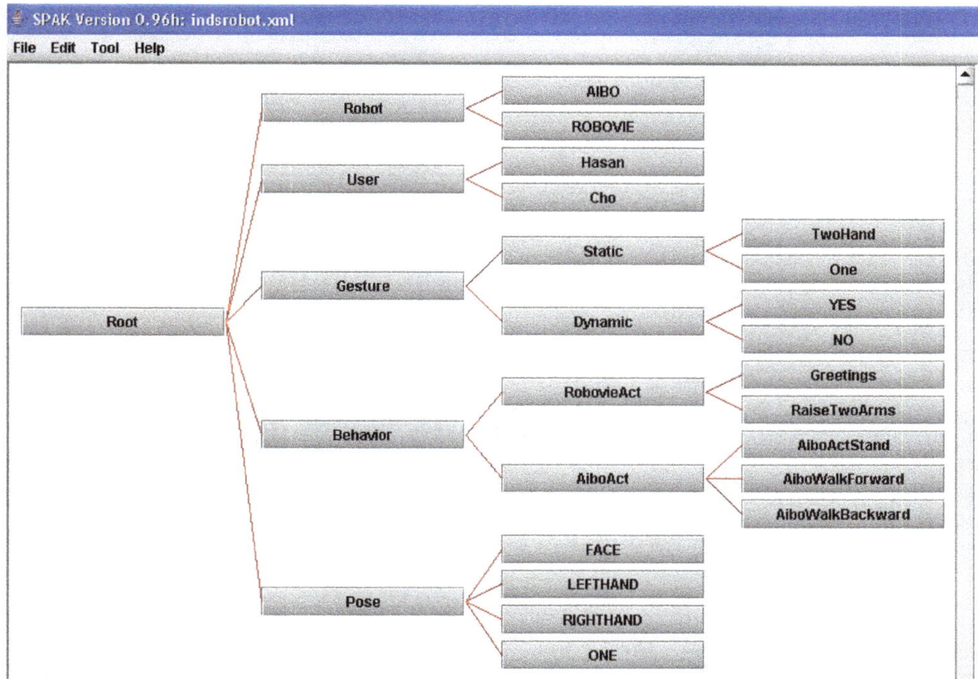

Fig. 2. Knowledge Editor showing the example gesture-based human-robot interaction

The frame system can be created in Graphical User Interface (GUI) and interpreted by the SPAK inference engine. Figure 2 shows the example knowledge Editor (part of the system) with the example of 'Robot', 'User', 'Gesture', 'Behaviour' (robot actions) and 'Pose' frames. The knowledge Editor Window displays the current frame hierarchy. Each frame is represented as click-able button. The buttons are connected with lines indicating IS_A relationships among the frames. Clicking on the frame button brings up its slot editor. Figure 3 shows an example of slot editor for the robot (Robovie) action-frame "RaiseTwoArms". The attributes for a Slot are defined by slot name, type, value, condition, argument, required, shared, etc. Figure 4 shows an example instance-frame 'Hasan' of the class-frame 'User' defined in SPAK. Figure 5 shows an example instance-frame 'Robovie' of class-frame 'Robot' defined in SPAK.

Property of RaiseTwoArms

File Edit

Name	Type	Value	Condition	Argument	Required	Shared	Unique
Name	String	RaiseTwo...	ANY		☐	☐	☐
mRobot	Instance		ANY	ROBOVIE	☑	☑	☑
mUser	Instance		ANY	Hasan	☑	☑	☑
mGesture	Instance		ANY	TwoHand	☑	☑	☑
OnInstanti...	String	roboposes...	ANY		☐	☑	☑

Fig. 3. Example of a robot action-frame 'RaiseTwoArms' in SPAK

Property of Hasan

File Edit

Name	Type	Value	Condition	Argument	Required	Shared
Name	String	Hasan	ANY		☑	☑
UserName	String		=	hasan	☑	☑
Y-High	Integer		<=	220	☐	☑
Y-Low	Integer		>=	60	☐	☑
I-High	Integer		<=	70	☐	☑
I-Low	Integer		>=	8	☐	☑
Q-High	Integer		<=	10	☐	☑
Q-Low	Integer		>=	-15	☐	☑

Fig. 4. Example of instance-frame 'Hasan' of class-frame 'User' in SPAK

Property of ROBOVIE

File Edit

Name	Type	Value	Condition	Argument	Required	Shared
Name	String	ROBOVIE	ANY		☑	☑
RobotName	String		=	robovie	☑	☑

Fig. 5. Example of instance-frame 'Robovie' of class-frame 'Robot' in SPAK

Property of FACE

File Edit

Name	Type	Value	Condition	Argument	Required	Shared
Name	String	FACE	ANY		☑	☑
mPose	String		=	face	☑	☑

Fig. 6. Example of instance-frame 'FACE' of class-frame 'Pose' in SPAK

Property of LEFTHAND

File Edit

Name	Type	Value	Condition	Argument	Required	Shared
Name	String	LEFTHAND	ANY		☑	☑
mPose	String		=	lefthand	☑	☑

Fig. 7. Example of instance-frame 'LEFTHAND' of class-frame 'Pose' in SPAK

Image analysis module classifies the hand and face poses and identifies the user. Image analysis module, sends user name (hasan, cho, etc.) and pose names (face, lefthand, righthand, etc.) to the SPAK knowledge module. According to pose name and user name corresponding pose frame and user frame will be activated. Figure 6, Figure 7, and Figure 8 shows the instance-frames 'FACE', 'LEFTHAND', 'RIGHTHNAD' of the class-frame 'Pose' respectively. If the required combination of the pose components is found then the corresponding gesture frame will be activated. Figure 9 shows the gesture frame 'TwoHand' in SPAK. It will be activated if pose fames 'FACE', 'LEFTHAND' and 'RIGHTHNAD' are activated. Using the received gesture and user information, SPAK processes the facts and activates the corresponding robot action frames to carry out predefined robot actions, which may include body movement and speech. The robot behaviour is user dependent and it is mapped based on user and gesture relationship (user-gesturer-robot-action) in the knowledge base. Figure 3 shows an example of 'Robovie' robot action-frame for the action "Raise Two Arms". This frame only activated if the identified user is 'Hasan', recognized gesture is "TwoHand" and the selected robot is 'Robovie'. User can add or edit the necessary knowledge frames for the users, face and hand poses, gestures and robot behaviors using SPAK knowledge Editor. The new robot behaviour frame can be included in the knowledge base according to generalization of multiple occurrences of the same gesture with user consent (first time only).

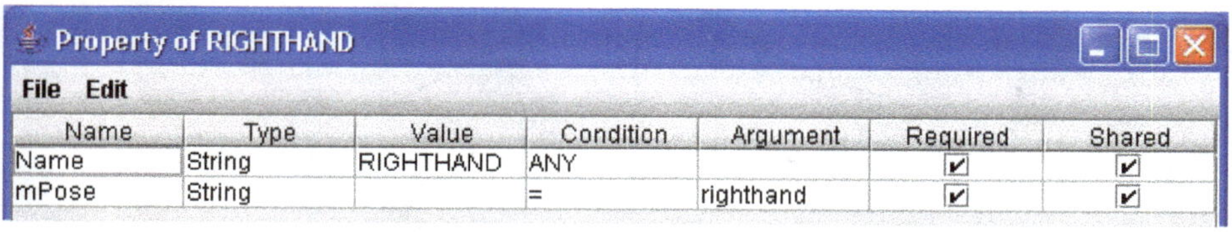

Property of RIGHTHAND

File Edit

Name	Type	Value	Condition	Argument	Required	Shared
Name	String	RIGHTHAND	ANY		✔	✔
mPose	String		=	righthand	✔	✔

Fig. 8. Example of instance-frame 'RIGHTHAND' of class-frame 'Pose' in SPAK

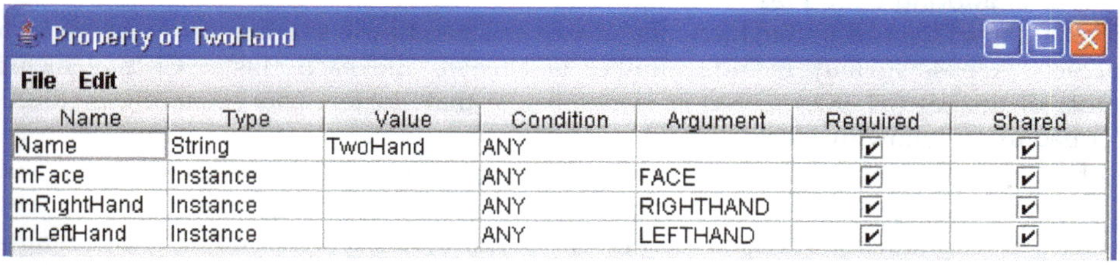

Property of TwoHand

File Edit

Name	Type	Value	Condition	Argument	Required	Shared
Name	String	TwoHand	ANY		✔	✔
mFace	Instance		ANY	FACE	✔	✔
mRightHand	Instance		ANY	RIGHTHAND	✔	✔
mLeftHand	Instance		ANY	LEFTHAND	✔	✔

Fig. 9. Example of instance-frame 'TwoHand' of class-frame 'Gesture' in SPAK

4. Users, poses, gesture and robot behavior adaptation

This section describes new users, poses, gestures and robot behaviours adaptation methods for implementing human-robot interaction. Suppose, the robot fixed with a same room with same lighting condition, in that case only the user skin color dominates on the color-based face and hands segmentation method. It is essential for the system to cope with the different persons. The new user may not be included in the system during training phase, so the person should be included using on-line registration process. The user may want to perform new gestures that is ever been used by others person or himself. In that case the system should include the new poses with minimal user interaction. The system learns new users, new poses using multi-clustering approach with minimum user interaction. To adapt to new users and new hand poses the system must be able to perceive and extract relevant properties from the unknown faces and hand poses, find common patterns among them and formulate discrimination criteria consistent with the goals of the recognition process. This form of learning is known as clustering and it is the first steps in any recognition process where discriminating features of the objects are not know in advance [Patterson, 1990]. Subsection 4.1 describes multi-clustering based learning method.

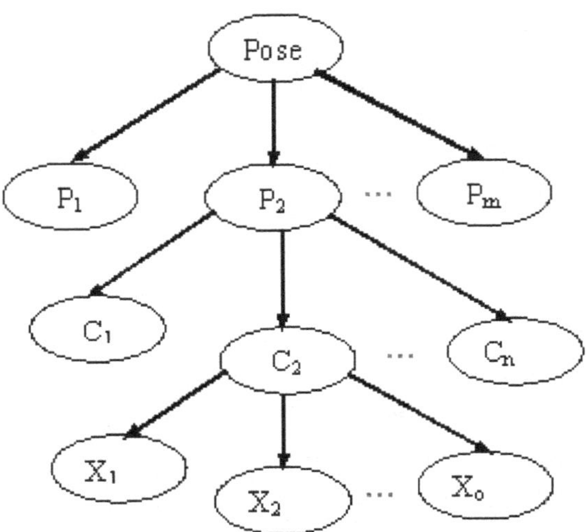

Fig. 10. Multi-cluster hierarchies for object classification and learning

4.1 Multi-clustering based learning method

Figure 10 shows the conceptual hierarchy of face and hand-pose learning using multi-cluster approach. A pose Pi may include number of clusters and each cluster Cj may include number of images (X1, X2, ..., Xo) as a member of that cluster. This clustering method is described using following steps:

Step 1. Generate eigenvectors from training images that includes all the known hand poses [Turk, 1991].

Step 2. Select m-number of eigenvectors corresponding to the higher order of eigenvalues. These selected eigenvectors are regarded as principal components. The eigenvalues are sorted from high to low values.

Step 3. Read the initial cluster image database (initialize with the known cluster images) and cluster information table that's hold the starting pointer of each cluster. Project each image onto the eigenspaces and form feature vectors using equation (1) and (2).

$$\omega_i^j = (u_m)^T (T_j) \tag{1}$$

$$\Omega_j = [\omega_1^j, \omega_2^j, \ldots, \omega_k^j] \tag{2}$$

Where, (u_m) is the *m-th* eigenvectors, T_j is the images (60×60) in the cluster database.

Step 4. Read the unlabeled images those should be clustered or labeled.

a. Project each unlabeled image onto the eigenspaces and form feature vectors (Ω) using equation (1) and (2).

b. Calculate Euclidean distance to each image in the known (clustered) dataset using equation (3) and (4),

$$\varepsilon_j = \| \Omega - \Omega_j \| \tag{3}$$

$$\varepsilon = \arg\min\{\varepsilon_j\} \tag{4}$$

Step 5. Find the nearest class,

a. If (Ti=< ε <=Tc) then add the image in the neighbor cluster; increment the insertion parameter.

b. If (ε <Ti), then the image is recognizable and no need to include it in the cluster database.

Step 6. If the insertion rate into the known cluster is grater than zero, then update the cluster information table that's holds the starting pointer of all clusters.

Step 7. Do the step 3 to step 6 until the insertion rate (α) into the known cluster (training) data set is zero (0).

Step 8. If insertion rate is zero, then cheek the unlabeled dataset, which follows the condition (Tc < ε <=Tf). Where, Tf is the threshold that defines for discarding the image.

Step 9. If maximum number of unlabeled data (for a class)>N (predefined), then select one image (based on minimum Euclidian distance) as a member of a new cluster. Then update the cluster information table.

Step 10. Repeat from step 3 to step 9 until the number of unlabeled data less than N.

Step 11. If height of the cluster (number of member images in the cluster) is>L, then add its as a new cluster.

Step 12. After clustering poses, the user defines the association of the clusters in the knowledge base. Each pose may be associated with multiple clusters. For undefined cluster there is no association link.

4.2 Face recognition and user adaptation

We have already mentioned that the robot should be able to recognize and remember the people and learn about them [Aryananda, 2002]. If the new user comes in front of the robot eyes camera or system camera the system identifies the person as unknown person and asks for registration. The face is first detected from the cluttered background using multiple feature-based approaches [Hasanuzaman, 2007]. The detected face is filtered in order to remove noises and normalized so that it matches with the size and type of the training image [Hasanuzzaman, 2006]. The detected face is scaled to be a square image with 60x60 dimensions and converted to be a gray image. The face pattern is classified using the eigenface method [Turk, 1991], whether it belongs to known person or unknown person. The eigenvectors are calculated from the known persons face images for all face class and m-number of eigenvectors corresponding to the highest eigenvalues are chosen to form principal components for each class. The Euclidean distance is determined between the weight vectors generated from the training images and the weight vectors generated from the detected face by projecting them onto the eigenspaces. If the minimal Euclidian distance is less than the predefined threshold value then person is known, otherwise unknown. For unknown person based on judge function learning process is activated and the system learned new user using multi-clustering approach [Section 4.1]. The judge function is based on the ratio of the number of unknown face to total number of detected faces for a specific time slot. The learning function develops new cluster/clusters corresponding to a new person. The user defines the person name and skin color information in the user profile knowledge base and associates with the corresponding cluster. For known user, person-centric skin color information (Y, I, Q components) is used to reduce the computational cost.

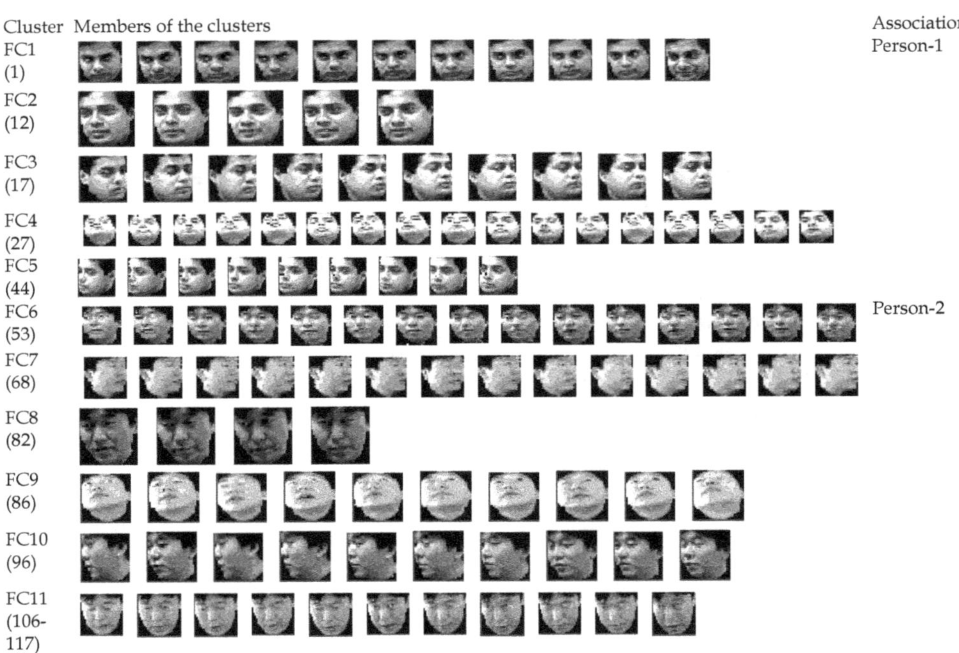

Fig. 11. Example output of multi-clustering algorithm for learning new user

Figure 11 shows the example output of multi-clustering approach for recognizing and learning new user. For example, the system is initially trained by the face images (100 face images with five directions) of person_1. The system learns and remembers this person using five clusters (FC1, FC2, FC3, FC4, FC5) as shown in top five rows of Figure 11 and clusters information table (that's hold the starting position of each cluster and end position of last cluster) contents are [1, 12, 17, 27, 44, 52]. For example, in the case of face classification, if any face image matches with the known member between 12 and 16 then classified as face cluster_2 (FC2). If the face is classified as any of these five clusters, the person is identified as person_1. Suppose, the system is deal with new person 100 face images and it could not identify the person and activate the learning function. Then the system develops new six clusters (FC6, FC7, FC8, FC9, FC10, FC11) and updates the cluster information table ([1, 12, 17, 27, 44, 53, 68, 82, 86, 96, 106, 116]) as shown in Figure 11 (rows 6 to rows 11). The new user is registered as person_2 and the associations with the clusters are defined. If any detected face images is classified as known cluster then corresponding person is identified.

4.3 Hand pose classification and adaptation

For machine it is difficult to understand the new poses without prior knowledge. It is essential to learn new poses based on specific judge function or predefined knowledge. The judge function determines the user intention, i.e., intention to create new gesture. The judge function is based on the ratio of the number of unknown hand poses to the total number hand poses for a specific time slot. For example, the user shows same hand pose continuously for 10 image frames that are unknown to the system that means he/she wants to use it as a gesture. In this proposed system the hand poses are classified using multi-cluster based learning method. For unknown pose, based on judge function learning function is activated and the system learns new pose. The learning function develops new cluster/clusters corresponding to new pose. The user defines the pose name in the knowledge base and associates with the corresponding cluster. If the pose is identified then corresponding frame will be activated. Figure 12 presents example output of learning new pose using multi-cluster approach. The system is first trained by pose 'ONE' and form one cluster for that pose. Then the system is trained by pose 'FIST' where formed another two clusters because the user uses two hands for that pose. Similarly other clusters are formed corresponding to new poses.

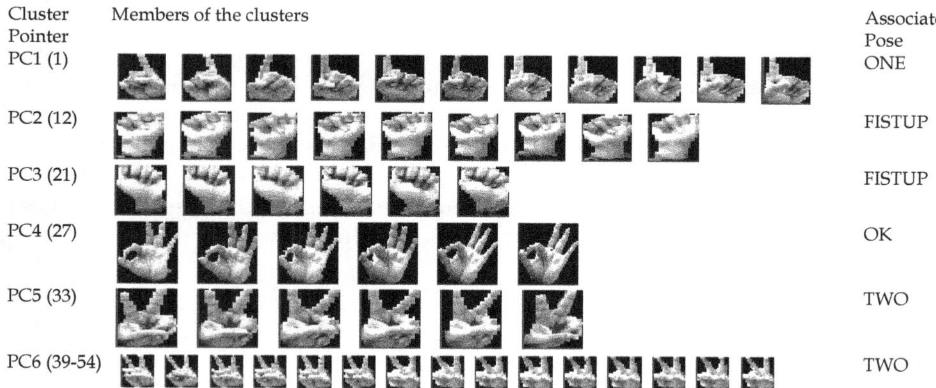

Fig. 12. Example output of multi-clustering approach for learning new pose

4.4 Gesture recognition and adaptation

The recognition of gesture is carried out in two phases. In the first phase, face and hand poses are classified from captured image frame using the method described in previous section. Then combinations of poses are analyzed to identify the occurrence of gesture. For example, if left hand palm, right hand palm and one face are present in the input image then it recognizes as "TwoHand" gesture [Figure 19(a)] and corresponding gesture frame will be activated. Interpretation of identified gesture is user-dependent since the meaning of the gesture may differ from person to person based on their culture. For example, when user'Hasan' comes in front of 'Robovie' eyes, 'Robovie' recognizes the person and says "Hi Hasan! How are you?", then 'Hasan' raises his 'Thumb up' and "Robovie" replies to 'Hasan'"Oh! You are not fine today". In the similar situation, for another user 'Cho', 'Robovie' says, "Hi, You are fine today?". That means 'Robovie' can understand the person-centric meaning of gesture. To accommodate different user's desires, our person-centric gesture interpretation is implemented using frame-based knowledge representation approach. The user predefines these frames into the knowledge base with necessary attributes (gesture components, gesture name) for all predefined gestures. Our current system recognizes 11 static gestures. These are: 'TwoHand' (raise left hand and right hand palms), 'LeftHand' (raise left hand palm), 'RightHand' (raise right hand palm), 'One' (raise index finger), 'Two' (form V sign using index and middle fingers), 'Three' (raise index, middle and ring fingers), 'ThumbUp' (thumb up), 'Ok' (make circle using thumb and index finger), 'FistUp' (fist up), 'PointLeft' (point left by index finger), 'PointRight' (point right by index finger).

It is possible to recognize more gestures including new poses and new rules for the gesture using this system. New poses can be included in the training image database using the interactive learning method and corresponding frame can be defined in the knowledge base to interpret the gesture. To teach the robot a new poses, the user should perform the poses several times (example 10 image frame times.). Then the learning method detects it as a new pose and creates cluster/clusters for that pose. Sequentially, it updates the knowledge base for the cluster information.

4.5 Robot behaviours adaptation

Robot behaviours or actions can be programmed or learned through experience. But it is difficult to perceive human or user intention to acts robot with his/her gestures. This system has proposed the experience based and user-interactive robot behaviour learning or adaptation method. In this method the history of the similar gesture-action map is stored in the knowledge base. According to maximum user desires action will be select for the gesture and ask for the user acknowledgement. If user consents or uses "YES" gesture (or types "Yes") the corresponding frame will be store permanently. For Example scenario:

Person_n: comes in front of Robovie.

Robovie: cannot recognize, and asks, "Who are you?"

Person_n: types "Person_n" (or says "Person_n").

Robovie: Says, "Hello Person_n, do you want to play with me?"

Person_n: shows "OK" hand gesture (make circle using thumb and index fingure).

Robovie: asks "Do you mean Yes"?

Person_n: again shows "OK" gesture.

Robovie: add "Ok ="Yes, for Person_n" into his knowledge base.

Suppose there are two users already use 'OK' gesture to mean Yes, so Robovie adds "OK=Yes for everybody" into his knowledge base.

5. Experimental results and discussions

This system uses a standard video camera and 'Robovie' eye's camera for data acquisition. Each captured image is digitized into a matrix of 320×240 pixels with 24-bit color. User and hand pose adaptation method is verified using real-time captured images as well as static images. The algorithm has also been tested with a real world human-robot interaction system using a humanoid robot, 'Robovie' developed by ATR .

5.1 Results of user recognition and adaptation

Seven individuals were asked to act for the predefined face poses in front of the camera and all the sequence of face images were saved as individual image frame. All the training and test images are 60x60 pixels gray face images. The adaptation algorithm is tested for 7 persons frontal face or normal face (NF) images, and five directional face images (normal face, left directed face, right directed face, up directed face, down directed face). Figure 13 shows the sample result of the user adaptation method for normal faces. In the first step, the system is trained using 60 face images of three persons and developed three clusters (top 3 rows of Figure 13) corresponding to three persons. The cluster information table contents are [1, 11, 23, 29]. For example, in this situation if any input face image matches with the known face image member between 1 and 10 then the person is identified as person 1.

Fig. 13. Sample outputs of the clustering method for frontal faces

In the second step, 20-face image sequences of another person are fed to the system as input. The minimum Euclidian distances (ED) from three known persons face images are shown using upper line graph (B_adap) in Figure 14. The system identifies these faces as unknown person based on predefined threshold value for the Euclidian distance and activates the user learning function. The user learning function developed new cluster (4th row of Figure 13) and updated the cluster information table as [1, 11, 23, 30, 37]. After adaptation, the minimum Euclidian distance distribution line (A_adap) in Figure 7.21 shows that, for 8 images, minimum ED is zero and those are included in the new cluster so that the system can recognize the person. This method is tested for 7 persons including 2 females, and as a result of learning, 7 clusters with different length (number of images per cluster) for different persons (as shown in Figure 13) were formed.

The users adaptation method is also tested for 700 five directional face images of 7 persons (sample output in Figure 11). Figure 15 shows the distribution of 41 clusters for the 700 face images of 7 persons. In the first step, the system is trained using 100 face images of person_1 and it formed 5 clusters based on 5-directional faces. At this time, the contents of the cluster information table (that holds staring pointer of each cluster in the training database) are [1, 12, 17, 27, 44, 52]. After learning person_2, the cluster information table contents are [1, 12, 17, 27, 44, 53, 68, 82, 86, 96, 106, 116]. Similarly, other persons are adapted. Figure 16 shows the example of errors in clustering process. In the cluster 26 up directed faces of person_6 and frontal face person_5 are overlapped and treated as one cluster (Figure 16 (a)). In the case of cluster 31, up directed faces of person_5 and normal (frontal) faces of person_6 are overlapped and grouped in the same cluster (Figure 16 (b)). This problem can be solved using narrow threshold, but in that case the number of iteration as well as discard rates of the images classification method will be increased.

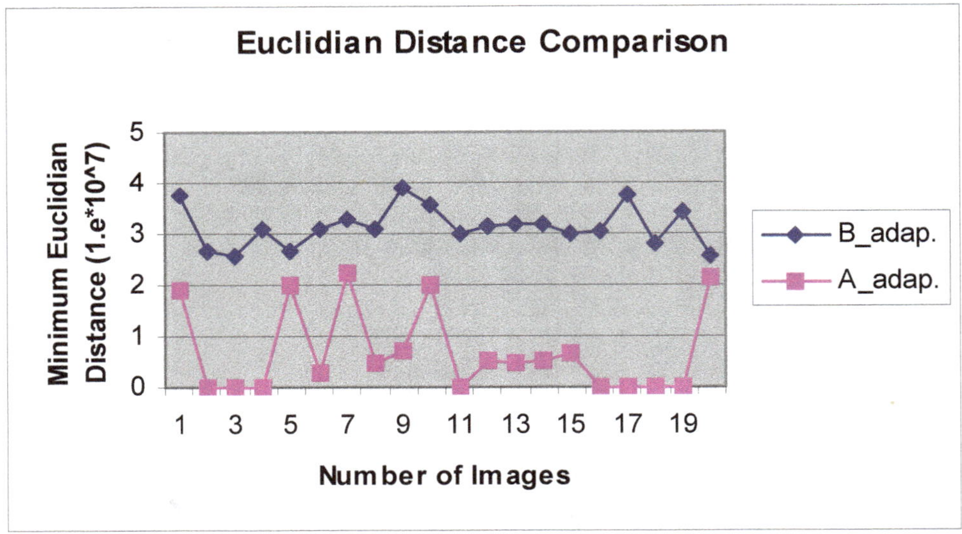

Fig. 14. Euclidian distances distribution of 20 frontal faces (before and after adaptation)

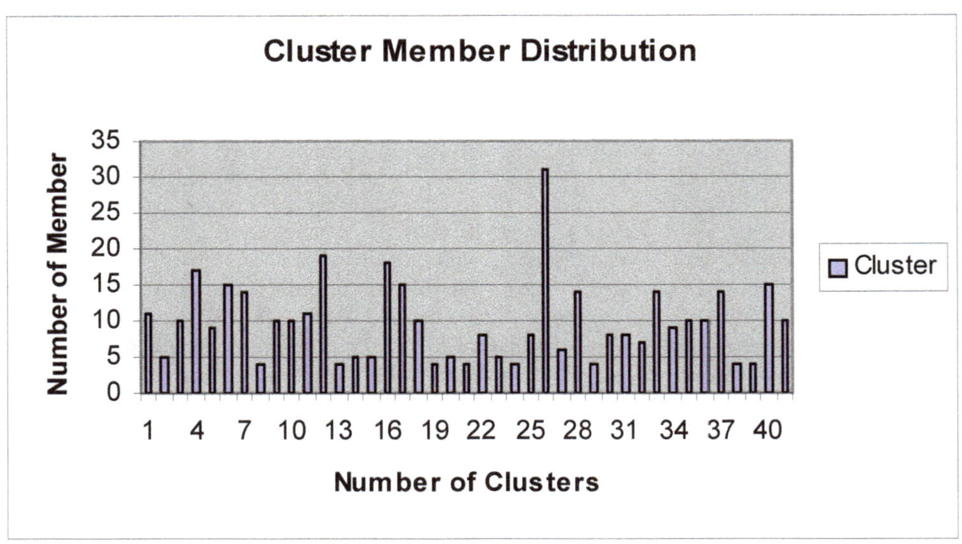

Fig. 15. Cluster member distributions for five directed face poses

(a) Cluster 26: Person_5 frontal face and person _6 up directed face overlapping (231-261)

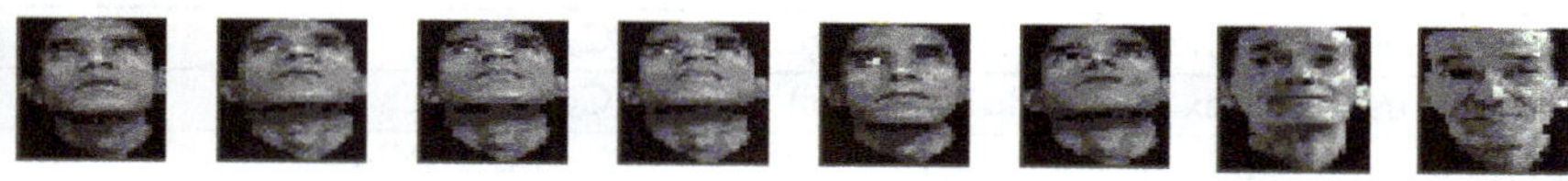

(b) Cluster 31: Person_5 up face and person_6 (from right 2 image) frontal face overlapping

Fig. 16. Example of errors in clustering process

5.2 Results of pose classification and adaptation

The system uses 10 hand poses of 7 persons for evaluating pose classification and adaptation method. All the training and test images are 60x60 pixels gray images. This system is first trained using 200 images of 10 poses of person_1 (20 images of each pose). It automatically clusters the images into 13 clusters. Figure 12 shows the sample outputs of hand poses learning method for person_1. If the user uses two hands to make the same pose then it forms two different clusters for the same pose. Different clusters can also be formed for the variation of orientation even the pose is same.

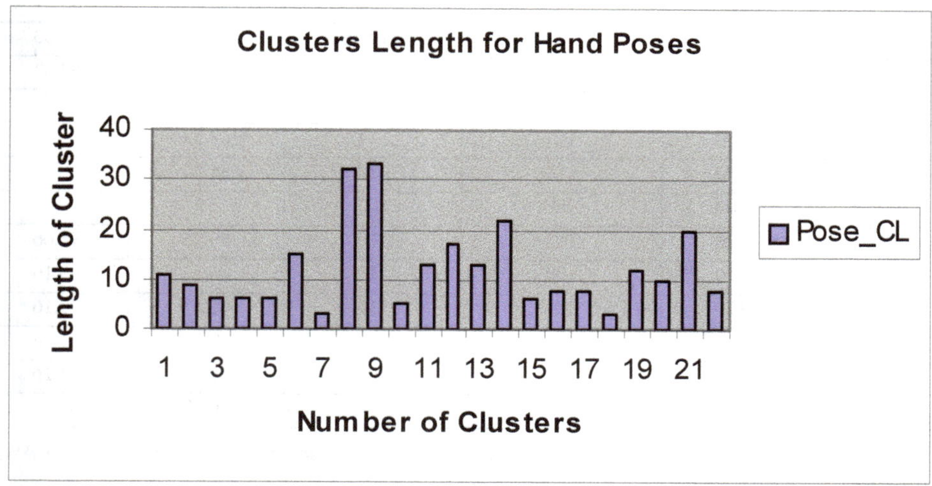

Fig. 17. Cluster member distributions for hand poses

If the person is change, then it may form different clusters for the same hand poses (gestures) due to the variation of hand shape and color. After trained with 10 hand poses of person_1, 200 images of 10 hand poses of person_2 are feed to the system. The system developed 9 more clusters for the person_2 corresponding to 8 hand poses. For, the 'LEFTHAND' and 'RIGHTHAND' palm it did not develop new clusters, rather inserted new members in those clusters. Table 1 shows the 22 clusters developed for 10 hand poses of 2 persons and their associations with hand poses. Figure 17 shows the distributions of the clusters of 10 hand poses for two persons.

Pose Name	Associated Clusters
ONE (Raise Index Finger)	PC_1, PC_{15}
FIST (Fist Up)	PC_2, PC_3, PC_{19}
OK (Make circle using thumb and index fingers)	PC_4, PC_{20}
TWO (V sign using index and middle fingers)	PC_5, PC_6, PC_{16}
THREE (Raise index, middle and ring fingers)	PC_7, PC_{17}, PC_{18}
LEFTHAND (Left hand palm)	PC_8
RIGHTHAND (Right hand palm)	PC_9
THUMB (Thumb Up)	PC_{10}, PC_{11}, PC_{14}
POINTL (Point Left)	PC_{12}, PC_{21}
POINTL (Point Right)	PC_{13}, PC_{22}

Table 1. List of hand poses and associated clusters

Input Images (ASL Char)	Number of Image	Correct Recognition		Accuracy (%)	
		Before_Adap	*After_Adap*	*B_Adap*	*A_Adap*
A	120	81	119	67.50	99.16
B	123	82	109	66.66	88.61
C	110	73	103	66.36	93.63
D	120	78	106	65	88.33
E	127	80	96	62.99	75.59
F	120	81	114	67.50	95
G	120	118	120	98.33	100
I	100	56	100	56	100
K	120	107	116	89.16	96.66
L	120	100	119	83.33	99.16
P	120	79	119	65.83	99.16
V	120	75	86	62.50	71.66
W	120	74	101	61.66	84.16
Y	120	85	98	70.83	81.66

Table 2. Comparison of pose classification accuracy (before and after adaptation) for 14 ASL Characters

In this study we have also compared pose classification accuracy (%) using two methods: one is multi-cluster based approach without adaptation, the other is multi-cluster based approach with adaptation. In this experiment we have used total 840 training images, (20 images of each pose of each person and 1660 test images of 14 ASL characters [ASL, 2004] of three persons. Table 2 presents the comparisons of two methods (B_Adap=before adaptation and A_Adap=after adaptation). This table shows that the accuracy of pose classification method which includes the adaptation or learning approach is better, because the learning function increments the clusters members or forms new clusters if necessary to classify the new images.

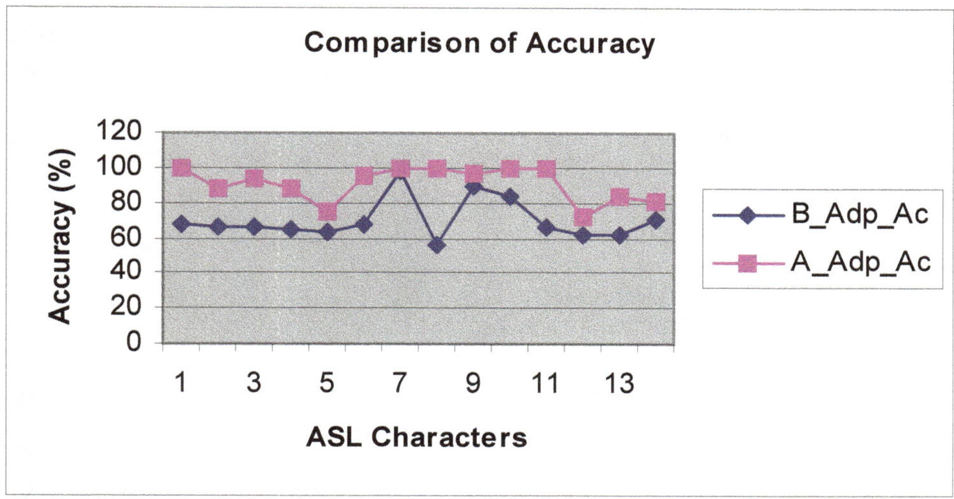

Fig. 18. Comparison of pose classification accuracy before after adaptation

Figure 18 depicts the graphical representations of 14-ASL characters classification accuracy using adaptation method (after adaptation) and without adaptation method (before adaptation). The comparison curves shows that if we include adaptation method then pose classification performance will be better, but needs user interaction that is bottleneck of this method.

6. Implementation of human-robot interaction

The real-time gesture based human-robot interaction is implemented as an application of this system. This approach has been implemented on a humanoid robot, name"Robovie". Since the same gestures can mean different tasks for different persons, we need to maintain the gesture with person-to-task knowledge. The robot and the gesture recognition PC are connected to SPAK knowledge server. From the image analysis and recognition PC, person identity and pose names are sent to the SPAK for decisions making and the robot activation. According to gesture and user identity, the knowledge module generates executable codes for robot actions. The robot then follows speech and body action commands. This method has been implemented for the following scenario:

User: "Person_1" comes in front of Robovie eyes camera and robot recognizes the user as "Person_1" .

Robot: "Hi Person_1, How are you?" (speech)

Person_1: uses the gesture "Ok"

Robot: " Oh Good! Do you want to play now?" (speech)

Person_1: uses the gesture "YES"

Robot: "Oh Thanks" (speech)

Person_1: uses the gesture "TwoHand"

Robot: imitates user's gesture "Raise Two Arms" as shown in Figure 19.

Person_1: uses the gesture "FistUp" (stop the interaction)
Robot: Bye-bye (speech).
User: "Person_2" comes in front of Robovie eyes camera, robot detects the face as unknown,
Robot: "Hi, What is your Name?" (speech)
Person_2: Types his name "Person_2"
Robot: " Oh, Good! Do you want to play now?" (speech)
Person_2: uses the gesture "OK"
Robot: "Thanks!" (speech)
Person_2: uses the gesture "LeftHand"
Robot: imitate user's gesture ("Raise Left Arm")
Person_2: uses the gesture "RightHand"
Robot: imitate user's gesture ("Raise Right Arm")
Person_2: uses the gesture "Three"
Robot: This is three (speech)
Person_2: uses the gesture "TwoHand"
Robot: Bye-bye (speech)

The above scenario shows that the same gesture can be used to represent different meanings and several gestures can be used to denote the same meaning for different persons. A user can design new actions according to his/her desires using 'Robovie' and can design corresponding knowledge frames using SPAK to implement their desired actions.

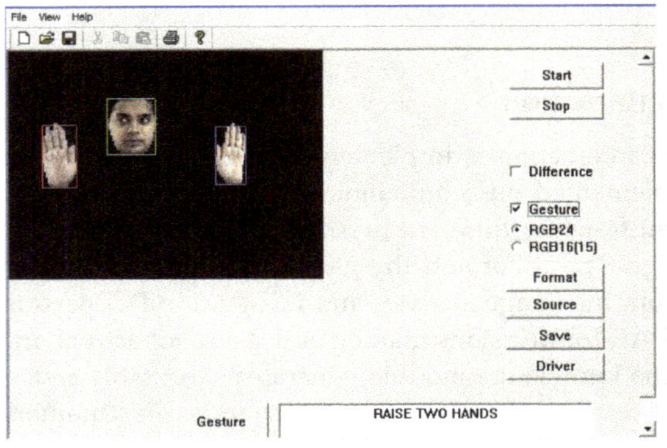

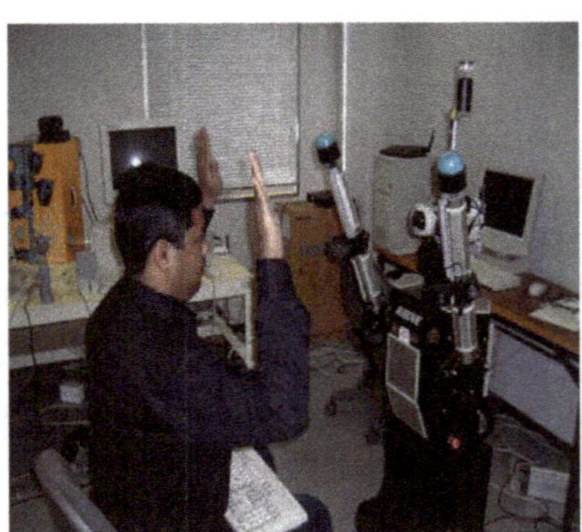

Fig. 19. Example human-robot (Robovie) interaction

7. Conclusion

This chapter describes users, gestures and robot behaviour adaptation method for human robot interaction by integrating computer vision and a knowledge-based software platform. In this method the user defines frames for users, poses, gestures, robots and robot behaviours. This chapter presents a multi-cluster based interactive learning approach for adapting new users and hand poses. However, if a large number of users use a large number of hand poses it is impossible to run this systems in real time. To overcome this

problem, in future we should maintain person-specific subspaces (individual PCA for each person of all hand poses) for pose classification and learning.
In this chapter we have also described how the system can adapt with new gestures and new robot behaviours using multiple occurrences of the same gesture with user interaction. The future aim is to make the system more robust, dynamically adaptable to new users and new gestures for interaction with different robots such as Aibo, Robovie, Scout, etc. Ultimate goal of this research is to establish a human-robot symbiotic society so that they can share their resources and work cooperatively with human beings.

8. Acknowledgment

I would like to express deep sense of gratitude and thanks to Dr. Haruki Ueno, Professor, Department of Informatics, National Institute of Informatics, Tokyo, Japan, for his sagacious guidance, encouragement and every possible help throughout this research work. I am grateful to Dr. Y. Shirai, Professor, Department of Human and Computer Intelligence, School of Information Science and Engineering, Ritsumeikan University, for his ingenious inspiration, suggestions and care in the whole research period. I would also like to express my sincere thanks to Professor H. Gotoda, Department of Informatics, National Institute of Informatics,Tokyo, Japan, for his valuable suggestions throughout the research. I must also thank to Dr. T. Zhang and Dr. V. Ampornaramveth for their assistance and suggestions during this research work.

9. References

[Ampornaramveth, 2004] V. Ampornaramveth, P. Kiatisevi, H. Ueno, "SPAK: Software Platform for Agents and Knowledge Systems in Symbiotic Robots", IEICE Transactions on Information and systems, Vol.E86-D, No.3, pp 1-10, 2004.

[Ampornaramveth, 2001] V. Ampornaramveth, H. Ueno, "Software Platform for Symbiotic Operations of Human and Networked Robots", NII Journal, Vol.3, No.1, pp 73-81, 2001.

[Aryananda, 2002] L. Aryananda, "Recognizing and Remembering Individuals: Online and Unsupervised Face Recognition for Humanoid Robot" in Proceeding of IEEE/RSJ International Conference on Intelligent Robots and Systems (IROS 2002), Vol. 2, pp. 1202-1207, 2002.

[ASL, 2004] "American Sign Language Browser" http://commtechlab.msu.edu/sites/aslweb/browser.htm, visited on April 2004.

[Augusteijn, 1993] M. F. Augusteijn, and T.L. Skujca, "Identification of Human Faces Through Texture-Based Feature Recognition and Neural Network Technology", in Proceeding of IEEE conference on Neural Networks, pp.392-398, 1993.

[Azoz, 1998] Y. Azoz, L. Devi, and R. Sharma, "Reliable Tracking of Human Arm Dynamics by Multiple Cue Integration and Constraint Fusion", in Proceeding of IEEE Computer Society Conference on Computer Vision and Pattern Recognition (CVPR'98), pp. 905-910, 1998.

[Bartlett, 1998] M. S Bartlett, H. M. Lades, and, T. Sejnowski, "Independent Component Representation for Face Recognition" in Proceedings of Symposium on Electronic Imaging (SPEI): Science and Technology, pp. 528-539, 1998.

[Belhumeur, 1997] P.N. Belhumeur, J. P. Hespanha, and D. J. Kriegman, " Eigenfaces vs. Fisherfaces: Recognition Using Class Specific Linear Projection" IEEE Transaction on Pattern Analysis and Machine Intelligence (PAMI), Vol. 19, pp. 711-720, 1997.

[Bhuiyan, 2004] M. A. Bhuiyan, V. Ampornaramveth, S. Muto, and H. Ueno, "On Tracking of Eye For Human-Robot Interface", International Journal of Robotics and Automation, Vol. 19, No. 1, pp. 42-54, 2004.

[Bhuiyan, 2003] M. A. Bhuiyan, V. Ampornaramveth, S. Muto, H. Ueno, "Face Detection and Facial Feature Localization for Human-machine Interface", NII Journal, Vol.5, No. 1, pp. 25-39, 2003.

[Birk, 1997] H. Birk, T. B. Moeslund, and C. B. Madsen, "Real-time Recognition of Hand Alphabet Gesture Using Principal Component Analysis", in Proceeding of 10th Scandinavian Conference on Image Analysis, Finland, 1997.

[Chellappa, 1995] R. Chellappa, C. L. Wilson, and S. Sirohey, "Human and Machine Recognition of faces: A survey", in Proceeding of IEEE, Vol. 83, No. 5, pp. 705-740, 1995.

[Crowley, 1997] J. L. Crowley and F. Berard, "Multi Modal Tracking of Faces for Video Communications", In proceedings of IEEE Conference on Computer Vision and Pattern Recognition (CVPR'97), pp. 640-645, 1997.

[Cutler, 1998] R. Cutler, M. Turk, "View-based Interpretation of Real-time Optical Flow for Gesture Recognition", in Proceedings of 3rd International Conference on Automatic Face and Gesture Recognition (AFGR'98), pp. 416-421, 1998.

[Darrel, 1993] T. Darrel and A. Pentland, "Space-time Gestures", in Proceedings of IEEE International Conference on Computer Vision and Pattern recognition (CVPR'93), pp. 335-340, 1993.

[Dai, 1996] Y. Dai and Y. Nakano, "Face-Texture Model Based on SGLD and Its Application in Face Detection in a Color Scene", Pattern Recognition, Vol. 29, No. 6, pp.1007-1017, 1996.

[Endres, 1998] H. Endres, W. Feiten, and G. Lawitzky, "Field Test of a Navigation System: Autonomous Cleaning in Supermarkets", in Proceeding of IEEE International Conference on Robotics & Automation (ICRA '98), 1998.

[Festival, 1999] "The Festival Speech Synthesis System developed by CSTR", University of Edinburgh, http://www.cstr.ed.ac.uk/project/festival

[Fong, 2003] T. Fong, I. Nourbakhsh and K. Dautenhahn, "A Survey of Socially Interactive Robots", Robotics and Autonomous System, Vol. 42(3-4), pp.143-166, 2003.

[Freeman, 1996] W.T. Freeman, K. Tanaka, J. Ohta, and K. Kyuma, "Computer Vision for Computer Games", in Proceedings of International Conference on Automatic Face and Gesture Recognition (AFGR'96), pp. 100-105, 1996.

[Hasanuzzaman, 2007a] Md. Hasanuzzaman, T. Zhang, V. Ampornaramveth, H. Gotoda, Y. Shirai, H. Ueno, "Adaptive Visual Gesture Recognition for Human-Robot Interaction Using Knowledge-based Software Platform", International Journal of Robotics and Autonomous Systems (RAS), Elsevier, Vol. 55(8), pp. 643-657, 2007.

[Hasanuzzaman, 2007b] Md. Hasanuzzaman, S. M. Tareeq, Tao Zhang, V. Ampornaramveth, H. Gotoda, Y. Shirai, H. Ueno, "Adaptive Visual Gesture Recognition for Human-Robot Interaction", Malaysian Journal of Computer Science, ISSN-0127-9084, Vol. 20(1), pp. 23-34, 2007.

[Hasanuzzaman, 2006] Md. Hasanuzzaman, T. Zhang, V. Ampornaramveth, and H. Ueno, "Gesture-Based Human-Robot Interaction Using a Knowledge-Based Software Platform", International Journal of Industrial Robot, Vol. 33 (1), pp. 37-49, 2006.

[Hasanuzzaman, 2004a] M. Hasanuzzaman, V. Ampornaramveth, T. Zhang, M. A. Bhuiyan, Y. Shirai, H. Ueno, "Real-time Vision-based Gesture Recognition for Human-Robot Interaction", in Proceeding of IEEE International Conference on Robotics and Biomimetics (ROBIO'2004), China, pp. 379-384, 2004.

[Hasanuzzaman, 2004b] M. Hasanuzzaman, T. Zhang, V. Ampornaramveth, M.A. Bhuiyan, Y. Shirai, H. Ueno, "Gesture Recognition for Human-Robot Interaction Through a Knowledge Based Software Platform", in Proceeding of IEEE International Conference on Image Analysis and Recognition (ICIAR 2004), LNCS 3211 (Springer-Verlag Berlin Heidelberg), Vol. 1, pp. 5300-537, Portugal, 2004.

[Hasanuzzaman, 2004c] M. Hasanuzzaman, T. Zhang, V. Ampornaramveth, P. Kiatisevi, Y. Shirai, H. Ueno, "Gesture-based Human-Robot Interaction Using a Frame-based Software Platform", in Proceeding of IEEE International Conference on Systems Man and Cybernetics (IEEE SMC'2004), Netherland, 2004.

[Kanda, 2002] T. Kanda, H. Ishiguro, T. Ono, M. Imai, and R. Nakatsu, "Development and Evaluation of an Interactive Humanoid Robot: Robovie", in Proceeding of IEEE International Conference on Robotics and Automation (ICRA 2002), pp. 1848-1855, 2002.

[Kanade, 1977] T. Kanade, "Computer Recognition of Human Faces", Brikhauser Verlag, Basel and Stuttgart, ISR-47, pp. 1-106, 1977.

[Kiatisevi, 2004] P. Kiatisevi, V. Ampornaramveth and H. Ueno, "A Ditributed Architecture for Knowledge-Based Interactive Robots", in Proceeding of 2nd International Conference on Information Technology for Application (ICITA' 2004), pp. 256-261, 2004.

[King, 1990] S. King and C. Weirman, "Helpmate Autonomous Mobile Robot Navigation System", in Proceeding of SPIE Conference on Mobile Robots, pp. 190-198,1990.

[Kjeldsen, 1996] R. Kjeldsen, and K. Kender, "Finding Skin in Color Images", in Proceedings of 2nd International Conference on Automatic Face and Gesture Recognition (AFGR'96), pp. 312-317, 1996.

[Kortenkamp, 1996] D. Kortenkamp, E. Hubber, and P. Bonasso, "Recognizing and Interpreting Gestures on a Mobile robot", in Proceeding of AAAI'96, pp. 915-921, 1996.

[Kotropoulos, 1997] C. Kotropoulos and I. Pitas, "Rule-based Face Detection in Frontal Views", in Proceeding of International Conference on Acoustics, Speech and Signal Processing, Vol. 4, pp. 2537-2540, 1997.

[Lin, 2002] J. Lin, Y. Wu, and T. S Huang, "Capturing Human Hand Motion in Image Sequences", in Proceeding of Workshop on Motion and Video Computing, Orlando, Florida, December, 2002.

[Miao, 1999] J. Miao, B. Yin, K. Wang, L. Shen, and X. Chen, "A Hierarchical Multiscale and Multiangle System for Human Face Detection in a Complex Background Using Gravity-Centre Template", Pattern Recognition, Vol. 32, No. 7, pp. 1237-1248, 1999.

[Minsky, 1974] M. Minsky "A Framework for Representing Knowledge", MIT-AI Laboratory Memo 306, 1974.

[Moghaddam, 1995] B. Moghaddam and A. Pentland, "Probabilistic Visual Learning for Object Detection", in Proceeding of 5th International Conference on Computer Vision, pp. 786-793, 1995.

[Nam, 1996] Y. Nam and K. Y. Wohn, "Recognition of Space-Time Hand-Gestures Using Hidden Markov Model", in Proceedings of ACM Symposium on Virtual Reality Software and Technology, pp. 51-58, 1996.

[Oka, 2002] K. Oka, Y. Sato, and H. Koike, "Real-Time Tracking of Multiple Finger-trips and Gesture Recognition for Augmented Desk Interface Systems", in Proceeding of International Conference in Automatic Face and Gesture Recognition (AFGR'02), pp. 423-428, Washington D.C, USA, 2002.

[Patterson, 1990] D. W. Patterson, "Introduction to Artificial Intelligence and Expert Systems", Prentice-Hall Inc., Englewood Cliffs, N.J, USA, 1990.

[Pavlovic, 1997] V. I. Pavlovic, R. Sharma and T. S. Huang, "Visual Interpretation of Hand Gestures for Human-Computer Interaction: A Review", IEEE Transactions on Pattern Analysis and Machine Intelligence (PAMI), Vol. 19, No. 7, pp. 677-695, 1997.

[Perzanowski, 2001] D. Perzanowski, A. C. Schultz, W. Adams, A. Marsh, and M. Bugajska, "Building a Multimodal Human-Robot Interface", IEEE Intelligent Systems, Vol. 16(1), pp. 16-21, 2001.

[Pineau, 2003] J. Pineau, M. Montemerlo, M. Pollack, N. Roy, S. Thrun, "Towards Robotic Assistants in Nursing Homes: Challenges and Results", Robotics and Autonomous Systems, Vol. 42, pp. 271-281, 2003.

[Rehg, 1994] J. M. Rehg and T. Kanade, "Digiteyes: Vision-based Hand Tracking for Human-Computer Interaction", in Proceeding of Workshop on Motion of Non-Rigid and Articulated Bodies, pp. 16-94, 1994.

[Rowley, 1998] H. A. Rowley, S. Baluja and T. Kanade, "Neural Network-Based Face Detection" IEEE Transactions on Pattern Analysis and Machine Intelligence (PAMI), Vol. 23, No. 1, pp. 23-38, 1998.

[Saber, 1998] E. Saber and A. M. Tekalp, "Frontal-view Face Detection and Facial Feature Extraction Using Color, Shape and Symmetry Based Cost Functions", Pattern Recognition Letters, Vol. 17(8) pp.669-680, 1998.

[Sakai, 1996] T. Sakai. M. Nagao and S. Fujibayashi, "Line Extraction and Pattern Detection in a Photograph", Pattern Recognition, Vol. 1, pp.233-248, 1996.

[Severinson-Eklundh, 2003] K. Severinson-Eklundh, A. Green, H. Huttenrauch, "Social and Collaborative Aspects of Interaction with a Service Robot", Robotics and Autonomous Systems, Vol. 42, pp.223-234, 2003.

[Shimada, 1996] N. Shimada, and Y. Shirai, "3-D Hand Pose Estimation and Shape Model Refinement from a Monocular Image Sequence", in Proceedings of VSMM'96 in GIFU, pp.23-428, 1996.

[Siegwart, 2003] R. Siegwart et. al., "Robox at Expo.02: A large-scale Installation of Personal Robots", Robotics and Autonomous Systems, Vol. 42, pp. 203-222, 2003.

[Sirohey, 1993] S. A. Sirohey, "Human Face Segmentation and Identification", *Technical Report* CS-TR-3176,University of Maryland, pp. 1-33, 1993.

[Sturman, 1994] D.J. Sturman and D. Zetler, "A Survey of Glove-Based Input" IEEE Computer Graphics and Applications, Vol. 14, pp-30-39, 1994.

[Tsukamoto, 1994] A. Tsukamoto, C.W. Lee, and S. Tsuji, "Detection and Pose Estimation of Human Face with Synthesized Image Models," in Proceeding of International Conference of Pattern Recognition, pp. 754-757,1994.

[Torras, 1995] C. Torras, "Robot Adaptivity", Robotics and Automation Systems, Vol. 15, pp.11-23, 1995.

[Torrance, 1994] M. C. Torrance, "Natural Communication with Robots" Master's thesis, MIT, Department of Electrical Engineering and Computer Science, Cambridge, MA, January 1994.

[Triesch, 2002] J. Triesch and C. V. Malsburg, "Classification of Hand Postures Against Complex Backgrounds Using Elastic Graph Matching", Image and Vision Computing, Vol. 20, pp. 937–943, 2002.

[Turk, 1991] M. Turk and A. Pentland, "Eigenface for Recognition" Journal of Cognitive Neuroscience, Vol. 3, No.1, pp. 71-86, 1991.

[Ueno, 2002] H. Ueno, "A Knowledge-Based Information Modeling for Autonomous Humanoid Service Robot", IEICE Transactions on Information & Systems, Vol. E85-D, No. 4, pp. 657-665, 2002.

[Ueno, 2000] H. Ueno, "A Cognitive Science-Based Knowledge Modeling for Autonomous Humanoid Service Robot-Towards a Human-Robot Symbiosis", in Proceeding of 10th European-Japanese Conference on Modeling and Knowledge Base, pp. 82-95, 2000.

[Waldherr, 2000] S. Waldherr, R. Romero, S. Thrun, "A Gesture Based Interface for Human-Robot Interaction", Journal of Autonomous Robots, Kluwer Academic Publishers, pp. 151-173, 2000.

[Weimer, 1989] D. Weimer and S. K. Ganapathy, "A Synthetic Virtual Environment with Hand Gesturing and Voice Input", in Proceedings of ACM CHI'89 Human Factors in Computing Systems, pp. 235-240, 1989.

[Wiskott, 1997] L. Wiskott, J. M. Fellous, N. Kruger, and C. V. Malsburg, "Face Recognition by Elastic Bunch Graph Matching", IEEE Transactions on Pattern Analysis and Machine Intelligence (PAMI), Vol. 19, No.7, pp. 775-779, 1997.

[Yang, 2002] M. H. Yang, D. J. Kriegman and N. Ahuja, "Detection Faces in Images: A survey", IEEE Transactions on Pattern Analysis and Machine Intelligence (PAMI), Vol. 24, No. 1, pp. 34-58, 2002.

[Yang, 2000] M. H. Yang, "Hand Gesture Recognition and Face Detection in Images", Ph.D Thesis, University of Illinois, Urbana-Champaign, 2000.

[Yang, 1998] J. Yang, R. Stiefelhagen, U. Meier and A. Waibel, "Visual Tracking for Multimodal Human Computer Interaction", in Proceedings of ACM CHI'98 Human Factors in Computing Systems, pp. 140-147, 1998.

[Yang, 1994] G. Yang and T. S. Huang, "Human Face Detection in Complex Background", Pattern Recognition, Vol. 27, No.1, pp.53-63, 1994.

[Yuille, 1992] A. Yuille, P. Hallinan and D. Cohen, "Feature Extraction from Faces Using Deformable Templates, International Journal of Computer Vision, Vol. 8, No. 2, pp 99-111, 1992.

[Zhang, 2004b] T. Zhang, M. Hasanuzzaman, V. Ampornaramveth, P. Kiatisevi, H. Ueno, "Human-Robot Interaction Control for Industrial Robot Arm Through Software Platform for Agents and Knowledge Management", in Proceedings of IEEE International Conference on Systems, Man and Cybernetics (IEEE SMC 2004), Netherlands, pp. 2865-2870, 2004.

[Zhao, 2003] W. Zhao, R. Chellappa, P. J. Phillips, and A. Rosenfeld, "Face Recognition: A Literature Survey" ACM Computing Surveys, Vol. 35, No. 4, pp. 399-458, 2003.

12

Learning Novel Objects for Domestic Service Robots

Muhammad Attamimi[1], Tomoaki Nakamura[1], Takayuki Nagai[1], Komei Sugiura[2] and Naoto Iwahashi[2]

[1]*The University of Electro-Communications*
[2]*National Institute of Information and Communications Technology*
Japan

1. Introduction

It is fair to say that robots which can interact with and serve humans especially in the domestic environment will spread widely in near future. A fundamental task called mobile manipulation is required for such domestic service robots. Therefore, many humanoid robots have been developed with the ability of mobile manipulation (1–5). Recently, competitions such as RoboCup@Home (6), Mobile Manipulation Challenge (7), and Semantic Robot Vision Challenge (8), have been proposed to evaluate such robots.

Since the tasks are implemented on domestic service robots, it stands to reason that natural interaction such as speech instruction should be used for the mobile manipulation. Here, we focus on the mobile manipulation using natural speech instruction such as "Bring me X" (X is an out-of-vocabulary (OOV) word). In order to realize this task, the integration of navigation, manipulation, speech recognition, and image recognition is required.

Image and speech recognition are difficult especially when novel objects are involved in the system. For example, there are objects specific to each home and new products can be brought into the home. It is impossible to register the names and images of all these objects with the robot in advance. Hence, we propose a method for learning novel objects with a simple procedure.

The robot, on which the proposed learning method is implemented, is intended to be used in a private domestic environment. Therefore, the procedure of teaching objects to the robot must be simple. For example, the user says, "This object is X" (X is the name of the object) and shows the object to the robot (Fig.1: Left). It is easy for a user to teach a robot many objects with this procedure. Then the user orders the robot to bring him/her something. For example, the user says, "Bring me X" (Fig.1: Right). As we mentioned earlier, such extended manipulation tasks are necessary for domestic service robots. However, there are three problems in teaching novel objects to the robots. The first problem is speech recognition of an object's name. In usual methods, phonemes of names must be registered in an internal dictionary. However, it is impossible to register all objects in advance. The second problem is the speech synthesis. A robot must utter the name of the recognized object for interaction with humans such as "Is it X?" However, conventional robot utterance systems cannot utter a word which is not registered in the dictionary. Even if the phoneme sequence of an OOV word can be recognized,

Fig. 1. Left: The user teaches the object to the robot. Right: The robot recognizes and utters the OOV word.

it cannot be used for speech synthesis since accuracy of phoneme recognition is less than 90%. The third problem is segmentation of the object region from a scene in the learning phase. When a robot learns an object, it must find where the object region is in the scene and segment it.

Methods for extracting OOV words in a speech have been proposed (9) for solving the first problem. Phonemes of OOV words can be obtained with these methods but they are not always correct. To solve the second problem, the user is required to restate the OOV word again and again so that correct phonemes are obtained (10). The robot can utter the word correctly but it is best that the robot learns the word from one user's utterance. There are also methods for situations in which the correct phonemes are not obtained. With such methods, the user utters the spelling of the OOV word to correct the phonemes (11). However, this requires a long time for the robot to learn OOV words by recognizing their spelling in Japanese or Chinese.

Considering these problems, we propose a system shown in Fig.2. We solve the first problem by extracting OOV words from a template sentence. The second problem may be solved by uttering phonemes of OOV words using a text-to-speech (TTS) system. However, it is difficult to recognize phonemes correctly. In the proposed method, the OOV part of the user's speech is converted to the robot's voice by voice conversion using Eigenvoice Gaussian mixture models (EGMMs) (12).

There has been research on the segmentation of images (13–15) for solving the third problem. The method developed by Rother *et al.* (13) requires a rough hand-drawn region of an object. Shi and Malik's method (14) can segment images automatically, but it cannot determine which segment is the object region. Mishra and Aloimonos's method (15) can segment the object accurately using color, 3D information, and motion. However, an initial point that locates inside the object region must be specified.

On the other hand, an object, which can be assumed as a human movements (e.g. hand movements), can be extracted from complicated scene because the proposed method is designed for a human to teach a robot an object. A color histogram and scale invariant feature transform (SIFT) are computed from extracted objects and registered in a database. This information is used for object recognition.

We implement the proposed method on a robot called "DiGORO". We believe it is important to evaluate the robot in a realistic domestic environment with a realistic task. When the robot moves to the object, it does not always arrive at an ideal position nor angle, and the illumination changes according to the position. A system is needed to work well in such

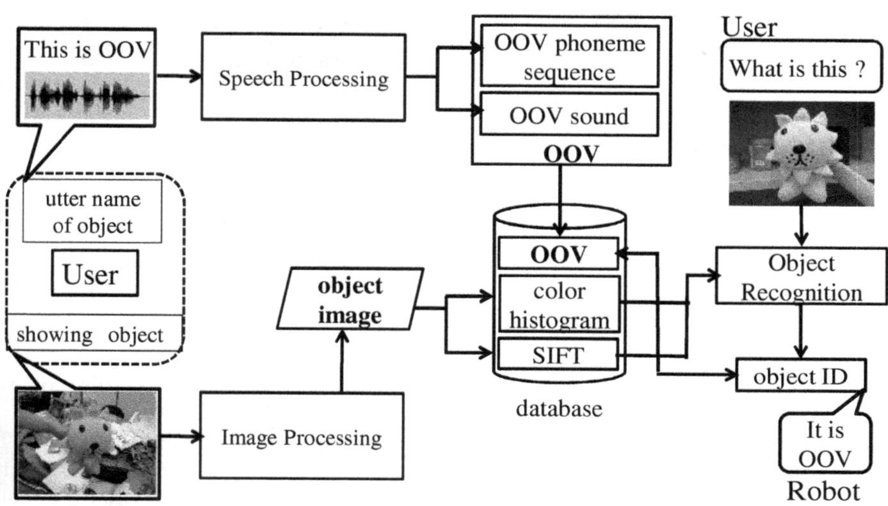

Fig. 2. Overview of learning novel objects.

an environment. In this research, we used the "Supermarket" task of RoboCup@Home (6) as the extended mobile manipulation task. RoboCup@Home is a competition that tests the ability of robots in a domestic environment. Supermarket is a standardized task based on the fetch-and-carry operation. There are also other tasks which can be used for evaluation (7; 8). The "Semantic Robot Vision Challenge" (8) evaluates the ability of a robot to find an object in a real environment. However, only three teams participated in the 2009 competition. Furthermore, Semantic Robot Vision Challenge is not for evaluating manipulation. The "Mobile Manipulation Challenge" was held at the 2010 International Conference on Robotics and Automation. Even this competition evaluates the mobile manipulation ability of robots, only four teams participated. It is difficult to determine what task should be used for evaluating robots, even though there are tasks (6–8) for it. We used one of the tasks of RoboCup@Home, which we believe is the most standard. RoboCup@Home has the largest number of participants [1] and has clearly-stated rules, which are open to the public. Besides, the rules are improved every year. From these reasons, such tasks are better than self-defined ones.

This chapter is organized as follows: the following section discusses a method for finding novel objects in cluttered scene. Then, the idea of pronouncing out-of-vocabulary words using voice conversion will be discussed in section 3. In section 4, the procedure of extended mobile manipulation task is described. Next section will discuss some experimental results to

[1] 24 teams participated in 2010 RoboCup@Home competition (6). On the other hand, a few teams participated in Mobile Manipulation Challenge (7), and Semantic Robot Vision Challenge (8).

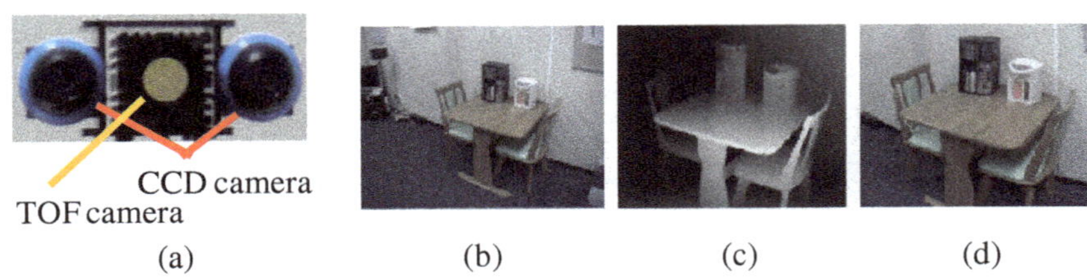

Fig. 3. (a)3D visual sensor. (b)color image (1024 × 768). (c)depth image (176 × 144). (d)mapped color image (176 × 144).

validate the proposed system. Discussion about proposed method will be done in section 6 followed by conclusion of this chapter in 7.

2. Finding novel objects in cluttered scene

2.1 3D visual sensor

Figure 3 shows the visual sensor used in this chapter. This sensor is able to acquire color and accurate depth information in real time by calibrating a TOF and two CCD cameras.

The distance measurement capability of TOF camera is based on the TOF principle. In TOF systems, the time taken for light to travel from an active illumination source to the objects in the field of view and return to the sensor is measured. A TOF camera SwissRanger SR4000 (23) is used as a part of 3D visual sensor. It emits a modulated near-infrared (NIR) and the CMOS/CCD imaging sensor measures the phase delay of the returned modulated signal at each pixel. These measurements in the sensor result in a 176 × 144 pixel depth map.

In the geometric camera calibration, the parameters that express camera pose and properties can be classified into extrinsic parameters (i.e. rotation and translation) and intrinsic ones (i.e. focal length, coefficient of lens distortion, optical center and pixel size). The extrinsic parameters represent camera position and pose in 3D space, while the intrinsic parameters are needed to project a 3D scene onto the 2D image plane. We use Zhang's calibration method in our proposed system, since the technique only requires the camera to observe a checkerboard pattern shown at a few different orientations. For the calibration of TOF camera, the reflected signal amplitude can be used to observe the checkerboard pattern. Therefore, it is straightforward to apply the same calibration method. Figure 3 (b), (c) and (d) show images captured from the visual sensor.

2.2 Motion attention based object segmentation

Assuming a user shows a target object to the robot, there may be people, objects, or furniture behind that object. The problem is object segmentation in such a complex background. Because the user has the object at hand, the object can be segmented out by taking into account the motion cue. This fact motivates us to use object segmentation based on motion attention.

Figure 4 shows an overview of motion attention. A motion detector first extracts the initial object region $M(x,y)$. Then, object information, such as color (hue) image $H(x,y)$ and depth image $D(x,y)$, is taken from the region. In particular, a hue histogram $f_H(h)$ and depth histogram $f_D(d)$ are taken from the region and normalized. Here, h and d represent the quantized value of hue and depth, respectively. Since these two histograms can be considered as probability density functions of the target object, the object probability map

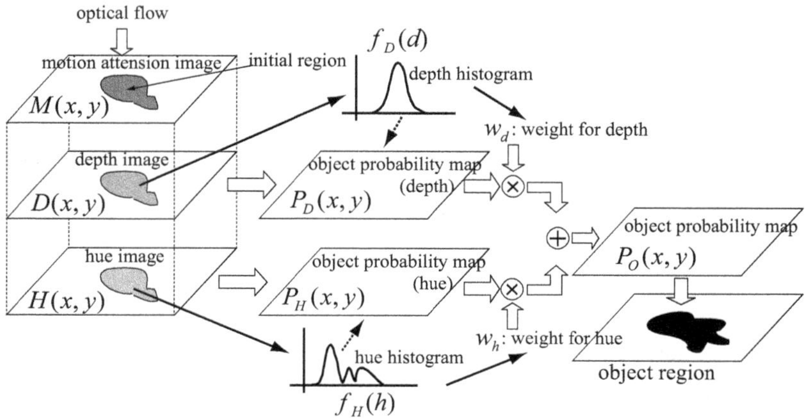

Fig. 4. Segmentation of object region using motion attention.

of each component ($P_D(x,y)$ and $P_H(x,y)$) at each pixel location can be easily obtained.

$$P_D(x,y) = f_D(D(x,y)), \tag{1}$$

$$P_H(x,y) = f_H(H(x,y)). \tag{2}$$

The weighted sum of these two object probability maps results in the object probability map $P_O(x,y)$.

$$P_O(x,y) = \text{LPF}[w_d P_D(x,y) + w_h P_H(x,y)], \tag{3}$$

The weights w_d and w_h are automatically assigned inversely proportional to the variance of each histogram. If the variance of the histogram is larger, its information is considered as inaccurate and the weight decreases. LPF represents a low pass filter, and we use a simple 3×3 averaging filter for it. The map is binarized, and then a final object mask is obtained using the connected component analysis. In the learning phase, object images are simply collected, then color histograms and SIFT features are extracted. These are used for object detection and recognition.

2.3 Object detection and identification in recognition phase

When the robot recognizes an object, the target object should be extracted from the scene. However, the same method in the learning phase is not applicable because the object is placed somewhere and it is not held by the user. Therefore, if objects are on the table, the plane detection technique is beneficial for detecting the objects. The 3D randomized Hough transform (24) is used for fast and accurate plane detection. This plane detection method is summarized below.

1. 3D information is captured in the scene.
2. Maximum plane is detected as table top using randomized Hough transform (24).
3. The plane is removed from 3D information.
4. The remaining point is projected on the plane.

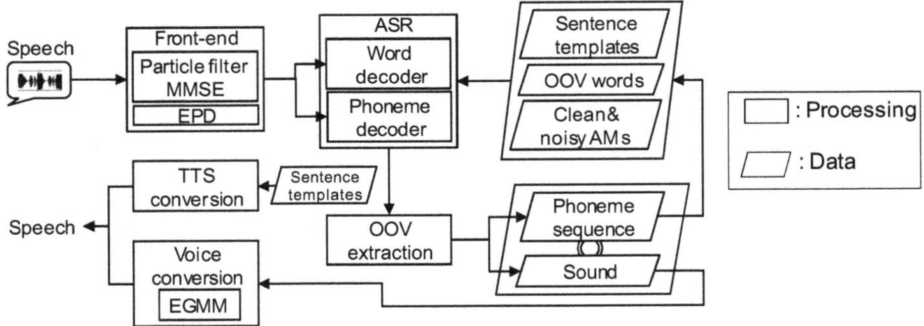

Fig. 5. Overview of the speech processing.

5. Connected components analysis is performed on the plane and each object is segmented out.

A SIFT descriptors is used for recognition. First, the candidates are narrowed down by using color information followed by the matching of SIFT descriptors, which are collected during the learning phase. It should be noted that the SIFT descriptors are extracted from multiple images taken from different viewpoints. Moreover, the number of object images is reduced for speeding up the SIFT matching process by matching among within-class object images and discarding similar ones. This process is also useful for deciding the threshold on the SIFT matching score.

3. Pronouncing out-of-vocabulary words using voice conversion

Figure 5 shows a schematic of the speech processing of the method, which uses automatic speech recognition (ASR) system called ATRASR (25). ATRASR is a hidden Markov model (HMM)-based speech recognition system, and it is used as a front-end and word/phoneme decoder. The phoneme decoder is used for obtaining the phoneme sequence of OOV words. Therefore, word- and phoneme-level speech recognition is possible.

To suppress noise, a particle filter is first applied to the online estimation of non-stationary noise, and then minimum mean square error (MMSE) estimation is used for noise reduction (26). Voice activity detection is conducted using endpoint detection (EPD) based on the frame's energy. This noise reduction part is of critical importance in RoboCup@Home tasks since the noise condition is severe.

Acoustic models (AMs) for the speech recognizer consist of "clean AMs" (male and female voices), which are trained using only clean voices, and "noisy AMs" (male and female voices), which are trained clean voices mixed with noise. This makes the speech recognition system robust in a noisy environment.

We use a template-based segmentation of words. To teach a robot an OOV word, the user is supposed to say template sentences such as "This is X". In terms of practical use, using a standard template sentence is reasonable since it is easy for users to understand how to teach a robot a word. A set of segmented voice and phoneme sequences is registered in a database. The phoneme sequence is used for utterance recognition of an OOV word.

For generating an utterance with an OOV word, the proposed method first converts the segmented voice recorded when the OOV word is learnt. The other part of the utterance is synthesized using XIMERA (27), which is a TTS conversion system. The OOV word part

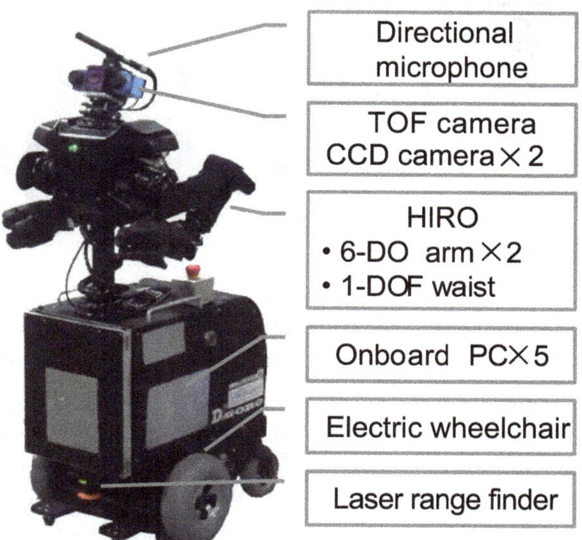

Fig. 6. The robot platform "DiGORO".

is converted into the robot's voice since the original sound is the user's voice, which is not naturally concatenated with a synthesized voice. The voice conversion is based on Eigenvoice Gaussian mixture models (EGMMs) (12). The recognized phoneme sequence of the OOV word is not used for synthesis since phoneme recognition accuracy is less than 90%, and the number of utterances for teaching an OOV word is virtually constrained to one owing to the time constraint of RoboCup@Home.

4. Procedure of extended mobile manipulation task

In this section, we describe the procedure of the mobile manipulation task called "Supermarket".

4.1 Robot platform: DiGORO

Figure 6 shows the robot "DiGORO" we previously developed (28). It is composed of the following hardware:

- Electric wheelchair
- HOKUYO laser range finder UTM-30LX
- KAWADA upper body humanoid robot
- Onboard PC (Intel Core2Duo processor) × 5
- Sanken directional microphone CS-3e
- YAMAHA loudspeaker NX-U10
- Mesa infrared TOF camera Swissranger
- Imaging Source CCD camera × 2

Fig. 7. Scenery of the learning phase. (a)The user gives a command for the robot to learn a new object. (b)The user shows the object from various directions.

4.2 Learning of novel object

Before the task, we need to teach objects to the robot. The procedure of teaching objects is summarized below.

1. The user shows the object to the robot and say "DiGORO, this object is X (X is an OOV word)" in Japanese [2]. The phoneme sequence and sound of the OOV word are extracted from the user's speech (Fig.7(a)).
2. The robot says "I start learning of X."
3. The user moves the object. Then 40 images of the object are segmented out and captured (Fig.7(b)).
4. Visual features (SIFT descriptors and color histogram) are calculated.
5. The phoneme sequence of the OOV word, sound of the OOV word and visual features of the object are registered in the object database. Then the robot says "I've memorized X."

If the user moves the object incorrectly, the motion attention cannot segment out the object. For example, if the user moves the object from outside the observed frame into it, the unintentional region is segmented out. To avoid such a situation, we instructed the user to say "DiGORO, this is X", while showing the object to the robot and to keep on showing it until the robot says, "I've memorized X."

4.3 Supermarket task

Supermarket is a task that the robot brings three specified objects. The detailed procedure of this task is summarized below.

1. The user says "DiGORO, bring me X (X is an OOV word)" (Fig.8(a)) .
2. The robot says "I'll bring you X. Is that correct?" Then if the user says "DiGORO, no", go to 1. If the user says "DiGORO, yes" go to next.
3. The robot says "Where is X?"
4. The user says "It is P (P is a name of the place.)".
5. The robot says "I'll go to P. Is that correct?" Then if the user says "DiGORO, no", go to 4. If the user says "DiGORO, yes" go to next.

[2] In this paper, the utterances are translated into English.

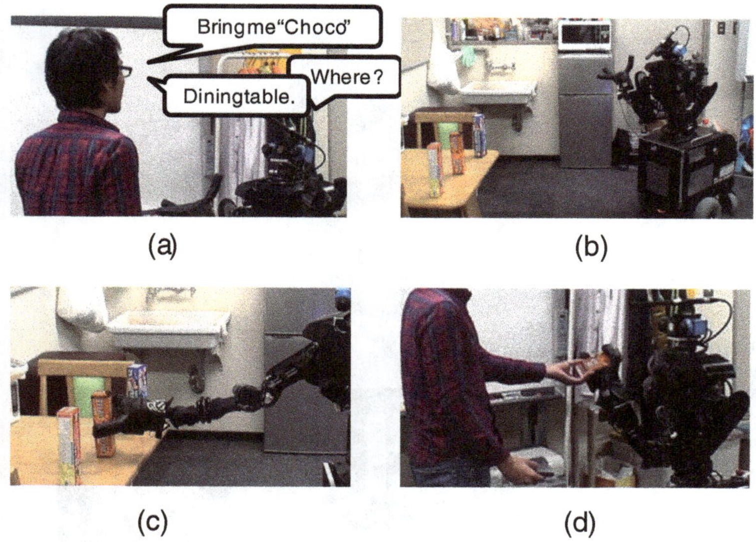

(a) (b) (c) (d)

Fig. 8. Scenery of the supermarket task. (a)The user ordered the robot to bring "choco", and the robot asked the user for needed information. (b)The robot navigated to the place where the user ordered. (c)The robot found the object and grasped it. (d)The robot returned to the start position and handed over the object to the user.

6. The robot goes to P (Fig.8(b)).
7. The robot looks around to find X using plane detection. If the robot can find X, the robot turns toward the object and says "I found X."
8. The robot grasps the object (Fig.8(c)) and returns to the start position (Fig.8(d)).
9. The task is completed when the robot brings three objects in total. Otherwise go to 1.

5. Experiments

In this section, experiments have been conducted to evaluate the proposed method and the applied system through the task called supermarket task.

5.1 Experiment 1: Evaluation of proposed method

5.1.1 Segmentation accuracy

In this experiment, we evaluated segmentation accuracy. It is necessary to segment the object region from a complex background which generally exists in the domestic environment. We used motion attention discussed in Section 2 because the object is held by a user in the learning phase.

The experiment was carried out in an ordinary living room, shown in Fig.9. We used 120 ordinary objects, as shown in Fig.10. A user taught the robot each object by showing and telling its name to the robot. The robot acquired 40 consecutive frames for each object and extracted the target object region from each image. Figure 11 shows examples of object segmentation.

Fig. 9. Experimental Environment.

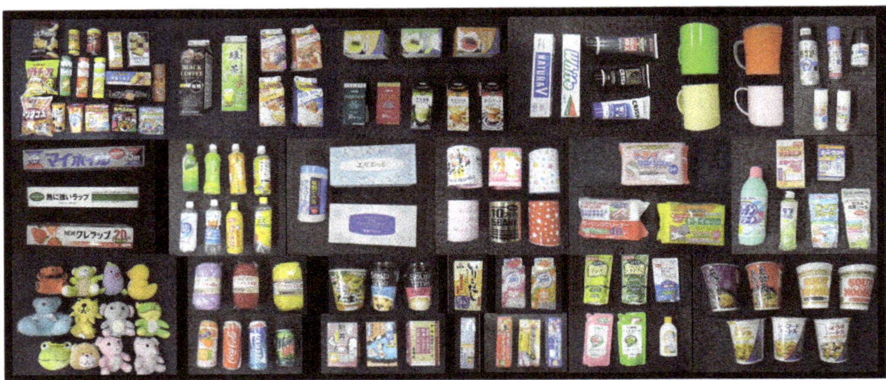

Fig. 10. 120 Objects used for experiments.

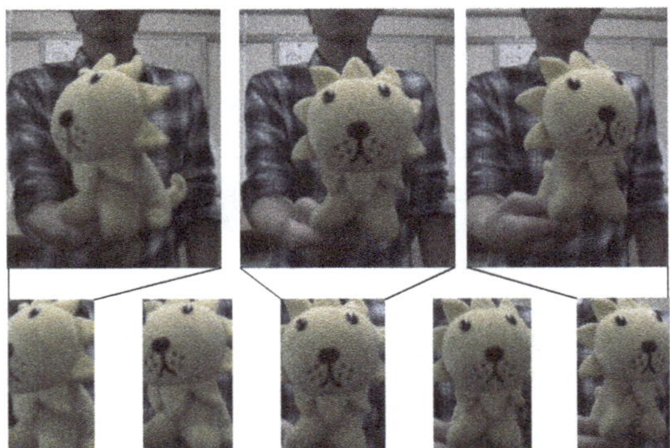

Fig. 11. Examples of object segmentation.

We extracted 10 out of the 40 frames for evaluating the segmentation accuracy of each object. Detection accuracy was measured using recall and precision rates, which are generically used for evaluation of classification, as shown in Fig.12(a), because it can be considered that pixels are classified into two classes, object region and non-object region. Here, "Object region"

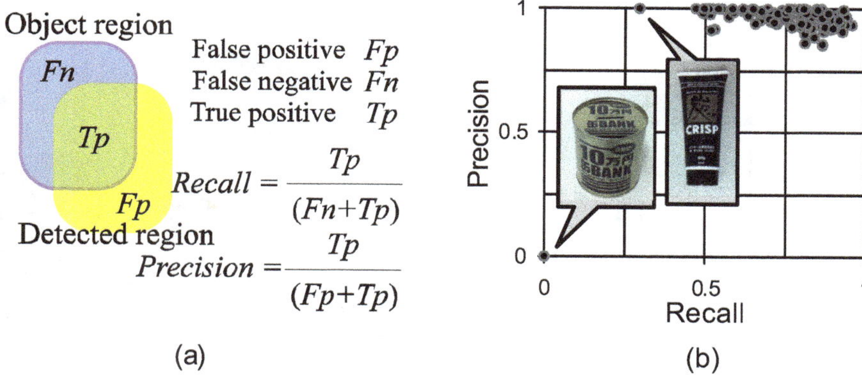

Fig. 12. (a) Definitions of recall and precision. (b) Results of object detection.

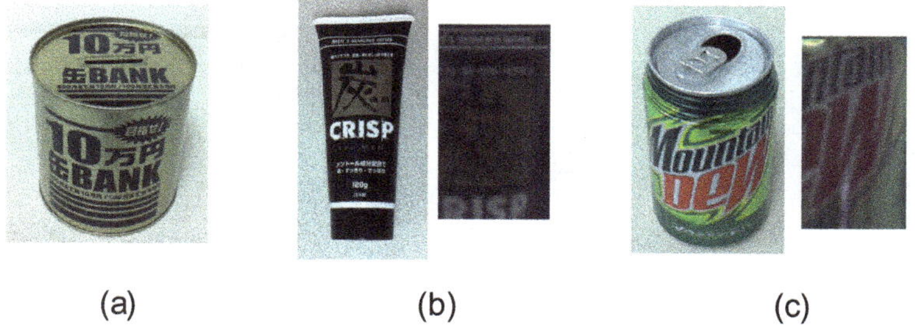

Fig. 13. Examples of object segmentation failure. (a) Object which could not be extracted. (b)Left: Object Right: Results of segmentation. (c)Left: Object Right: Results of segmentation.

	Place 1	Place 2	Place 3
Recognition rate	91%	89%	89%

Table 1. Object recognition rates.

indicates the manually labeled object region. Figure 12(b) shows a 2D plot of recall vs. precision. Each point represents an average of a single object (10 frames). As a result, the averages of all objects were 76.2% for recall and 95.8% for precision. This result indicates that the inside of object regions is extracted correctly because the precision was high. Therefore, it will not negatively affect object recognition.

Figure 13 shows examples of segmentation failure. The object in Fig.13(a) was not segmented at all because object's entire surface reflected near infrared rays which lead to fail on measuring 3D information. A part of object in Fig.13(b) was segmented because the black region absorbed near infrared rays. Moreover, a part of the object in Fig.13(c) was segmented because near infrared rays are reflected partially. We can see that black and metallic objects tend to cause low recall.

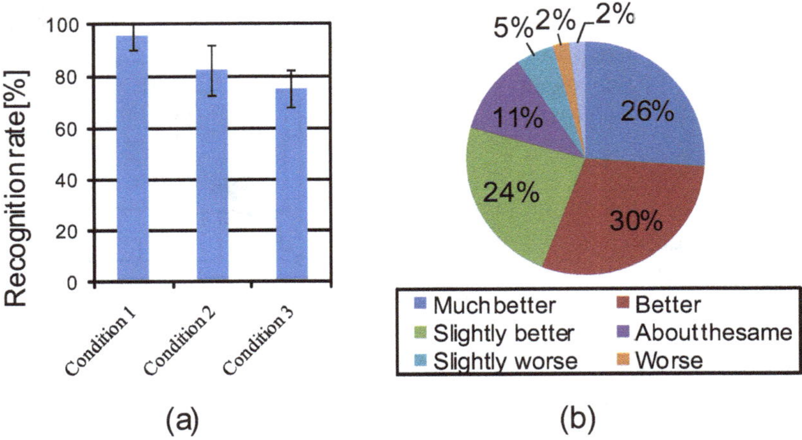

Fig. 14. (a) Recognition results. (Condition 1: Recognition with correct phonemes. Condition 2: Teacher and user are same person. Condition 3: Teachers taught the OOV words.) (b) Evaluation of voice conversion. The CMOS of VC was 1.45.

Quality	Score
Much better	3
Better	2
Slightly better	1
About the same	0
Slightly worse	-1
Worse	-2
Much worse	-3

Table 2. CMOS evaluation and scores.

5.1.2 Object recognition accuracy

We used 120 common objects, which had been learnt by the robot, as mentioned in the previous subsection. Three different locations with different lighting conditions in the living room were selected, and each object, which was segmented out using motion attention, was recognized. The results are listed in Table 1. The average recognition rate was about 90%. A major problem was false recognition between similar kinds of object such as cup noodles with different taste, because those objects have similar texture.

Next, we evaluated the proposed recognition method with the COIL100 database (29). COIL100 consist of 100 objects and 72 images per object. 36 images of each object were used for learning and the other 36 images were used for recognition. The recognition rate was 97.6%.

5.1.3 Recognition accuracy of out-of-vocabulary words

We evaluated the recognition accuracy of OOV words. The experimental procedure is described as follows. The teacher taught the robot OOV words such as "This is X". In a domestic environment, the teacher may not be only one person but also his/her family or friends. Therefore, we conducted the experiment under the condition that OOV words are

taught by several teachers including the user who asked the robot to bring something. For comparison, we conducted the experiment under simpler conditions. In each condition, volunteers uttered sentences "Bring me X" which consist of 120 words and the robot recognized X. There were eight volunteers and 960 utterances were recognized in total. The distance between the volunteer and the microphone was 50 cm. The ambient noise level in the experiment was set as 55dBA, which simulated the standard noise level in the RoboCup@Home competition when there is no other noise source such as announcement. If the speech recognition system can work in 55dBA noise, it can also work in a domestic environment. The recognition rate was calculated from these utterances.
Figure 14(a) shows the recognition rate in each condition, and the details of each condition are as follows:

1. **Recognition with correct phonemes:** Correct phonemes of the 120 words were manually registered in the dictionary. Each volunteer uttered "Bring me X" (X is the object name) and the robot recognized the object name.
2. **Teacher and user is the same person:** Each volunteer uttered 120 sentences "This is X" (X is the object name) and the robot learned the 120 OOV words. The robot recognized the 120 OOV words spoken by the volunteer who was the same as the teacher.
3. **Teachers taught OOV words:** First, 120 words were randomly assigned to eight teachers and these words were taught to the robot by them. Then, the robot recognized the 120 OOV words spoken by the user who is one of teachers. Therefore, the words were not always taught by the user. 118 out of the 960 were spoken by the teacher, i.e. teacher was the same as the user, and 842 out of the 960 utterances were spoken by others, i.e. teacher was not the same as the user.

The recognition rate was 95.2% in Condition 1, as shown in Fig.14(a). On the other hand, the accuracy of phonemes was 69.3% and the recognition rate was 82.4% in Condition 2. This indicates that the recognition rate was over 80%, which is satisfactory in a practical situation. In Condition 3, the recognition rate was 75.2%, as shown in Fig.14(a). The recognition rate was 83.4% when the teacher was the same as the user and 74.1% when the teacher was not the same as the user. Note that the speech files used in the training and those used in the test were different, even if the trainer and the tester was the same person. We can see that the recognition rate was lower than that in Condition 2. However, this is not a problem if restating is allowed.

5.1.4 Quality evaluation of robot's utterances

The objective of this experiment is to evaluate the quality of the robot's utterances. The experimental procedure is described below.
First, we made a database that included 960 utterances. It had 120 unique words and each word was uttered by eight volunteers. The ambient noise level was 55 dBA and the distance between the volunteer and microphone was 50 cm. Next, robot's utterances were generated using the proposed method. Utterances were also generated using a baseline method for comparison. These two methods are summarized as follows:

Voice Conversion (VC) (proposed): The utterances in the database are converted to robot voice by using EGMMs (12) (details of the proposed method were explained in Section 3).

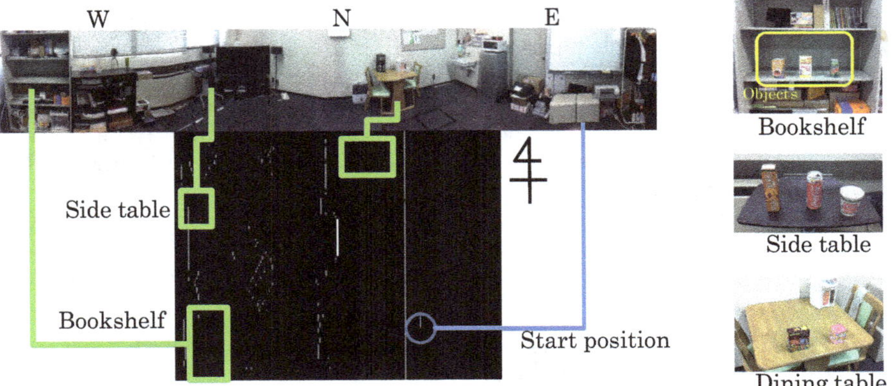

Fig. 15. The map and location of the tables/shelves.

Text-To-Speech (TTS) (baseline): The phoneme sequences obtained by phoneme recognition were used for generating robot utterances.

We then formed another group of six volunteers to evaluate the quality of generated utterances. Each volunteer listened to the utterances generated using TTS and VC. These utterances were composed of 120 unique words. The order of words was chosen at random. The order of TTS and VC samples was also chosen at random for each trial.
The comparison mean opinion score (CMOS) was used for evaluation. CMOS is specified by ITU-T recommendation P.800 (30). In the field of speech synthesis, CMOS is used for comparing voices synthesized with two methods. Specifically, the evaluation was conducted using the following questionnaire.

> (Volunteer listens to two robot's utterances.) Do you think the former is more accurate than the latter in terms of pronunciation?

The evaluation and its scores are listed in Table 2.
The evaluation results are shown in Fig.14(b). The CMOS of VC was 1.45, which suggests that VC is preferred. We can see that the proposed method, which utilizes VC, is efficient if the word which has been learnt is uttered once.

5.2 Experiment 2: Evaluation of applied system in mobile manipulation

We implemented an integrated audio-visual processing system on DiGORO and performed an experiment in a living room. The purpose of this experiment is to evaluate the robot in which the proposed method has been applied in mobile manipulation. We chose a task called "Supermarket" in the RoboCup@Home league. RoboCup@Home has several advantages, that is competition that has large number of participants, and clearly-stated rules, which are open to the public. In addition, improvements on the rules are done annually.

5.2.1 Experimental setup

Figure 15 illustrates a map generated from DiGORO's own on board SLAM mapping module. The location of the tables/shelf is also shown.
We designed the task module according to the flow in Section 4.3. A volunteer first interacted with the robot at the start position. Then the robot navigated to a table/shelf, recognized the

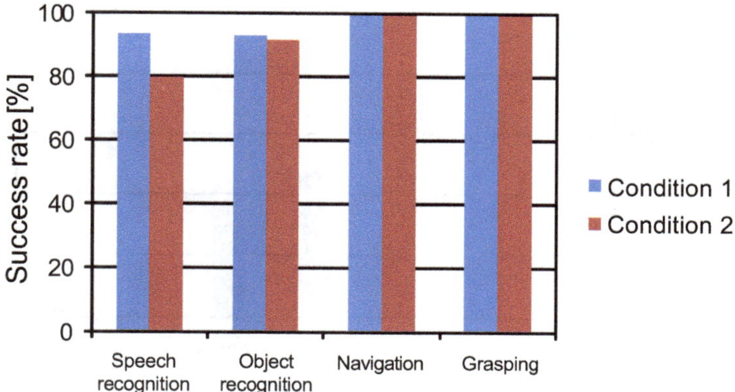

Fig. 16. Success rates. (Condition 1: words are taught by the same as requester. Condition 2: words are taught by different volunteers.)

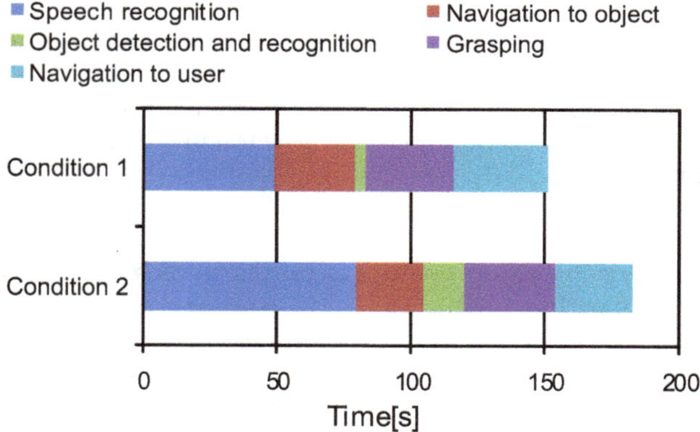

Fig. 17. Elapsed time of each process. (Condition 1: words are taught by the same as requester. Condition 2: words are taught by different volunteers.)

specified object, grasped it, and came back to the volunteer. This process was repeated for three objects.

We conducted the task under two conditions. One was similar to a real competition and the other was a more difficult condition. In each condition, we changed the dictionary of speech recognition because the user who teaches the object to the robot may not be the only person. The details of the conditions are as follows.

Condition 1: In the learning phase, each volunteer taught the robot the objects' names. The same volunteer asked the robot to bring the objects in the execution phase.

Condition 2: In the learning phase, 120 words were randomly assigned to eight volunteers and they taught these words to the robot. Each volunteer asked the robot to bring objects in the execution phase. Therefore, the names of the objects to bring were not always taught by the same volunteer who commanded the robot in the execution phase.

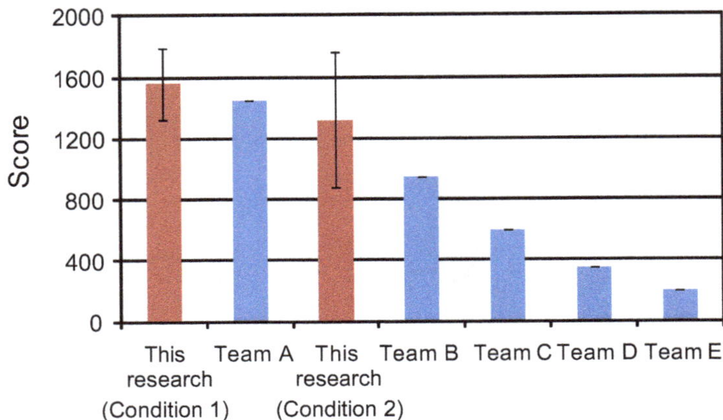

Fig. 18. Score comparison. (Condition 1: words are taught by the same as requester. Condition 2: words are taught by different volunteers.)

In the two different experimental setups, five volunteers who don't have prior knowledge of the robot conducted the task.
Therefore, the robot was supposed to bring 30 objects throughout this experiment. In each task, 30 out of the 120 objects were randomly chosen. The training data for these objects were obtained in Experiment 1.

5.2.2 Experimental results

We evaluated the results from three view points, success rate of each process, process elapsed time, and the score as a total performance.
Figure 16 shows the success rate of each process. We can see that high success rates over 90% were obtained, except for speech recognition. The speech recognition rate was 93% in Condition 1. On the other hand, it was 80% in Condition 2. This is because the phoneme sequences in the lexicon were not accurate.
Figure 17 depicts the average elapsed time for each process (per object). The results suggest that the trial can be completed within 10 min (elapsed time should be tripled and added 60 sec for the robot's instruction). The phase of instruction to the robot took a long time. The confirmation from the robot such as "I will bring X. Is this correct?" or restating the instruction to the robot such as "Bring me X" by the volunteer when the robot could not recognize the object name, were responsible for it. The instruction phase in Condition 1 was shorter than that in Condition 2 because false recognition in Condition 1 was less than that in Condition 2. This figure also shows that the time of the object recognition phase in Condition 2 was longer than that in Condition 1 because the object location was chosen randomly in both conditions. It accidently took a long time to find objects in Condition 2, depending on their location.
Next, we evaluated the task scores as a reference. Note that the comparison of the scores may be unfair because there are differences between a laboratory and competition environments. However, we used the scores since they can be the only source for comparison among different robots through the same realistic task.
Figure 18 shows a comparison of scores among teams that participated in an actual competition in 2009. The average score in Condition 1 was 1560. From this score, DiGORO would outperform the best team in the competition. Furthermore, the average score in

Condition 2 was 1320, which was comparable to Team A. It should be noted that we used the average scores. Although, a team could performed a task only once in the competition. In that respect, this comparison may be unfair.
In an actual competition, three objects which the robot brings were selected from ten common objects whose names are listed and was given to the teams. Therefore, it was possible to manually register the names of all the objects in the dictionary. On the other hand, objects which the robot brings were chosen from 120 objects in our experiment. Moreover, no manual process was included in the learning process. Considering these conditions, we can see that our robot obtained promising results even though the environment was different from the competition.

6. Discussion

6.1 Image processing

In this section, we will discuss the results from the evaluation of segmentation accuracy. Precision was 95.8%, which indicates that the inside of the object region was extracted correctly. On the other hand, recall was 76.2%, which was less than the precision. This indicates that sometimes only part of the object region was segmented out. This is because the TOF camera could not capture 3D information due to the material of the object. For example, 3D information cannot be captured from black or metallic objects because these objects reflect or absorb near infrared rays. We believe this will be improved by using a stereo vision. DiGORO (Fig.6) has two CCD cameras and can compute stereo disparity with them.
We now discuss the results of object recognition. The object recognition rate was about 90%. We used color and SIFT features for object recognition. Generally, it is difficult to recognize objects that have the same color and/or with no textures. For future work, we plan to use an object recognition method that integrates 3D shape information (31), which can significantly improve object recognition performance.

6.2 Learning and recognition of OOV words

For this research, the robot learned OOV words from one user's utterance and it is possible for the robot to recognize and utter them. The recognition rate was 82.4% and utterance was judged as better than the baseline method, which means a practical system is constructed. Failure in recognition was because false phonemes were learnt in the learning phase. The recognition rate can be improved by a user confirmation which phonemes were learnt correctly or not after learning. For example, a user utters "This is X" and the robot learns the object. Then the user confirms which "X" can be recognized or not by asking "Did you memorize X?" If the robot utters "Yes, I memorized X'" ($X = X'$), then the OOV word is registered correctly. Otherwise, the OOV word may not be registered correctly and the user can teach the object name again to the robot.

6.3 Evaluation in domestic environment

We evaluated the system in a domestic environment using the Supermarket task, which is one of the tasks in the RoboCup@Home league. Here, let us briefly discuss the evaluation task. As we mentioned earlier, it is difficult to determine what task should be used for evaluation, and there is no global standardized tasks for this. This situation makes it very difficult to evaluate robots, which were developed by different groups, through a same realistic task. We cannot compare our robot with others by using a self-defined task, since it is almost impossible to build their robots from scratch. Therefore, we think global standardized tasks are needed.

In this research, we propose to utilize the format of the task of RoboCup@Home, since we strongly believe that the tasks are the most standard tasks for evaluating robots for the following reasons:

1. The rules are open to the public.
2. Many teams from around the world participate, i.e. the task has already been performed by many robots.
3. The rules have been improved by many robotics researchers.

Unfortunately, the comparison of the scores in the current form is not fair enough. Hence the score should be treated as reference. Although the score is just for a reference, DiGORO outperforms the best team who participated in the competition, and it shows DiGORO can perform well in a domestic environment. Any deduction in points was a result of the robot not recognizing what a user wanted it to bring. This can be improved by user confirmation in the learning phase, as mentioned above.

The learning and recognition of OOV words can be applied to other tasks. For example the "Who is who?" task, which is one of the tasks in RoboCup@Home, involves the learning of human faces and names. In this task, a user utters "My name is X" and the robot learns "X" as his/her name. With this method, we can deal with a vast number of names.

Furthermore, DiGORO has many other abilities, and it can carry out eight other tasks. For example the robot can carry out the command "Follow Me", which is for following humans, and "Shopping Mall", which is for learning the location in an unknown place. These advanced features led our team to the 1st place at RoboCup@Home 2010. This suggests that DiGORO can stably work in a domestic environment.

7. Conclusion

We proposed a practical learning method of novel objects. With this method a robot can learn a word from one utterance. It is possible to utter an OOV word using the segmentation of the word from a template sentence and voice conversion. The object region is extracted from a complicated scene through a user moving the object. We implemented them all in a robot as an object learning system and evaluate it by conducting the Supermarket task. The experimental results show that our robot, DiGORO, can stably work in a real environment.

8. References

[1] T. Inamura, K. Okada, S. Tokutsu, N. Hatao, M. Inaba, and H. Inoue, "HRP-2W: A humanoid platform for research on support behavior in daily life environments," Robotics and Autonomous Systems, vol.57, no.2, pp.145–154, 2009.

[2] K. Wyrobek, E. Berger, H. Van der Loos, and J. Salisbury, "Towards a personal robotics development platform: Rationale and design of an intrinsically safe personal robot," IEEE International Conference on Robotics and Automation, pp.2165–2170, 2008.

[3] F. Weisshardt, U. Reiser, C. Parlitz, and A. Verl, "Making High-Tech Service Robot Platforms Available," Proceedings-ISR/ROBOTIK 2010, pp.1115–1120, 2010.

[4] J. Stückler, and S. Behnke, "Integrating indoor mobility, object manipulation, and intuitive interaction for domestic service tasks," IEEE-RAS International Conference on Humanoid Robots, pp.506–513, 2009.

[5] D. Holz, J. Paulus, T. Breuer, G. Giorgana, M. Reckhaus, F. Hegger, C. Müller, Z. Jin, R. Hartanto, P. Ploeger, et al., "The b-it-bots RoboCup@ Home 2009 team description paper," RoboCup 2009@ Home League Team Descriptions, Graz, Austria, 2009.
[6] "RoboCup@Home," `http://www.ai.rug.nl/robocupathome/`, 2010.
[7] "2010 Mobile Manipulation Challenge," `http://www.willowgarage.com/mmc10`, 2010.
[8] "Semantic Robot Vision Challenge," `http://www.semantic-robot-vision-challenge.org/`, 2009.
[9] I. Bazzi, and J. Glass, "A multi-class approach for modelling out-of-vocabulary words," Seventh International Conference on Spoken Language Processing, pp.1613–1616, 2002.
[10] M. Nakano, N. Iwahashi, T. Nagai, T. Sumii, X. Zuo, R. Taguchi, T. Nose, A. Mizutani, T. Nakamura, M. Attamim, et al., "Grounding New Words on the Physical World in Multi-Domain Human-Robot Dialogues," 2010 AAAI Fall Symposium Series, pp.74–79, 2010.
[11] H. Holzapfel, D. Neubig, and A. Waibel, "A dialogue approach to learning object descriptions and semantic categories," Robotics and Autonomous Systems, vol.56, no.11, pp.1004–1013, 2008.
[12] T. Toda, Y. Ohtani, and K. Shikano, "One-to-Many and Many-to-One Voice Conversion Based on Eigenvoices," IEEE International Conference on Acoustics, Speech and Signal Processing, vol.4, pp.1249–1252, 2007.
[13] C. Rother, V. Kolmogorov, and A. Blake, "Grabcut: Interactive foreground extraction using iterated graph cuts," ACM Transactions on Graphics (TOG), vol.23, no.3, pp.309–314, 2004.
[14] J. Shi, and J. Malik, "Normalized cuts and image segmentation," IEEE Transactions on Pattern Analysis and Machine Intelligence, vol.22, no.8, pp.888–905, 2002.
[15] A.K. Mishra, and Y. Aloimonos, "Active Segmentation," International Journal of Humanoid Robotics, vol.6, pp.361–386, 2009.
[16] S. Hasler, H. Wersing, S. Kirstein, and E. Körner, "Large-Scale Real-Time Object Identification Based on Analytic Features," Artificial Neural Networks–ICANN 2009, pp.663–672, 2009.
[17] H. Kim, E. Murphy-Chutorian, and J. Triesch, "Semi-autonomous learning of objects," Computer Vision and Pattern Recognition Workshop, p.145, 2006.
[18] H. Wersing, S. Kirstein, M. Gotting, H. Brandl, M. Dunn, I. Mikhailova, C. Goerick, J. Steil, H. Ritter, and E. Korner, "Online learning of objects in a biologically motivated visual architecture," International Journal of Neural Systems, vol.17, no.4, pp.219–230, 2007.
[19] N. Iwahashi, "Robots that learn language: Developmental approach to human-machine conversations," Symbol Grounding and Beyond, pp.143–167, 2006.
[20] D. Roy, "Grounding words in perception and action: computational insights," Trends in Cognitive Sciences, vol.9, no.8, pp.389–396, 2005.
[21] M. Fujita, R. Hasegawa, T. Takagi, J. Yokono, and H. Shimomura, "An autonomous robot that eats information via interaction with humans and environments," IEEE International Workshop on Robot and Human Interactive Communication, pp.383–389, 2002.
[22] M. Johnson-Roberson, G. Skantze, J. Bohg, J. Gustafson, R. Carlson, and D. Kragic, "Enhanced Visual Scene Understanding through Human-Robot Dialog," 2010 AAAI Fall Symposium on Dialog with Robots, pp.143–144, 2010.

[23] "Mesa imaging," http://www.mesa-imaging.ch/index.php.
[24] K. Okada, S. Kagami, M. Inaba, and H. Inoue, "Plane segment finder: Algorithm, implementation and applications," IEEE International Conference on Robotics and Automation, vol.2, pp.2120–2125, 2005.
[25] S. Nakamura, K. Markov, H. Nakaiwa, G. Kikui, H. Kawai, T. Jitsuhiro, J. Zhang, H. Yamamoto, E. Sumita, and S. Yamamoto, "The ATR multilingual speech-to-speech translation system," IEEE Transactions on Audio, Speech, and Language Processing, vol.14, no.2, pp.365–376, 2006.
[26] M. Fujimoto, and S. Nakamura, "Sequential non-stationary noise tracking using particle filtering with switching dynamical system," IEEE International Conference on Acoustics, Speech and Signal Processing, vol.1, pp.769–772, 2006.
[27] H. Kawai, T. Toda, J. Ni, M. Tsuzaki, and K. Tokuda, "XIMERA: A new TTS from ATR based on corpus-based technologies," Fifth ISCA Workshop on Speech Synthesis, pp.179–184, 2004.
[28] H. Okada, T. Omori, N. Iwahashi, K. Sugiura, T. Nagai, N. Watanabe, A. Mizutani, T. Nakamura, and M. Attamimi, "Team eR@sers 2009 in the @home league team description paper," , 2009.
[29] S.A. Nene, S.K. Nayar, and H. Murase, "Columbia Object Image Library (COIL-100)," Technical report, Feb. 1996.
[30] International Telecommunication Union, "ITU-T P.800," http://www.itu.int/rec/T-REC-P.800/en.
[31] M. Attamimi, A. Mizutani, T. Nakamura, T. Nagai, K. Funakoshi, , and M. Nakano, "Real-Time 3D Visual Sensor for Robust Object Recognition," IEEE/RSJ International Conference on Intelligent Robots and Systems, pp.4560–4565, 2010.
[32] RoboCup@Home league committee, "RoboCup@Home Rules & Regulations," http://www.ai.rug.nl/robocupathome/documents/rulebook2009_FINAL.pdf, 2009.

13

A Multi-Modal Panoramic Attentional Model for Robots and Applications

Ravi Sarvadevabhatla and Victor Ng-Thow-Hing
Honda Research Institute USA, 425 National Ave Suite 100,Mountain View CA 94043 USA

1. Introduction

Humanoid robots are becoming increasingly competent in perception of their surroundings and in providing intelligent responses to worldly events. A popular paradigm to realize such responses is the idea of attention itself. There are two important aspects of attention in the context of humanoid robots. First, *perception* describes how to design the sensory system to filter out useful salient features in the sensory field and perform subsequent higher level processing to perform tasks such as face recognition. Second, the *behavioral response* defines how the humanoid should act when it encounters the salient features. A model of attention enables the humanoid to achieve a semblance of liveliness that goes beyond exhibiting a mechanized repertoire of responses. It also facilitates progress in realizing models of higher-level cognitive processes such as having people direct the robot's attention to a specific target stimulus(Cynthia et al., 2001).

Studies indicate that humans employ attention as a mechanism for preventing sensory overload(Tsotsos et al., 2005),(Komatsu, 1994) – a finding which is relevant to robotics given that information bandwidth is often a concern. The neurobiologically inspired models of Itti (Tsotsos et al., 2005), initially developed for modeling visual attention have been improved(Dhavale et al., 2003) and their scope has been broadened to include even auditory modes of attention(Kayser et al., 2008). Such models have formed the basis of multi-modal attention mechanisms in (humanoid) robots(Maragos, 2008),(Rapantzikos, 2007).

Typical implementations of visual attention mechanisms employ a bottom-up processing of camera images to arrive at the so-called "saliency map", which encodes the unconstrained salience of the scene. Salient regions identified from saliency map are processed by higher-level modules such as object and face recognition. The results of these modules are then used as referential entities for the task at hand(e.g. acknowledging a familiar face, noting the location of a recognized object). Building upon the recent additions to Itti's original model(Tsotsos et al., 2005), some implementations also use top-down control mechanisms to constrain the salience(Cynthia et al., 2001),(Navalpakkam and Itti, 2005),(Moren et al., 2008).

In most of the implementations, the cameras are held fixed, simplifying processing and consequent attention mechanism modeling. However, this restricts the visual scope of attention, particularly in situations when the robot has to interact with multiple people who may be spread beyond its limited field-of-view. Moreover, they may choose to advertise their presence through a non-visual modality such as speech utterances.

Attempts to overcome this situation lead naturally to the idea of widening the visual scope and therefore, to the idea of a *panoramic* attention. In most of the implementations which utilize a panorama-like model (Kayama et al., 1998),(Nickel and Stiefelhagen, 2007), the panoramic region is discretized into addressable regions. While this ensures a complete coverage of the humanoid's field of view, it imposes high storage requirements. Most regions of the panorama remain unattended, particularly when the scene is static in nature. Even when dynamic elements are present(e.g. moving people), the corresponding regions require attention for a limited amount of time before higher-level tasks compel the system to direct its attention to other parts of panorama(Nickel and Stiefelhagen, 2007).

These limitations can be addressed by employing a *multi-modal panoramic attention model* – the topic of this chapter. In its basic mode, it operates on an egocentric panorama which spans the pan and tilt ranges of the humanoid's head cameras(Nickel and Stiefelhagen, 2007). As a baseline characteristic, the model naturally selects regions which can be deemed interesting for cognitively higher-level modules performing face detection, object recognition, sudden motion estimation etc. However, saliencies are maintained only for cognitively prominent entities (e.g. faces, objects, interesting or unusual motion phenomena). This frees us from considerations of storage structures and associated processing that are present in image pixel-based panoramic extensions of the traditional attention model(Kayama et al., 1998),(Ruesch et al., 2008). Also, the emphasis is not merely on obtaining a sparse representation in terms of storage as has been done in previous work. One of the objectives is also to assign and manipulate the semantics of sparsely represented entities in an entity-specific fashion. This chapter describes how this can be achieved with the panoramic attention model.

The panoramic attention model has an idle mode driven by an idle-motion policy which creates a superficial impression that the humanoid is idly looking about its surroundings. Internally, in the course of these idle motions, the humanoid's cameras span the entire panorama and register *incidental observations* such as identities of people it comes across or objects present in gaze directions it looks at. Thus, the idle-motion behavior imparts the humanoid with a human-like liveliness while it concurrently notes details of surroundings. The associated information may be later accessed when needed for a future task involving references to such entities i.e. the humanoid can immediately attend to the task bypassing the preparatory search for them. The active mode of the panoramic attention model is triggered by top-level tasks and triggers. In this mode, it responds in a task-specific manner (e.g. tracking a known person). Another significant contribution from the model is the notion of *cognitive panoramic habituation*. Entities registered in the panorama do not enjoy a permanent existence. Instead, their lifetimes are regulated by entity-specific persistence models(e.g. isolated objects tend to be more persistent than people who are likely to move about). This habituation mechanism enables the memories of entities in the panorama to fade away, thereby creating a human-like attentional effect. The memories associated with a panoramically registered entity are refreshed when the entity is referenced by top-down commands.

With the panoramic attention model, out-of-scene speakers can also be handled. The humanoid robot employed(Honda, 2000) uses a 2-microphone array which records audio from the environment. The audio signals are processed to perform localization, thus determining which direction speech is coming from, as well as source-specific attributes such as pitch and amplitude. In particular, the localization information is mapped onto the panoramic framework. Subsequent sections shall describe how audio information is utilized.

At this point, it is pertinent to point out that the panoramic attention model was designed for applicability across a variety of interaction situations and multi-humanoid

platforms. Therefore, having the model components,particularly the interfaces with sensors and actuators, operate in a modular fashion becomes important. Apart from technical considerations, one of the design goals was also to provide a certain level of transparency and user-friendly interface to non-technical users, who may not wish or need to understand the working details of the model. As described later, the model is augmented by an intuitive panoramic graphical user interface(GUI) which mirrors the model helps avoid cognitive dissonance that arises when dealing with the traditional, flat 2-d displays – a feature that was particularly valuable for operators using the interface in Wizard-Of-Oz like fashion.
The chapter will first review previous and related work in attention, particularly the panoramic incarnations. The details of the panoramic attention model are provided next, followed by a description of experimental results highlighting the effectiveness of the cognitive filtering aspect in the model. Interactive application scenarios (multi-tasking attentive behavior, Wizard-of-Oz interface, personalized human-robot interaction) describing the utility of panoramic attention model are presented next. The chapter concludes by discussing the implications of this model and planned extensions for future.

2. Related work

The basic premise of an attention model is the ability to filter out salient information from the raw sensor stream (camera frames, auditory input recording) of the robot(Tsotsos et al., 2005). In most implementations, the model introduced by Itti et al. (Itti et al., 1998) forms the basic framework for the bottom-up attention mechanism where low-level features such as intensity, color, and edge orientations are combined to create a *saliency map*. The areas of high saliency are targeted as candidates for gaze shifts (Cynthia et al., 2001; Dhavale et al., 2003; Ruesch et al., 2008; Koene et al., 2007). In some cases, these gaze points are attended in a biologically-inspired fashion using foveated vision systems(Sobh et al., 2008). In order to decide how to combine the various weights of features to create the saliency map, a top-down attention modulation process is prescribed that assigns weights generally based on task-preference(Cynthia et al., 2001). This top-down weighting can be flexibly adjusted as needed(Moren et al., 2008) according to the robot's current task mode.
The term panoramic attention has also been used to describe cognitive awareness that is both non-selective and peripheral, without a specific focus(Shear and Varela, 1999). There is evidence that the brain partially maintains an internal model of panoramic attention, the impairment of which results in individuals showing neglect of attention-arousing activities occurring in their surrounding environment(Halligan and Wade, 2005).
Although mentioned as a useful capability(Cynthia et al., 2001; Kayama et al., 1998), there have been relatively few forays into panorama-like representations that represent the scene beyond the immediate field of view. The term panoramic attention is usually confined to using visual panorama as an extension of the traditional fixed field-of-view representations (Kayama et al., 1998) or extension of the saliency map(Bur et al., 2006; Stiehl et al., 2008). The work of (Ruesch et al., 2008) uses an egocentric saliency map to represent, fuse and display multi-modal saliencies. Similar to the work described in this chapter, they ensure that salient regions decay. However, this mechanism operates directly on raw saliency values of discretized regions mapped onto the ego-sphere. In contrast, the cognitive decay mechanism (described in a subsequent section) works on sparsely represented higher-level entities such as faces. (Nickel and Stiefelhagen, 2007) employ a multi-modal attention map that spans the discretized space of pan/tilt positions of the camera head in order to store particles derived from visual and acoustic sensor events. Similar to the model presented in the chapter,

they exploit the information in their panoramic attention map to decide on gaze behavior for tracking existing people or for looking at novel stimuli to discover new people. The idea of a 'stale panorama' from (Stiehl et al., 2008) served as a motivating aspect for the present work.However, in their implementation only the raw video frames are used to create a panoramic image for a teleoperation implementation. The model presented in the chapter requires less storage because only the high-level semantic information from the visual field is extracted and stored such as identification and location of entities in the visual field. This is also important in that these semantic features can be used by the robot for decision-making, whereas further processing would need to be done with a raw image panorama. As part of their humanoid active vision system, (Koene et al., 2007) refer to a short term memory module that stores the location of perceived objects when they are outside the current field of view. They use this mechanism to associate sound with previous visual events, especially since audio may come from a source not in view.

In many of the panorama-like models mentioned above, the user interface does not mirror the world as perceived by the humanoid robot. The benefits of matching interface displays and controls to human mental models include reductions in mental transformations of information, faster learning and reduced cognitive load (Macedo et al., 1999) – a factor which inspired the design of the application interfaces described along with the model. As described in Section 5.1, the user interface mirrors the panoramic portion of the attention model, thereby minimizing cognitive dissonance. In the interest of focus and space, the numerous references to various Wizard-of-Oz systems and robot control interfaces will not be discussed.

In general, the aim for panoramic attention model is to fulfill a broader role than some of the aforementioned approaches. In particular, the sensor filtering characteristics of the bottom-up attention modules are combined with higher level spatial and semantic information that can then be exploited to by a behavior module for the robot. The semantic knowledge of an object allows refined modules to model the spatio-temporal behavior of objects perceived.

3. The panoramic attention model

A simple three-layer attention model forms the backbone of the panoramic attention model with each successive layer processing the outputs of its predecessor. Higher layers process data with increasing cognitive demand than the lower ones (Figure 1). To begin with, the visual portion of the attention model is described, followed by the audio portion.

3.1 Visual attention

Low-level visual attention: The traditional role of low-level visual attention is to rapidly determine the most salience-worthy saccade[1] locations for a given scene in a context-independent fashion.

The lowest among the layers, this processes input sensor data, i.e. color video frames streaming in from the humanoid's cameras. The processing for low-level layer happens as outlined in (Itti et al., 1998). Each input color frame is processed to arrive at the target ROI(Region of Interest)s. To improve the color contrast of incoming frames, Contrast-Limited Adaptive Histogram Equalization(Pizer et al., 1990) is applied on them. The contrast-adjusted color frame and its gray-scale version are used to generate various feature maps. The generic

[1] According to `http://en.wikipedia.org/wiki/Saccade`, a saccade is a fast movement of an eye, *head* or other part of an animal's body or device. The term saccade in this chapter refers to head movement since the humanoid used does not have a foveated vision control mechanism.

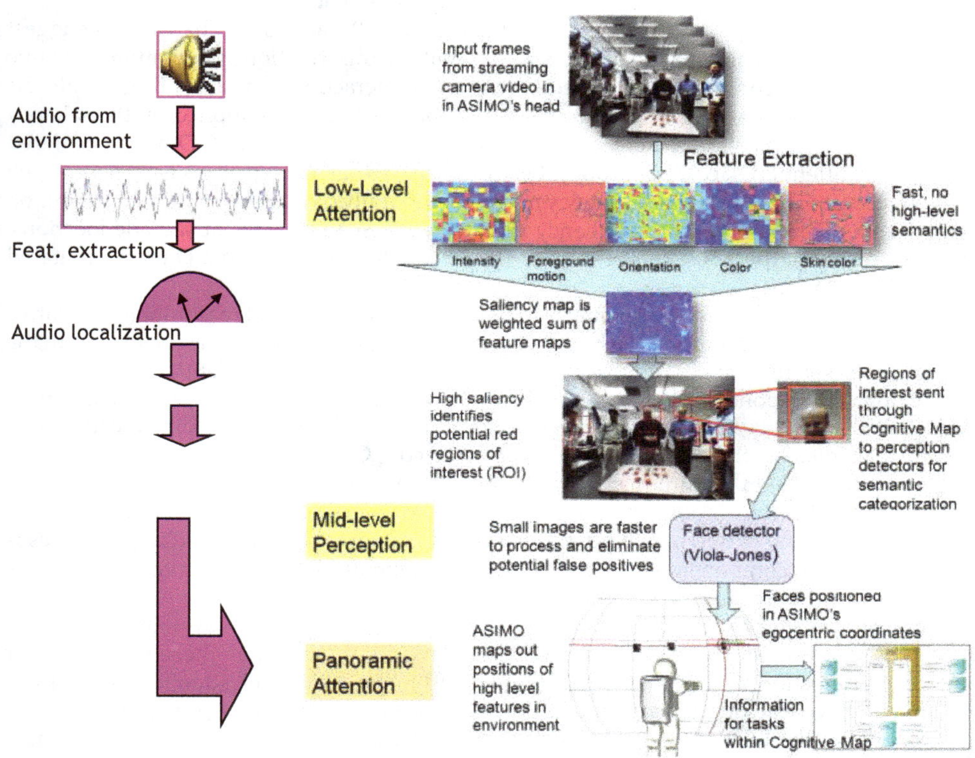

Fig. 1. Three layer panoramic attention model

procedure is to generate a Gaussian pyramid from input frame, perform center-surround differencing on select pyramid level pairs and add up the differences and finally normalize them – for details on feature map generation, refer to (Itti et al., 1998)[2]. In the current implementation, Color, Intensity, Orientation, Skin, and Motion features are used.

The base-level intensity map is derived from the corresponding channel of the HSI color space. The Color and Orientation feature maps are again constructed as suggested in (Itti et al., 1998). For Motion map, an adaptive foreground estimation method(McFarlane and Schofield, 1995) is used, which works well even in the presence of variable rate background motion. For Skin map, samples of skin segments were collected from training images and estimate the probability of skin pixels(Kakumanua et al., 2006). Another option for Skin map is to threshold the ratio of red and green channels of the RGB image and perform connected-component analysis on the result to obtain the feature-map level ROIs. The current implementation bypasses the pyramid reconstruction and interpolates the Motion and Skin feature maps to the level of other feature maps before forming the final saliency map.

[2] (Itti et al., 1998) refer to feature maps as conspicuity maps.

The saliency map is computed as a weighted addition of Feature Maps, with the weight proportional to the importance of the corresponding map. The following weights: $w_{color} = 0.1, w_{intensity} = 0.1, w_{orientation} = 0.1, w_{skin} = 0.2, w_{motion} = 0.5$ are used. The weights suggest relative importance to motion and skin color cues making the attention system more sensitive to regions that may contain faces as needed in human interaction scenarios. To generate the attentional saccade locations, a habituation/inhibition mechanism is applied on the saliency map(Ruesch et al., 2008). The Region of Interest (ROI)s are obtained as 2-D masks centered around each attention saccade location. Feature maps are generated from features of each frame and combined to produce a saliency map. A temporally dynamic habituation-inhibition mechanism is then applied to the saliency map to produce a sequence of saccade locations. The layer finally outputs Regions of Interest (ROI), which correspond to 2-D masks centered on eye-gaze locations.

Human-robot interaction scenarios may also require proximity-related reasoning(e.g. relative positions of people surrounding the robot). For this reason, the depth map associated with the RGB image is also processed. However, instead of processing the entire depth map, the depth map regions corresponding to the ROIs obtained are processed and a depth measurement relative to the robot's location is estimated for each ROI. Since the proximity concerning people is of interest, the Skin feature map is utilized. Connected component analysis is performed on the Skin feature map. For each component, the depth values corresponding to the location of each component's pixels are thresholded (with a lower and upper threshold to eliminate outliers) to obtain valid depth values. A histogram of such valid depth values is estimated for each component. The location of the histogram bin with the largest bin count is assigned as the depth value of the ROI.

Mid-level perception: This layer serves as a filter for the outputs of low-level attention. This layer forms the proper location for detectors and recognizers such as those employed for face/object detection and localization. Outputs from the low-level attention – the ROIs mentioned above – are independent of context. Effectively, the mid-level perception grounds and categorizes the ROIs into higher-level entities such as faces and objects, which are then passed on to the panoramic attention layer.

The mid-level perception contains face/object localization and recognition modules. Indeed, it is only at mid-level perception that some semblance of identity arises(Tsotsos et al., 2005). As mentioned before, one role of mid-level perception layer is to serve as a filter for incoming ROIs. ROIs provide candidate regions from the original image for the face module. In Section 4, the performance of mid-level perception layer is examined in its filtering role with an analysis of the reasons behind the achieved performance.

Panoramic attention layer: This layer manages the spatial and temporal evolution of entities detected by the robot's visual attention over its entire field of view, including the additional range obtained from the pan and tilt angle ranges in the camera head. To accomplish this, it registers entities from the mid-level perception onto a *cognitive*[3] panorama. It is also at this layer that auditory information is utilized to perform the aforementioned entity registration in a multi-modal fashion. The persistence of registered entities is managed in the course of updates to the panorama. External modules can communicate with the panoramic attention module and the information for their specialized purposes. In addition, an idle-motion policy

[3] The word *cognitive* here denotes the semantic information stored in this panorama and also serves to distinguish from the usual definition of panorama as a large field of view composite image stitched from smaller images.

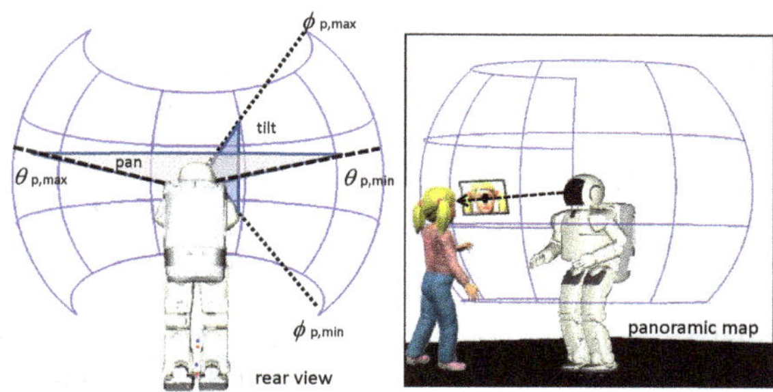

Fig. 2. Left: Panoramic map centered on ego-centric view from camera. Right: Given knowledge of the robot's head position and orientation and the 2-D image, it can be deduced that the person is somewhere along the line of sight connecting the camera origin through the entity's position in the 2-D image.

determines the current portion of the panorama that should be attended to, in the absence of any top-level controls.

The panoramic attention layer places the information about each entity obtained from the mid-level perception layer into the robot's ego-centric spherical coordinate system centered about its head (Figure 2). This information can then be accessible by behavior modules performing tasks within the robot system. In contrast to the lower layers which receive low-level sensor or image information, the input to the panoramic attention layer consists of higher level semantic features, such as face or object positions.

The main purpose of the panoramic attention layer is to expand environmental awareness beyond the robot's immediate field of view. It encompasses the total range of vision given the robot's head motion and the field of view of its cameras. Although in the current implementation only 2-D information is stored, the panoramic map can be easily augmented with three dimensional information given depth information obtained through stereo images or time-of-flight sensors. Nevertheless, given 2-D positions and the current joint angles of the head, some spatially-aware tasks are achievable. For example, in Figure 2, the position of the girl's face lies somewhere along the line of sight cast from the camera origin to the image's 2-D projection of face on the panoramic surface. If depth information is available, the distance along this ray can be esolved. Although items in the panoramic map cannot be localized for manipulation or navigation tasks without depth information, there is enough information to conduct many useful interaction tasks such as directed gaze because only relative directional information from the robot is required. If entities are beyond arm's reach of the robot, pointing operations can also be performed.

To accomplish these useful interaction tasks, an egocentric-based coordinate system (Figure 2, Left) is used, where the origin is localized at the robot's center of rotation of its head. Since all entities in the panoramic map are stored in the camera coordinate system and the configuration of the camera on the robot's body is known, the relative directional information of an entity with respect to its head can be reconstructed. If depth information is known, the relative position can also be estimated. Raw image data from the cameras are rectified to eliminate spherical distortion before being sent to the mid-level perception detectors. Consequently, there is no need to compensate for intrinsic camera parameters for the incoming image coordinates sent to the panoramic attention layer.

As the robot detects new entity locations, they are stored in the corresponding positions in the panorama. The positions are updated depending on the information provided by the mid-level perception modules. For example, if the mid-level modules perform tracking, the new positions simply replace the existing entries in the panoramic map. If no tracking is done, tracking behavior is built into the panoramic map using nearest-neighbour updates by treating the observation events as a time series of samples.

As the observations made by the panoramic map are made in an opportunistic manner (no explicit instructions to seek out or observe the entity), this allows certain commands to be performed with minimal preparation time. For example, when an instruction to look or point is given for an object already stored in the map, the action is carried out immediately without first needing to find the object, even in situations where the entity is off-screen.

3.2 Auditory attention

The audio signal from the environment is captured via the microphone array situated on the robot. Using the difference in time taken for the audio signal to travel to the microphones in the array, the azimuthal location of sounds(i.e. direction of sound as referenced in a horizontal plane) is localized(see Figure 1). This process of localization can be considered as selection of salient sounds(speech, specifically), analogous to the task of determining ROIs in low-level visual attention processing. In addition, pitch and amplitude features are extracted from the audio signal. The localization information, along with these extracted features, is combined at the panoramic layer of the model thereby making the process of entity registration a multi-modal affair. At the panoramic level, this incoming information is analyzed to identify the number of active speakers in the environment. The design choice of combining audio information with the visual in order to perform entity registration at the panoramic level can be considered as a late-binding model,wherein the modalities(audio and visual) are processed independently and in parallel to eventually combine at cognitively high levels where modality-specific information tends to be fairly grounded.

At the panoramic level, the audio-related information is received in the form of *sound sample messages*. Each message is associated with an angle corresponding to the azimuthal direction of the origin of the sound from the environment. In order to identify the number of speakers, the messages are clustered in real-time using the azimuthal angle as the clustering attribute, each cluster corresponding to a speaker and the center of the cluster corresponding to the azimuthal directional location of the speaker(Guedalia et al., 1998). During interaction, it is quite natural for the speakers to move about in the environment and this includes entirely disappearing from the zone of interaction. Therefore, the created clusters, which correspond to *active* speakers, do not persist indefinitely. Each cluster C_i is associated with an activation α_i modelled as an exponential decay function:

$$\alpha_i = \exp(-t/\tau_i), \tag{1}$$

where t represents the elapsed time since the last sample was added to the cluster and τ_i is a time-constant used to control the rate of decay.

As the cluster C_i is created, it is initially assigned an activation $\alpha_i = 1$. The activation value then decays according to the above equation. If a new sound sample falls into the cluster, the activation value is reset to 1. Otherwise, if the activation value falls below a threshold, the cluster is removed.

4. Experimental assessment of using ROIs

An immediate consequence of the panoramic attention model is that it doubles up as a filter for part-based, localized detection and recognition modules which reside at the mid-level perceptual layer. As mentioned previously, this mechanism decreases both the processing time and false positive rate for these modules. To verify this, these two quantities were measured in videos captured by the humanoid. The video data was collected with varying number of people and their activities (e.g. standing still, gesturing, conversing with each other, walking). The results in Table 1 validate the filtering role of the model[4]. The explanation for decrease in processing time is obvious. The consequence of the decreased processing time becomes significant in that additional time can be devoted for detailed processing of ROIs(e.g. determining facial expressions) without loss in throughput. It must also be noted that the decrease in false-positive rate is not simply a consequence of decrease in size of processing region. The low-level attention model described in the previous section provides good candidate image regions(ROIs) which have been selected using multiple cues such as color, intensity, skin and motion-regions. In particular, the skin and motion related feature maps offer valuable information about human presence. This, combined with the appropriate weighting of feature maps, enables the selection of ROIs which tend to contain good face candidates. On the other hand, the inability to exploit the aforementioned human-related cues contributes to larger false positive rates when face detection is performed on entire image by face detection algorithms which cannot utilize real-time human presence cues. Arguably, the model could fail to attend all of the entities present in the time-span of processing a single frame(thereby resulting in a non-trivial false negative rate), but this is a trade-off between the number of interest regions considered (consequently, processing time) and the rate at which the inhibition-habituation occurs over time.

Region size	Processing time	False positive rate
Full-frame (1024x768)	1204.71 milli-sec	4.74%
ROIs (200x200)	96.26 milli-sec	1.24%

Table 1. A quantitative measurement of filtering role played by the attention model

5. Applications

Once the panoramic attention framework is established, its information can be used to model various behaviors. In particular, models to direct the robot's gaze under various situations such as idle moments, encountering a new individual and following conversations between humans will be described in this chapter. Studies have shown that a robot's gaze cues can manipulate people's behavior and the roles they see themselves in during interaction with robots and other humans. For example, gaze can indicate strong turn-yielding cues to participants in a conversation (Mutlu et al., 2009). In idle moments when the robot is not conducting any primary tasks, the lack of visible activity can lead people to believe the robot is no longer functional. For this reason, small head motions are sometimes introduced in the robot's behavior to indicate liveliness. In (Nozawa et al., 2004), head motions were randomly perturbed to give it visual expression while it talked to compensate for the robot's lack of a mouth.

[4] The full-frame results were obtained using the OpenCV implementation of the Viola-Jones face detector(Viola and Jones, 2004)

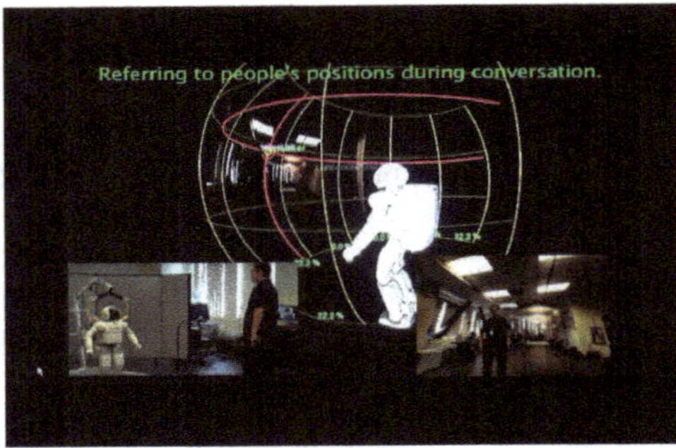

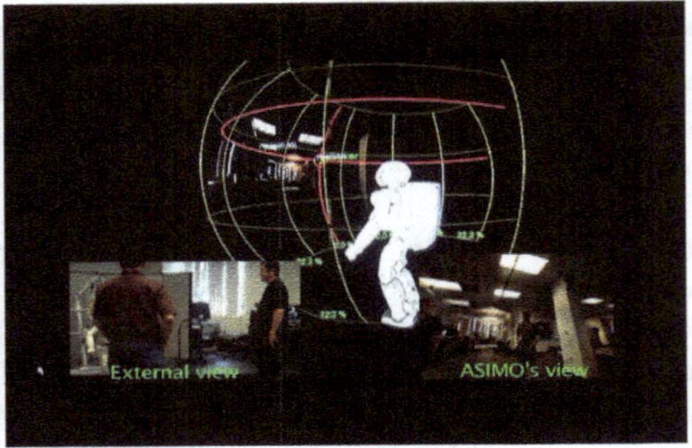

Fig. 3. Panoramic attention in action: In the absence of attention-triggers, the robot looks about idly with the gaze locations generated by idle-policy model. The associated gaze direction probabilities can be seen at the bottom of the panorama in green as percentages. In the left figure, the robot has just seen a familiar face so it registers identity and location on panorama. When it observes a second person(right figure), it repeats the registration process.

Rather than play pre-recorded or purely random motions, the panoramic attention framework is used in combination with gaze control to let the robot actively map out its surrounding environment during its idle moments. This forms the basis of the panoramic layer's idle-motion policy. The panorama is first subdivided along its horizontal camera pan angle range $[\theta_{min}, \theta_{max}]$ into n bins of width w where:

$$w = \frac{\theta_{max} - \theta_{min}}{n} \tag{2}$$

and the bounds of bin i are defined as:

$$(\theta_{min} + i * w, \theta_{min} + (i+1) * w), i = 0, \cdots, n-1. \tag{3}$$

A frequency histogram can be built from the number of gaze occurrences that fall into each bin with f_i corresponding to the frequency of bin i. It is desirable to have the robot select new gaze directions based on areas it has visited infrequently to ensure it is adequately sampling its entire viewing range. Consequently, the probability of visiting bin i is defined as:

$$p_i = \frac{\max_j f_j - f_i}{\max_j f_j * n - f_{total}} \tag{4}$$

where f_{total} is the total number of gaze hits. From Equation 4, a bin that receives many gaze hits is less likely to be selected while those bins that have no current hits are very likely to be seen next. This strategy eventually allows the robot to map out entities in its surrounding environment while it is idle. (Refer to Figures 3 and 5 where the probabilities are marked at the bottom of the panorama in green). A lively-looking humanoid also results as a convenient by-product of this idle-policy.

In situations where entities appear in the robot's view (such as faces), their positions are noted within the panoramic map (Figure 4). If the face is recognized as a new person, the robot can respond to the person's presence(e.g. by greeting him/her) and subsequently tracks the changing position of the person via updates from the mid-level attention layer. This can be done for more than one person in a surreptitious manner while the robot is performing other tasks.

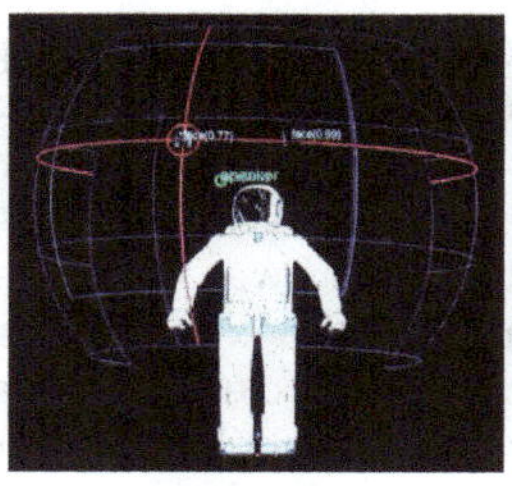
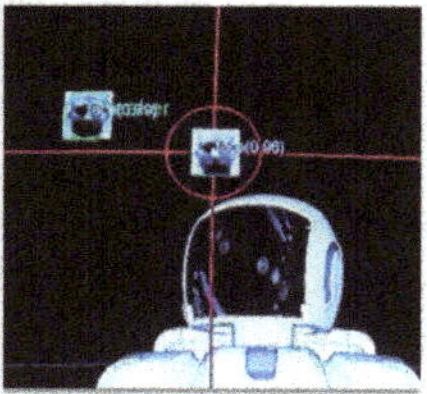

Fig. 4. Multiple faces appear in panoramic attention system (red cross-hairs represent current focus)

In situations where the robot hears the name of a familiar person during a conversation, if the person's location is already in the panoramic map, the robot can redirect its gaze towards that person, providing a strong visible reminder to the person that conversation was directed at him or her. For example, in Figure 3(left), the robot first recognizes a person and registers his location, then its attention is directed to a second person(Figure 3(right)). When the latter asks the former a question(Figure 5(left)), the robot immediately knows the location of the person being referenced and can look back at him without searching(Figure 5(right)). In case the person being referenced has moved in the meantime, the robot commences the search from the last known stable position. If the person can be tracked within the field of view of the first person, the robot skips the search phase. Either way, this strategy is more efficient than a full-blown panorama-wide search.

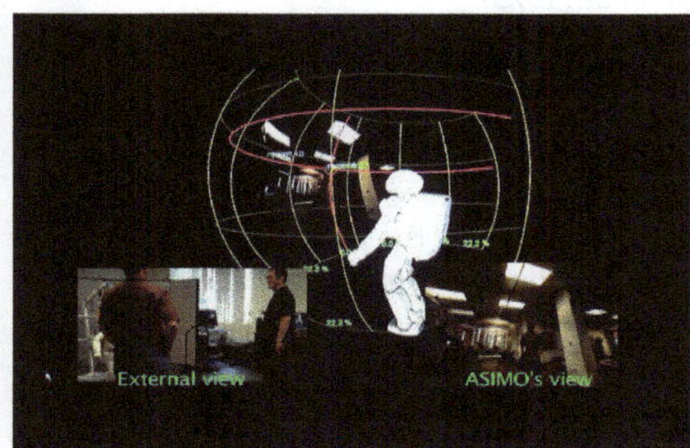

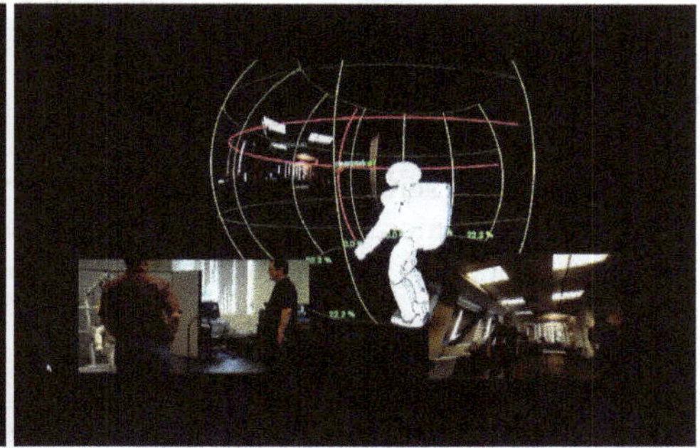

Fig. 5. The robot follows the conversation between them. When it hears a referential phrase (e.g. name being called)(left), it proceeds to look at the referent and participate in the conversation appropriately(right).

When observations of faces appear in the system, they are assigned an activation signal initially set to $\alpha = 1$. As time progresses, the signal decays according to the exponential decay function:

$$\alpha = \exp(-t/\tau), \tag{5}$$

where t refers to time passed since the initial observation and the decay rate is set to τ, with smaller values of τ causing faster decay.

This activation signal is used to provide a finite bound on the time the observation stays in the panoramic map. If a new observation of the entity is made, the activation signal is reset to 1. The decay rate can be assigned different values based on entity type. For example, static and isolated objects can be given a slow decay rate while moving people can be assigned faster decay rates. The activation signal can also indicate how stable an observation is. For example, a rapidly moving face will have its activation constantly being reset to one. When encountering people in the scenario mentioned previously, they were often observed in motion before coming to a stop. When the robot's head was initially set to look at the person on the first observation event, the robot would often look at the position where the person was just moments before. However, if the robot is directed to look at the person only after the activation signal has reduced to a lower value threshold, the robot will look at a person only after his or her position stays stable long enough for the activation signal to decay to the threshold set. This behavior is called *cognitive panoramic habituation* – a panoramic analogue of the habituation-inhibition model in traditional attention architectures.

When the robot moves its head, a *head-motion-pending* message is broadcast to temporarily disable the low-level attention system and prevent incorrect motion-detected events caused by the robot's own head motion. When head motions have stopped, a *head-motion-stopped* message is broadcast to reactivate the low-level attention system. This mechanism also informs the high level panoramic attention layer to ignore incoming face position events since locating its position within the 3-D panoramic surface would be difficult because the robot's view direction is in flux. This strategy is supported by studies related to saccade suppression in the presence of large scene changes(Bridgeman et al., 1975).

5.1 Other applications

When used simultaneously with multiple modalities, this combined information can create a better estimate of the robot's environment. Since the panoramic attention map encompasses the robot's entire sensory range of motion beyond the immediate visual field, entities not immediately in camera range can still be tracked. The ability for the robot to obtain a greater sensory awareness of its environment is utilized in the following applications:

5.1.1 Wizard-of-Oz interface

In addition to providing monitoring functions, the panoramic map can be used an interactive interface for allowing a robot operator to directly select targets in the robot's field of view for specifying pointing or gaze directions (Figure 6). The two-dimensional image coordinates of entities within the active camera view are mapped onto the three-dimensional panoramic view surface, producing three dimensional directional information. This information enables a robot or any other device with a pan/tilt mechanism to direct their gaze or point their arm at the entity in its environment using its egocentric frame of reference(Sarvadevabhatla et al., 2010). Two important features of the interface are evident from Figure 6 – the interface mirrors the world as seen by the robot and the semi-transparent, movable operator GUI controls ensure a satisfactory trade-off between viewing the world and operating the interface simultaneously.

5.1.2 Multi-tasking attentive behavior

The panoramic attention model can be extended to manage other modalities such as motion. This allows a person to get a robot's attention by waving their hands. This ability is important in multi-tasking situations where a robot may be focused on a task. In Figure 7, a person waves his hand to get the robot's attention while the robot is actively playing a card game with a different person. The attention-seeking person may either shout a greeting (possibly off-camera) or wave their hands in the robot's visual field. Once the robot directs its gaze to

Fig. 6. Wizard-of-OZ application for direct selection of gaze and pointing targets

the source of the sensory disturbance, face detectors automatically locate the face to allow the robot to shift its gaze directly to the person's face.

5.1.3 Multi-speaker logging

The panoramic map can be used to log information and assign these as properties to entities in its environment. For example, we have actively kept track of the amount of speaker participation in group scenarios and attributed spoken utterances to the likely speaker based on the location of the sound utterance (Figure 8). Utilizing multi-modal information, it can be confirmed if a sound utterance coincides with a face appearing in the same location to avoid false positive identification of people by non-human sound sources in the environment. This mechanism has been used to allow the robot to monitor relative activity from a group of participants in a conversation.

6. Discussion and future work

The low-level attention layer utilized in the multi-modal panoramic attention model described in this chapter is commonly used(Itti et al., 1998), therefore the other components of the model can be readily layered upon existing implementations of the same. An important benefit of the three-layer attention model is the ability of its mid-level layer to act as a spatiotemporal filter for sensory input. The panoramic map provides a high-level environmental assessment of entities around the robot that can be used to develop behavioral models or aid in task completion. The idle gaze behavior model combines exploratory gaze behavior, active gaze directed at new observations and subsequent gaze tracking behavior. Significantly, these benefits are obtained for free while providing the impression of liveliness. Finally, the

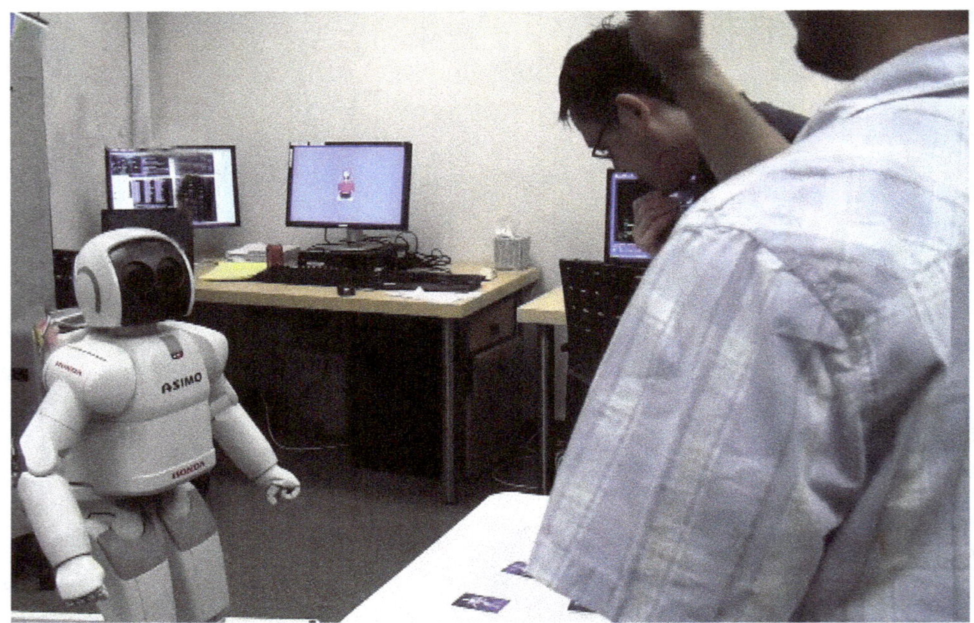

Fig. 7. Hand-waving to get robot's attention while it is playing card game with another person

incidental information obtained through casual observation can be quickly recalled when location-based queries are performed.

6.1 Limitations

The current attention pipeline can be accelerated significantly if the processing related to producing regions of interest is parallelized. This can either be done during low-level feature computation by utilizing GPU acceleration techniques or at higher levels by allowing regions of interest to be processed simultaneously by multiple types of detectors or the regions of interest could be paritioned spatially to be analyzed in parallel by the same type of detector. In a distributed system, the communication between the different levels of the attention model is handled using network socket communication. If there are numerous communication links from the raw sensory input to the highest panoramic attention layer, the cumulative latency will create a delayed response in perception and any subsequent behavioral reponse. By judiciously allow increased computational load on a single computer, network communication can be replaced by very rapid memory-based messaging schemes which will allow the robot to respond quicker to environmental changes.

To reduce excessive computational overhead, the use of context can be applied to adjust weighting of different features when determining regions of interest. For example, a highly visual task may not require frequent sound processing. The attention system can choose to either adjust the sampling rate of the sound detector or disable processing completely. Increased processing can then be allocated to relevant features to produce more accurate measurements and subsequent state estimates. At the middle layers, introducing a top-down mechanism of modulating different types of detectors can create efficient resource allocation.

Fig. 8. Automatic logging of speaker activity and locations for multi-speaker applications: (inset) Panoramic attention locates speakers and logs amount of speaker activity for each participant. White circle regions represent past clusters of sound activity labeled with current number of utterances and cumulative time spoken.

For example, if it is discovered that certain objects are not present in an environment or not relevant to a task, the corresponding detectors can be suspended to allow relevant detectors to perform at faster rates, improving overall system performance.

The mechanism for top-down modulation of the attention system should be handled by a behavior system that is driven by the set of active applications running on the robot. The behaviors are dictated by the current task and there should be a clear layer of separation between the behavior and perception system. Behaviors are triggered by changes of the environmental state which is inferred from the panoramic attention system. Therefore, it should be the behavior system that configures which detectors and low-level features the perception system should perform to allow proper decision-making for behaviors to be made. The behavior system can consist of either low-level reactive actions that may be triggered directly by low-level features, or high-level deliberative behaviors that may spawn multiple sub-tasks themselves. Since reactive behaviors have fewer intervening communication links, generated responses will automatically occur quicker in response to changes in sensory input.

Since the locations in the panoramic map are in egocentric coordinates, they need to be updated whenever the robot moves to a new location. Although these locations can be completely re-sensed and re-calculated once the robot moves to its new position, the

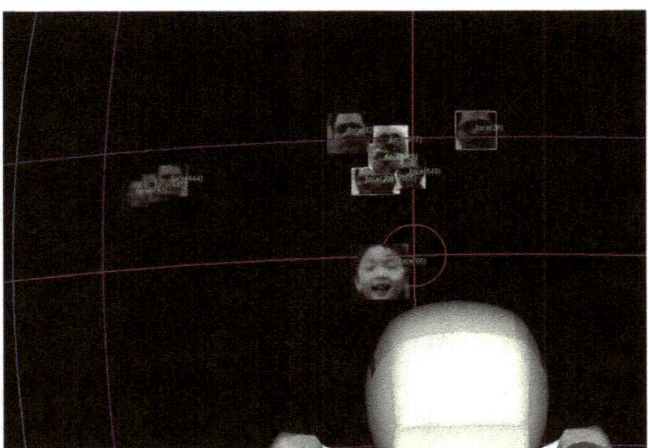

Fig. 9. Child faces appears lower in panorama.

panoramic map would be invalid during actual motion, limiting the robot's perception while it is moving. Continuously tracking the position of environmental objects during motion and incorporating the known ego-motion of the robot could be employed to create a continuously updated panoramic map even while the robot is moving.

6.2 Future work

To make the low-level attention model more consistent, it would be preferable to replace the skin and motion feature maps with their multi-resolution versions. By adding more cues to the low-level model such as depth information and audio source localization, the panoramic map can be extended to aid in navigation and manipulation tasks, as well as identification of speaker roles to follow conversations or identify the source of a query. This would provide a multi-modal basis to the existing model. If an object-recognition module is also integrated into the mid-level layer, it would widen the scope to scenarios involving objects and human-object interactions.

The semantic information stored in the panoramic attention map can be exploited to produce better motion models. For example, it is expected that a face belonging to a person is more likely to be mobile compared to an inanimate object resting on a table. Consequently, the dynamic motion models can be tuned to the entity type.

Spatial relationships in the panoramic map can also be used to deduce information about entities in the system. In Figure 9, one of the faces appears in a lower vertical position than the others. Given that the bounding boxes for both faces are in the same order of magnitude, the robot can infer that one person could be a child or sitting down. If the robot was not able to detect an inanimate chair in the region of the lower face, it could further deduce the face belongs to a child.

7. References

C. Breazeal & A. Edsinger & P. Fitzpatrick & B. Scassellati(2001). Active vision for sociable robots, In: *IEEE Transactions on Systems, Man, and Cybernetics, A*, vol. 31, pp 443-453.

L. Itti & C. Koch & E. Niebur(1998). A model of Saliency-based Visual Attention for Rapid Scene Analysis, In: *IEEE Transactions on Pattern Analysis and Machine Intelligence*, vol. 28, pp 1254-1259.

L.Itti & G. Rees & J.K. Tsotsos(2005). In: *Neurobiology of Attention*, Academic Press, 2005, ISBN 0123757312, 9780123757319.

L. Komatsu(1994). In: *Experimenting With the Mind: Readings in Cognitive Psychology*, Brooks-Cole Publishing Company.

L. Itti & N. Dhavale & F. Pighin(2003). Realistic Avatar Eye and Head Animation Using a Neurobiological Model of Visual Attention, In: *Proc. SPIE 48th Annual International Symposium on Optical Science and Technology*, vol. 5200, pp 64-78.

C.Kayser & C. I. Petkov & M. Lippert & N.K. Logothetis(2008). Mechanisms for Allocating Auditory Attention: An Auditory Saliency Map, In: *Current Biology*, Vol. 15

P. Maragos(2008). In: *Multimodal Processing and Interaction: Audio, Video, Text*, pp. 181-184, Springer, ISBN 0387763155, 9780387763156.

K.Rapantzikos & G. Evangelopoulos & P.Maragos & Y. Avrithis(2007). An Audio-Visual Saliency Model for Movie Summarization, In: *IEEE 9th Workshop on Multimedia Signal Processing(MMSP)*, pp.320-323.

V. Navalpakkam and L. Itti(2005). Modeling the influence of task on attention, In:*Journal of Vision Research*, vol. 45, no. 2, pp. 205-231.

J.Moren & A.Ude & A.Koene(2008). Biologically based top-down attention modulation for humanoid interactions, In: *International Journal of Humanoid Robotics (IJHR)*, vol. 5, pp 3-24

B. Mutlu & T. Shiwa & T. Kanda & H. Ishiguro & N. Hagita(2009). Footing in Human-Robot Conversations: How Robots Might Shape Participants Roles Using Gaze Cues. In: *Proceedings of the 4th International Conference on Human-Robot Interaction (HRI).*

Y. Nozawa & J. Mori & M. Ishizuka(2004). Humanoid Robot Presentation Controlled by Multimodal Presentation Markup Language MPML, In:*Proc. 13th IEEE International Workshop on Robot and Human Interactive Communication (RO-MAN).*

T. Sobh & A. Yavari & H.R. Pourreza(2008). Visual Attention in Foveated Images, In:*Advances in Computer and Information Sciences and Engineering*, Springer, ISBN 9781402087417, pp. 17-20.

J. Shear & F.J. Varela(1999). In: *The View from Within: First-person Approaches to the Study of Consciousness*, Imprint Academic, ISBN 0907845258, 9780907845256

P. W. Halligan & D. T. Wade(2005). In: *The Effectiveness of Rehabilitation for Cognitive Deficits*, Oxford University Press, ISBN 0198526547, 9780198526544

A. Bur & A. Tapus & N. Ouerhani & R. Siegwar & H.Hiigli(2006), Robot Navigation by Panoramic Vision and Attention Guided Features, In: *International Conference on Pattern Recognition(ICPR)*, vol. 1, pp. 695-698.

K. Kayama & K. Nagashima & A. Konno & M.Inaba & H. Inoue(1998). Panoramic-environmental description as robots visual short-term memory, In:*IEEE International Conference on Robotics and Automation(ICRA)*, pp. 3253-3258.

K. Nickel and R.Stiefelhagen(2007). Fast Audio-Visual Multi-Person Tracking for a Humanoid Stereo Camera Head, In:*IEEE-RAS 7th International Conference on Humanoid Robots - Humanoids'07*, Pittsburgh, PA.

J. Ruesch & M. Lopes & A. Bernardino & J. Hörnstein & J. Santos-Victor & R. Pfeifer(2008), Multimodal saliency-based bottom-up attention framework for the humanoid robot iCub, In:*IEEE International Conference on Robotics and Automation (ICRA)*, pp. 962-967

A. Koene & J. Moren & V. Trifa & G. Cheng(2007). Gaze shift reflex in a humanoid active vision system, In:*Proceedings of the 5th International Conference on Computer Vision Systems (ICVS).*

W. D. Stiehl & J. K. Lee & R. Toscano & C. Breazeal(2008). The Huggable: A Platform for Research in Robotic Companions for Elder care, In:*AAAI Fall Symposium, Technical Report FS-08-02*, The AAAI Press, Menlo Park, California.

J. Macedo & D. Kaber & M. Endsley & P. Powanusorn & S. Myung(1999). The effects of automated compensation for incongruent axes on teleoperator performance, In:*Human Factors*, vol 40, pp. 541-553.

S.M.Pizer & R.E. Johnston & J.P. Ericksen & B.C. Yankaskas & K.E. Muller(1990). Contrast-limited adaptive histogram equalization: speed and effectiveness, In:*Proceedings of the First Conference on Visualization in Biomedical Computing*, pp. 337-345.

N. J. B. McFarlane & C. P. Schofield(1995). Segmentation and tracking of piglets in images, In:*Journal of Machine Vision and Applications*, vol. 8, no. 3, pp. 187-193,Springer.

P. Kakumanua & S. Makrogiannisa & N. Bourbakis(2006). A survey of skin-color modeling and detection methods, In:*Pattern Recognition*, vol. 40(3), 2006

G. Bridgeman & D. Hendry & L. Start(1975). Failure to detect displacement of visual world during saccadic eye movements. In: *Vision Research*, 15, pp. 719-722.

I.D. Guedalia & M. London & M.Werman (1998). An On-Line Agglomerative Clustering Method for Nonstationary Data. In: *Neural Computation*, vol. 11(2), pp.521-540.

ASIMO humanoid robot (2000). Honda Motor Co. Ltd.

P.A.Viola & M.J.Jones (2004). Robust Real-Time Face Detection. In: *International Journal of Computer Vision*, vol. 57(2), pp.137-154.

S. Ravi Kiran & V. Ng-Thow-Hing & S.Okita, In: *The 19th IEEE International Symposium on Robot and Human Interactive Communication(RO-MAN)*. pp.7-14.

Part 4

Current and Future Challenges for Humanoid Robots

14

Rob's Robot: Current and Future Challenges for Humanoid Robots

Boris Durán and Serge Thill
University of Skövde
Sweden

1. Introduction

The purpose of this chapter is both to review the current state of the art in robotics and to identify some of the challenges that the field has yet to face. This is done and illustrated by following an imaginary roboticist, Rob, who is currently designing a new humanoid robot from scratch. Rob's ultimate vision is that this robot will be humanoid in the sense that it possesses the abilities of a human being. Therefore, his initial aim is to identify relevant work as well as areas in which innovations are still needed.

Research in robotics has come a long way in the last few decades, whether one considers humanoids specifically or just any type of architecture. The world is increasingly moving away from being confined to industrial robots with precisely programmed movement plans and is beginning to include adaptive robots that can robustly operate in a variety of scenarios as well as robots that show increasing levels of cognitive abilities and intelligence. We have witnessed an increasing collaboration between research areas that, previously, had very little in common. Nowadays, it is not difficult to find research groups uniting neuroscientists, engineers, psychologists, linguists, even philosophers and musicians, all coming together to study the human body as well as the human mind. There is thus plenty of previous work that Rob can build upon when designing his robot. At the same time, it is clear that the field is still a long way from creating truly human-like machines and that there are thus significant challenges to overcome.

In this chapter, we first provide a brief history of humanoid robotics research since it sets the background to Rob's work. We then present the state of the art in the different components of a robot, hardware as well as software, discussed in contrast with Rob's requirements and highlighting particular areas in which further research is needed before Rob's robot can become reality.

2. A brief history of humanoid robotics research

Our fellow roboticist Rob goes back in time in order to trace the roots of his passion. He finds out that attempts to create automated machines can be traced back to the first century of our era with the inventions of Hero of Alexandria, a Greek mathematician and engineer from Roman Egypt. Among the first human-like automata he found one created by another polymath inventor from Mesopotamia in the 12th century known as Al-Jazari. His group of "robots" used water to drive their mechanisms and play four different instruments.

Among others who designed and/or created human-like machines he discovers that:

- Leonardo Da Vinci designed a mechanical knight at the end of the 15th century. This machine was able to sit up, wave its arms and move its head and jaw.
- Hisashige Tanaka, a Japanese craftsman in the 19th century, created several extremely complex toys able to serve tea, fire arrows and paint kanji characters.
- In 1929, Japanese biologist Makoto Nishimura designed and created *Gakutensoku*, a robot driven by air pressure that could move its head and hands.

Rob also comes across some interesting information about the word "robot". It has its origins in the 1921 play "R.U.R." (Rossum's Universal Robots) created by the Czech writer Karel Capek, although he later credited his brother Josef Capek as the actual person who coined the word. In Czech, the word "robota" refers to "forced labor"; in this sense a robot is a machine created to perform repetitive and expensive labor. In 1942 Isaac Asimov introduces the word "robotics" in the short story "Runaround". His robot stories also introduced the idea of a "positronic brain" (used by the character "Data" in Star Trek) and the "three laws of robotics" (later he added the "zeroth" law).

However it wasn't until the second half of the 20th century that fully autonomous robots started being developed and used in greater numbers. Following the rhythm of a demanding industrial revolution the first programmable (autonomous) robot was born. Unimate was used to lift and stack hot pieces of die-cast metal and spot welding at a General Motor plant. From then on, other single arm robots were created to cope with larger production requirements. Nowadays industrial robots are widely used in manufacturing, assembly, packing, transport, surgery, etc.

Humanoid robotics, as a formal line of research was born in the 1970s with the creation of Wabot at Waseda University in Japan. Several projects around the world have been developed since then. The next section will briefly describe the current state-of-the-art in anthropomorphic robots.

2.1 State of the art

This section starts by presenting HONDA's project ASIMO, Fig. 1(a). HONDA's research on humanoid robots began in 1986 with its E-series, a collection of biped structures that developed, without doubt, the most representative example among state of the art humanoid robots (Hirose et al., 2001). Today's ASIMO has 34 degrees-of-freedom (DOF), weights 54kg and has a total height of 130cm. This robot is able to walk and run in straight or circular paths, go up or down stairs and perform an always increasing set of computer vision routines like object tracking, identification, and recognition, localization, obstacle avoidance, etc.

A joint project between the University of Tokyo and Kawada Industries gave birth to the Humanoid Robot "H6" and ended up with HRP-3 in 2007 (Kaneko et al., 2008). The National Institute of Advanced Industrial Science and Technology (AIST) continued this project and developed today's HRP-4, Fig. 1(c), a 34 DOF walking robot weighting 39kg and measuring 151cm in height. This light-weight full size robot was designed with a reduction of production costs in mind, thus allowing access to the average robotics research group around the world to a pretty advanced research tool.

Korea's Advanced Institute of Science and Technology (KAIST) started working on humanoid platforms in 2002 with KHR-1. Today's KAIST humanoid is known as HUBO2 (KHR-4) and has a total of 40 DOF, a height of 130cm and weight of 45kg, Fig. 1(b). HUBO2 is also able to walk, run and perform an increasing set of sensor-motor tasks (Park et al., 2005).

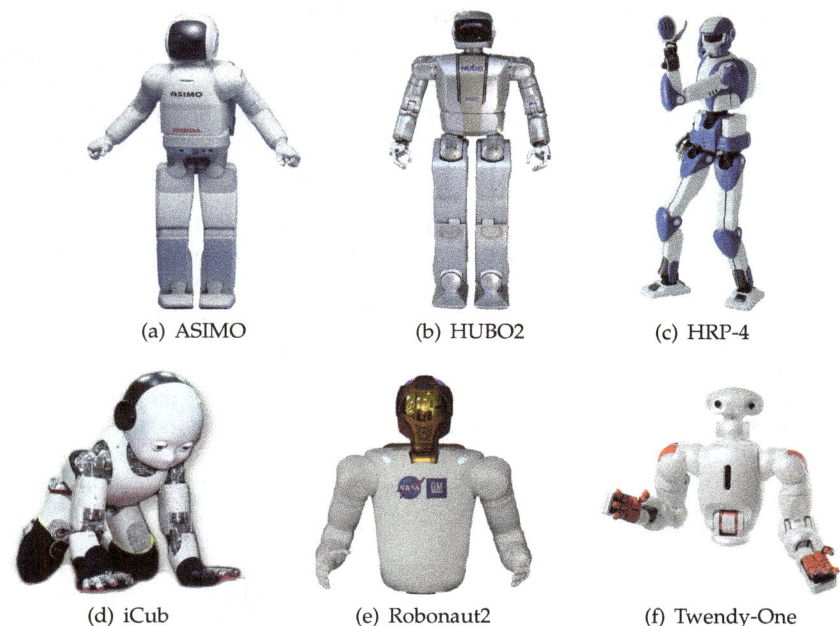

(a) ASIMO (b) HUBO2 (c) HRP-4

(d) iCub (e) Robonaut2 (f) Twendy-One

Fig. 1. Some state of the art humanoid platforms.

The RobotCub Consortium formed by 11 European institutions created iCub, a child sized humanoid robot which until 2010 was only able to crawl, Fig. 1(d). This robot has 53 DOF, weights 22kg and has a height of 104cm (Tsakarakis et al., 2007). A remarkable feature of this project is that everything developed within it was made available as open source, including the mechanical design and software products. This move has created a growing community that actively collaborates at all levels of research.

General Motors joined NASA for the second version of NASA's project Robonaut (Ambrose et al., 2000). This upper body humanoid robot (Fig. 1(e)) is a very dexterous robotic platform with 42 DOF, 150kg and 100cm (waist to head). Another upper-body robotic platform worth including as state-of-the-art is Waseda University's TwendyOne, a very dexterous and high-power robot with 47 DOF, 111kg and 147cm height, Fig. 1(f).

One common characteristic of all the above mentioned platforms is the use of electric torque motors in their joints. There have been, however, new attempts to use hydraulic and pneumatic actuators such as the ones in Anybots' biped Dexter which is also able to jump (Anybots, 2008); or Boston Dynamics' PetMan which introduces a much faster and more human-like gait (Petman, 2011). These two examples make use of pneumatic and hydraulic actuators respectively but there is also a growing interest around the use of pneumatic artificial muscles or air muscles. Shadow Robot's pneumatic hand (Shadow Robot Company, 2003) performing highly dexterous movements and Festo's use of artificial muscles (Festo, 2011) are also worth mentioning as state-of-the-art due to the high accuracy of these otherwise difficult to control kind of actuators.

3. Mechanical requirements and engineering challenges

It is a matter of philosophical discussion how far it is possible to discuss hardware and software separately. Section 4.2 in particular will discuss theories from cognitive science which indicate that in the human, body and mind are intrinsically intertwined and that some of the hardware can in fact be considered to fulfill certain roles of the software too. However, entering into such a debate in detail is beyond the scope of this chapter. Therefore, when considering the different components of Rob's robot, we will largely separate hardware and software in the sense that we will discuss purely mechanical requirements first and computational requirements next. Since hardware can fulfill computational requirements and mechanical requirements can impose software requirements as well, there may be some natural overlap, which in a sense already illustrates the validity of some of the ideas from embodied cognition (to be discussed in section 4.2.1).

First then, Rob needs to identify mechanical and engineering requirements for the creation of his humanoid robot. He is interested in particular in the state of the art of components like sensors and actuators, and must consider how close these elements are to his ambitious requirements and what major challenges still need to be addressed before his vision can become reality. The remainder of this section is dedicated to answering these questions.

3.1 Sensors

Throughout the history of humanoid robotics research there has been an uneven study of the different human senses. Visual, auditory and tactile modalities have received more attention than olfactory and gustatory modalities. It may not be necessary for a robot to eat something but the information contained in taste and odor becomes important when developing higher levels of cognitive processes. Thinking about the future of humanoid robots we should be careful not to leave behind information that may be helpful for these machines to successfully interact with humans.

In this section Rob studies some of the devices that are either currently being used or are under development in different robotic platforms; they are categorized according to the human sense they relate to the most.

3.1.1 Vision

Humanoid robots arguably pose the largest challenge for the field of computer vision due to the unconstrained nature of this application. Rob's plans for his autonomous humanoid involve it coping with an unlimited stream of information changing always in space, time, intensity, color, etc. It makes sense for Rob that among all sensory modalities within robotic applications, vision is (so far) the most computationally expensive. Just to have an idea of how challenging and broad the area of computer vision is, Rob remembers that in 1966 Marvin Minsky, considered by many to be the father of artificial intelligence (AI), thought one master project could "solve the problem of computer vision". More than 40 years have passed and the problem is far from solved.

In robotic applications, a fine balance between hardware and software is always used when working with vision. During the last decade, Rob witnessed an increasing interest on transferring some software tasks to the hardware side of this modality. Inspired by its biological counterpart, several designs of log-polar CCD or CMOS image sensors (Traver & Bernardino, 2010) (Fig. 2(b)) and hemispherical eye-like cameras (Ko et al., 2008) (Fig. 2(c)) have been proposed. However, the always increasing processing power of today's computers and specialized graphical processing units (GPU) have allowed many researchers to successfully tackle different areas of vision while keeping their low cost Cartesian cameras (Fig. 2(a)) and solve their algorithms with software. Nonetheless, Rob welcomes the attempts to reproduce the high resolution area at the center of the visual field, fovea, with hardware and/or software. He thinks that nature developed this specific structure for our eyes throughout evolution and it may be wise to take advantage of that development.

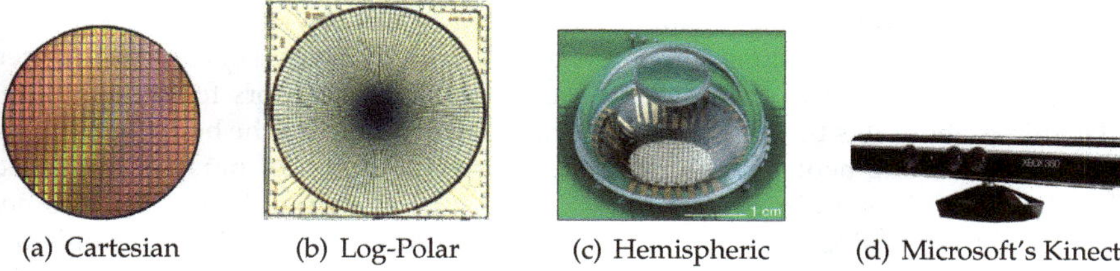

(a) Cartesian (b) Log-Polar (c) Hemispheric (d) Microsoft's Kinect

Fig. 2. Samples of different configurations for image sensors.

Rob identifies several challenges for the hardware component of vision sensors. Potentially the most important is the need to widen the field of view of current cameras. Normal limits for the human field of view are approximately 160 degrees in the horizontal direction and 135 degrees in the vertical direction. A typical camera used in robotic applications has a field of view around 60 degrees for the horizontal direction and 45 degrees for the vertical direction. Important information is lost in those areas where current cameras can not reach in a single stimulus. Adaptation to changes in light conditions is also a complex and difficult process for artificial vision, especially when working within a dynamic environment. Rob is considering a few options from the video security market based on infra-red and Day/Night technology that could be adapted for use in humanoid robots. Finally, the robotics community has experienced a growing interest on using RGB-D (Red-Green-Blue-Depth) cameras after the release of the low-cost motion capture system Kinect by Microsoft (Shotton & Sharp, 2011), Fig. 2(d). This technology combines visual information and high-resolution depth, opening therefore new possibilities for overcoming the challenges of 3D mapping and localization, object identification and recognition, tracking, manipulation, etc.

3.1.2 Tactile

Even though touch sensors are being used on few specific points of humanoid platforms (e.g. tip of fingers, feet, head), Rob thinks that an efficient solution to acquiring information from large surfaces is needed in order to be able to exploit the richness of textures, shapes, temperatures, firmness, etc. Therefore he argues the importance of developing skin-like sensors as another future challenge for humanoid robotics.

In robotic applications, Rob has used his imagination to create solutions for detecting objects in the path of an arm, finger or leg. One of these solutions has been the detection of peaks of current in their electric motors or using torque and force sensors. This approach could be considered tactile sensing as well but within the area of proprioception since measurements are made as relative positions of neighbouring parts in static and dynamic situations. In humanoid robots, exteroceptive sensors include dedicated pressure sensors placed on points of interest. The technology used in most pressure sensors so far include: resistive, tunnel-effect, capacitive, optical, ultrasonic, magnetic, and piezoelectric. Resistive and capacitive sensors are certainly the most common technology used not only in robotic but in many consumer electronic applications (Dahiya et al., 2010).

Advances in nano-materials are making it possible to integrate a large amount of sensors in flexible and stretchable surfaces (Bao et al., 2009; Peratech, 2011; Takei et al., 2010). Rob finds the approach taken by Prof. Zhenan Bao very interesting since it includes not only a sensitive resolution that surpasses that of human skin, but also the possibility to sense chemicals or biological materials (Bao et al., 2009). In addition, they are also working on embedding solar cells within the same films (Lipomi et al., 2011), which is an outstanding innovation since it would help to solve another major challenge for humanoid robots, i.e. energy.

Rob remembers that human skin is composed of two primary layers: epidermis and dermis. The dermis is located beneath the epidermis and contains all sensors for temperature and touch. He thinks about this because the dermis helps also to cushion the body from stress and strain, a task that is augmented by the morphological interaction of muscles, water, bones, etc. This dynamic plays an important role in tasks such as manipulation and locomotion. Therefore the design of Rob's future humanoid robot's skin should include not only a large amount and various types of sensors, but also an inner mechanism that provides similar dynamics like those found in human bodies.

3.1.3 Sound

Rob knows that microphones are the standard tools for acquiring sound information from the environment in a robotic platform. They can be installed in different configurations and detect/regulate sound in ways that surpass human abilities. But detecting a sound is just the starting point of a more complex challenge. Once a sound has been detected, it is time to localize the origin or source of that sound. Humans' remarkable ability to detect, localize and recognize sound can be exemplified using the *cocktail-party effect*. This term describes the ability of a person to direct his/her attention on the source of a sound amidst other sound sources which makes possible, for example, to talk to another person in a noisy, crowded party (Haykin & Chen, 2005).

It has been shown in humans that a continuous interaction between auditory and visual cues take place in the brain (Cui et al., 2008). In humanoid robots the localization of a sound source is mainly done by computing the difference in times and levels of signals arriving to the different sensors, in this case microphones (Keyrouz, 2008). Attempts to integrate sensor information from visual with auditory modalities has also been done in humanoid platforms (Ruesch et al., 2008).

Rob trusts that at least from the hardware point of view, he will not have many problems. The real challenge for the future is to improve the way different sensor modalities can be merged to replicate or surpass the abilities found in humans. However this point belongs to the *mindware* of his platform, so he will focus on working in optimized configurations of microphones.

3.1.4 Odor and taste

Rob gets a strange feeling at this point and thinks, "Should I even consider spending time thinking about a robot with these two senses? After all, robots do not need to eat anything as humans do, they just need a power cable and will have all the energy they need. And electricity has no smell, right?" But then he starts thinking in more holistic terms– maybe robots will need to somehow recognize smells and flavors; moreover, they will need to associate that information with visual, auditory and tactile cues. Otherwise it is going to be difficult, if not impossible, to interact with humans using more or less the same language.

Artificial noses and "tongues" have already been developed and they are also overmatching those of humans (Mahmoudi, 2009). Artificial noses are able to detect and classify thousands of chemical molecules ranging from biological agents to wine and perfume. As in the case of sound, it does not seem difficult to trust that in the near future Rob will have access to a device that gives his robot information about odor and taste, at the very minimum at human levels.

3.2 Actuators

In the case of actuators, Rob illustrates the current state of the art of active components, although he is lately more biased towards approaches which make use of spring-damper components. He knows that the right choice of these components forms one of the critical challenges for his future humanoid robot. Rob's preference for non-linear components comes from the fact that the human body is made, from an engineering perspective, almost exclusively of spring-damper type elements: cartilages, tendons, muscles, fat, skin, etc. whose interaction results in stability, energy efficiency and adaptability.

3.2.1 Whole-body motion

Humans have a large repertoire of motor actions to move the whole body from one point to another. Walking could be considered the most representative behavior in this repertoire but we learn to adapt and use our bodies depending on the circumstances. Humans are also able to run, crawl, jump, climb or descend stairs, and if using the arms, then we can squeeze our bodies between narrow spaces or if in water we can learn to swim.

Rob realizes that researchers in whole body motion for humanoid robots have focused most, if not all their attention in walking only. The most common approach to control the walking behavior and balance of a humanoid robot is called Zero Moment Point (ZMP) and it has been used since the beginning of humanoid robotics research (Vukobratovic & Borovac, 2004). This approach computes the point where the whole foot needs to be placed in order to have no moment in the horizontal direction. In other words, ZMP is the point where the vertical inertia and gravity force add to zero. Most state of the art humanoid platforms like ASIMO, HRP-4, and HUBO2 make use of this approach. The main drawbacks of ZMP arise from the need to have the whole foot in contact with a flat surface, and it assumes that this surface has enough friction to neglect horizontal forces.

Passive-dynamic walkers are an alternate approach to humanoid locomotion (Collins et al., 2005). These platforms try to exploit not only the non-linear properties of passive spring-damper components but the interaction of the whole body with the environment. The result is, in most cases, a more human-like gait with a heel-toe step in contrast to the flat steps seen in ZMP-based platforms. State of the art passive-dynamic walkers use dynamic balance control (Collette et al., 2007) which provides them with robustness to external disturbances and more human-like whole-body motor reactions. Examples of this approach can be found in platforms such as Dexter (Anybots, 2008), PetMan (Petman, 2011), and Flame (Hobbelen et al., 2008).

Rob also found out that running and jumping have already been implemented in a few of the current humanoid platforms (Anybots, 2008; Niiyama & Kuniyoshi, 2010), although there is still much work to do to reach human-like levels. Once humanoid robotics started to become the meeting point of different scientific groups such as developmental psychologists, neuroscientists and engineers, interesting topics emerged. One of them was the study of infant crawling, its implementation in a humanoid platform was done using the iCub robot (Righetti & Ijspeert, 2006), Fig. 1(d).

For our roboticist Rob, the challenge of making humanoid robots replicate the different types of motor behaviors found in humans is just one part of a larger challenge. An equally interesting project will be the design of a decision making control that allows the agent to switch between the different motor behaviors. Switching between walking to crawling and vice versa, from walking to trotting or running and back. Those changes will need to be generated autonomously and dynamically as a response to the needs of the environment. The traditional way of programming a robot by following a set of rules in reaction to external stimuli does not work and will not work in dynamic, unconstrained environments. Rob is thinking about Asimo tripping over a step and hitting the floor with his face first. Rob knows that a more dynamic, autonomous and adaptive approach is needed for the future of his humanoid robot.

3.2.2 Manipulation

It is now time for Rob to think about manipulation. He looks at his hands and starts thinking about the quasi-perfection found in this mechanism. He realizes that there is no other being in the animal kingdom that has reached similar levels of dexterity. His hands are the result of an evolutionary process of thousands of years which got them to the sophisticated level of manipulation the enjoy today. This process could have gone "hand-in-hand" with the evolution of higher cognitive abilities (Faisal et al., 2010). Considering human hands only, there is an even larger repertoire of motor actions that range from simple waving and hand signals to complex interactions such as manipulating a pen or spinning a top or shooting a marble (see Bullock & Dollar (2011) for a more detailed classification of human hand behaviors).

Rob is aware of the particular engineering challenge presented by interacting with objects in the real world. Trying to replicate as close as possible the human counterpart, current state of the art humanoid hands (Fig. 3) have four (Iwata et al., 2005; Liu et al., 2007) and five (Davis et al., 2008; Pons et al., 2004; Shadow Robot Company, 2003; Ueno & Oda, 2009) fingers, sometimes under-actuated through a set of tendons/cables as in the case of iCub's hands (Fig. 1(d)). However three-finger hands are also being used for manipulation, the most remarkable case being the work done at the Ishikawa-Oku Lab at the University of Tokyo (Ishikawa & Oku, 2011),Fig. 3(d). Through the use of high-speed cameras and actuators they have showed that it is possible to perform complex actions like throwing and catching or batting a ball, knoting a flexible rope with one manipulator, throwing and catching any type of objects with the same hand, or manipulating tools of different shapes.

The interaction between skin and bones in a human hand creates a dynamic balance between sensing and acting capabilities. By including skin-like materials into the design of robotic end-effectors it will be possible to re-create the optimized functionality of human hands. Skin that includes soft materials in its construction will add spring-damping properties to grasping behaviors. Accurate control of the hardware is transferred to the physical properties of the materials thus saving energy, time and resources. The future humanoid manipulator should be capable of interacting at a minimum with the same objects and displaying similar active and passive properties as human hands.

3.3 Interim summary 1

Rob realizes that the technology necessary to design and build a humanoid platform seems to be reaching a very mature level. In many cases the biggest obstacle is the lack of resources to put together state of the art sensors and actuators in the same project. We have seen remarkable innovations in the area of nano-materials which can help to overcome current obstacles in the acquisition of tactile, visual, and olfactory/taste information. At the same

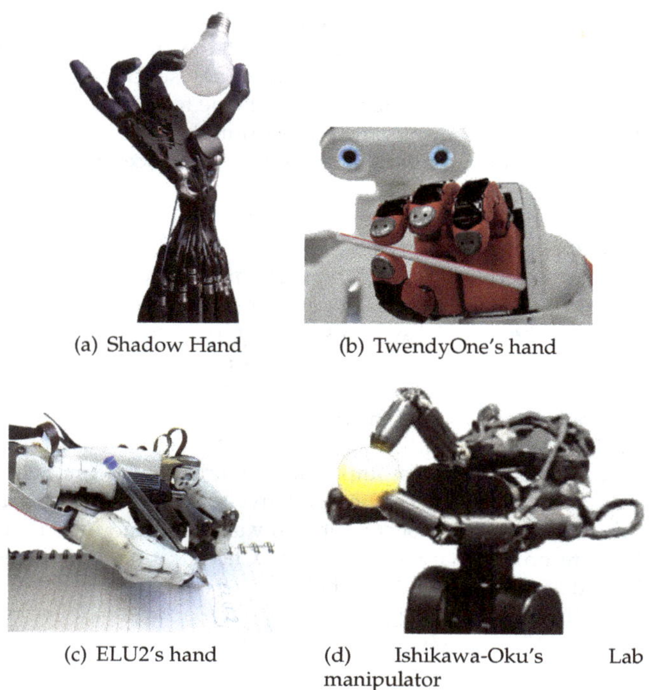

(a) Shadow Hand (b) TwendyOne's hand

(c) ELU2's hand (d) Ishikawa-Oku's Lab manipulator

Fig. 3. Samples of different state of the art manipulators.

time, the exponential growth of computational power gives enough freedom to integrate all this information.

In terms of actuators, Rob is impressed by the results from the use of spring-damper components. Previous attempts to control this kind of materials gave many researchers (especially to those working with traditional control theory) several headaches. However current implementations for whole-body motion and manipulation have shown the feasibility of this approach, thanks most likely to the use of alternative methodologies, e.g. nonlinear dynamical systems.

4. Computational requirements

Having considered his mechanical requirements, Rob now turns to computational requirements for his robot. In his mind, this means anything that could be needed to make the hardware perform interesting behaviors. This includes, but is not limited to, learning how to solve (not necessarily pre-defined) tasks in uncertain environments under changing or undefined conditions and interacting with humans in a non-trivial manner.

Within robotics, Rob has distinguished two main approaches of interest to him: one which we shall call the traditional symbolic approach (also known as *Cognitivism*) and one that we will name embodied approach (also known as *Emergent Systems*). Although both approaches can sometimes use the same technologies (neural networks are a prime example of that), they differ philosophically in the sense that symbolic is first and foremost an AI approach with its roots in computer science, whereas the embodied approach has its roots in the cognitive sciences. An important consequence is that the traditional symbolic approach sees computations as fundamentally disassociated from any specific hardware and therefore operating in an abstract, amodel level while this is not true for the embodied approach. This section therefore begins by briefly discussing some examples of the symbolic approach and its limitations before moving on to embodiment and its relevance to robotics in general and Rob's robot in particular.

4.1 Traditional symbolic approaches

The symbolic paradigm sees cognition as a type of computation between any information that can be represented in code and the outcome of the operations carried out on those codes (Vernon, 2008). In other words, symbols are manipulated through a set of rules. This determines that the way an artificial system sees the world around it is only through the eyes of the designer/programmer of that system. All its freedom and limitations will be dependent on the amount of symbols and rules that the person who created this model wrote in code.

Symbolic approaches make use of mathematical tools that help them select the best response within a specific goal. Among the most important are probabilistic modeling and machine learning techniques. Both of these have been relatively successful in solving task specific problems. Hundreds if not thousands of commercially available products for character, voice, and face recognition are on the market right now. Classification and regression problems are easily solved by this approach. Nonetheless, the dependency of the system to the programmer's view of the world challenges symbolic approaches in solving problems such as the symbol grounding problem, the frame problem, and the combinatorial problem (Vernon, 2008).

Rob has seen Asimo tripping and falling a couple of times without any reaction to protect itself from the fall. This gives Rob a clue about the limitations of the symbolic approach for controlling unexpected situations. At some point a group of engineers devised the most likely situations that Asimo would encounter when climbing or descending a number of stairs given that they would be flat, non slippery and nobody else would be around. So they knew about the symbols present in an specific sequence of events, they knew the constraints needed and imposed and finally, they wrote enough rules to fulfill the task.

Rob starts thinking about those events that cannot be foreseen and therefore cannot be calculated. What if a pebble is in the wrong place or at the wrong time? What if a cat crosses my robot's path? What if a person pushes my robot? After all, my robot will be moving around in places where any other human moves around, and we encounter these types of situations all the time. Rob has already considered the importance of a good design for both actuators and their wrapping materials. This would help by transferring some of the problems from the digital control part into the physical interactions with the environment. Yet a symbolic approach does not seem to be as open as needed in the unconstrained environments that humans work and live.

4.2 Embodied and cognitively inspired approaches

Rob thus comes to the realization that the traditional symbolic approaches do not fulfill his requirements. In this part, he therefore discusses embodied approaches. In a nutshell, this class of approaches sees the particular body of an agent as intertwined with its mind: the cognitive abilities of an agent is a consequence of both and cannot be understood by studying one in the absence of the other.

In the simplest form, the embodiment can provide a solution to the *symbol grounding problem* (Harnad, 1990), namely the problem of attaching a real meaning to symbols. For example, the sequence of symbols *to grasp* has no meaning by itself; it only becomes meaningful when an agent can associate it with the corresponding sensory perceptions and motor actions. In a more advanced form, the embodiment of a particular agent can actually reduce the computations necessary by a controller in the traditional sense by what is called morphological computing (Pfeifer et al., 2006).

The remainder of this section thus first briefly introduces the core concept of embodied cognition as relevant to roboticists such as Rob. We then consider examples of research which uses embodiment in the sensory-motor grounding sense as well as examples of morphological computing as such. Finally, we briefly discuss *dynamic field theory*, a particular modeling approach explicitly built on ideas from embodied cognition.

4.2.1 Embodied cognition

Thill (2011) provides a brief introduction of embodied cognition as relevant to the design of artificial cognitive systems in general (rather than the specific case of robots). The brief introduction here follows this discussion, albeit adapted to suit the needs of roboticists in particular.

The basic claim within embodied cognition (Anderson, 2003; Chrisley & Ziemke, 2003; Gallagher, 2005), as mentioned before, is that the body intrinsically shapes the mind. A simple illustration of this influence comes from an early study by Strack et al. (1988), who showed that people rate cartoons as more funny when holding a pen between their teeth (activating smiling muscles) than when holding a pen between their lips (activating frowning muscles). Another example is the SNARC (Spatial-Numeric Association of Response Codes) effect (Fischer, 2008, see Pezzulo et al. 2011). In essence, people respond to smaller numbers faster with the left hand than with the right hand and vice versa for large numbers. Similarly, when asked to produce random numbers while simultaneously shaking their heads left to right, people are biased towards smaller numbers during left turns than during right ones. A further illustration of the body's influence on mental processes can be seen in language processing, particularly when involving action verbs. Such processing (for instance while reading a sentence) can be shown to activate motor regions within the brain and lead to either facilitation of or interference with executing an action (involving the same end-effector as the sentence, see Chersi et al., 2010, for a review and a more thorough discussion).

Although examples such as those above and several more not discussed in detail here (but see Pezzulo et al., 2011, for additional ones) clearly show that body and mind are intertwined, it is still an open debate how intricate the relationship is. While researchers like Pfeifer & Bongard (2007) or Pezzulo et al. (2011) argue strongly in favor of such an intertwined relationship, Mahon & Caramazza (2008), for instance, are amongst those who favour a view that sees mental processes operating at an abstract symbolic representation, with concepts that are merely grounded by sensory-motor information. In other words, in this view (also related to Harnad, 1990), cognition does not require a body as such, although the latter may be necessary to ground the symbols used.

The relevance of embodied cognition to robotics in general is thus clear: when designing the controller for a robot, one faces decisions as to how much the higher-level cognitive abilities of the machine need to involve the particular embodiment and sensory features. Thus the relationship between robotics and the study of embodied cognition is mutually informative: on one hand, a robot provides cognitive scientists with a real body in which to test their theories and computational models thereof (Pezzulo et al., 2011) while on the other hand, insights from embodied cognition can allow the better design of robots. For Rob's robot, the latter is clearly the most interesting aspect and hence the focus of the remainder of this section.

4.2.2 Mirror neurons and the social robot

One of the key requirements for Rob's robot is the ability to be social in the sense that it can interact with human beings in a sensible, autonomous manner. Such an interaction can take many forms. One example is *learning by imitation*, which would allow the robot to learn novel tasks by observing a human demonstrating the desired behavior. It could also take the form of *cooperation*, in which Rob's robot has to solve a task together with a human. Or it could be *verbal interaction*, which may be required in all social scenarios involving robots.

What is common to all these scenarios is first and foremost the requirement that Rob's robot be able to *understand* the actions of a human and related concepts, whether shown to the robot by demonstration or given via verbal labels. In other words, these scenarios can benefit from the embodied theories on sensory-motor grounding in order to facilitate this understanding.

Within this context, the discovery of mirror neurons –neurons that are active both when an agent executes a (goal-directed) action and when he observes the action being executed by another (Gallese et al., 1996) –is attracting Rob's attention. Indeed, one of the most prominent hypotheses on the functional role of mirror neurons is that they form the link between action perception and action understanding Bonini et al. (2010); Cattaneo et al. (2007); Fogassi et al. (2005); Rizzolatti & Sinigaglia (2010); Umiltà et al. (2008), which appears highly relevant for Rob's robot. Even so, it is always worthwhile to remember that this claim remains a *hypothesis*, not a proven fact, and does not come without criticism (Hickok, 2008).

It is further thought that mirror neurons may play a role in learning by imitation (Oztop et al., 2006) as well as in the sensory-motor grounding of language (see Chersi et al. (2010) for a discussion). Although these are again hypotheses rather than facts (, for instance rightly points out that most of the neurophysiologic data supporting the theories come from macaque monkeys, which neither imitate nor use language), they have inspired some robotics research (see for instance Yamashita & Tani (2008) for an example of learning by imitation and Wermter et al. (2005) for an example of language use). Overall, it thus comes as no surprise that mirror neurons are also of high interest to the field of humanoid robotics, since they may provide the key to grounding cognition in sensory-motor experiences. For this reason, Rob is interested in some of the work within the field of robotics that is based on insights learned from mirror neurons.

4.2.3 Mirror neuron based robots

Oztop et al. (2006) provides a general review of computational mirror neuron models (including models that can be seen to have components which function similarly to mirror neurons rather than being explicitly inspired by mirror neuron research), together with a proposed taxonomy. Not all of these are relevant to robotics, but we will briefly mention a few controllers that are examples of early mirror-neuron related work in robotics.

One of the first such examples is the recurrent neural network with parametric bias (RNNPB) by Tani et al. (2004), in which "parametric bias" units of a recurrent neural network are associated with actions, encoding each action with a specific vector. Then the system may then either be given a PB vector to generate the associated action, or it may be used to predict the PB vector associated with a given sensory-motor input. As such, these PB units can be understood as replicating some of the mirror functionality. Tani et al. (2004) demonstrate the utility of the overall architecture in three tasks. First, they show that the architecture enables a robot to learn hand movements by imitation. Second, they demonstrate that it can learn both end-point and cyclic movements. Finally, they illustrate the ability of the architecture to associate word sequences with the corresponding sensory-motor behaviors.

Yamashita & Tani (2008) similarly use a recurrent neural network at the heart of their robot controller but endow it with neurons that have two different timescales. The robot then learns repetitive movements and it is shown that the neurons with the faster timescale encode so-called movement primitives while the neurons with the slower timescale encode the sequencing of these primitives. This enables the robot to create novel behavior sequences by merely adapting the slower neurons. The encoding of different movement primitives within the neural structure also replicates the organization of parietal mirror neurons (Fogassi et al., 2005), which is at the core of other computational models of the mirror system (Chersi et al., 2010; Thill et al., In Press; Thill & Ziemke, 2010).

While the RNNPB architecture encodes behavior as different parametric bias vectors, Demiris & Hayes (2002); Demiris & Johnson (2003) propose an architecture in which every behavior is encoded by a separate module. This architecture combines inverse and forward models, leading to the ability to both recognize and execute actions with the same architecture. Learning is done by imitation, where the current state of the demonstrator is received and fed to all forward modules. These forward modules each predict the next state of the demonstrator based on the behavior they encode. The predicted states are compared with the actual states, resulting in confidence values that a certain behavior is being executed. If the behavior is known (a module produces a high confidence value), the motors are then actuated accordingly. If not, a new behavioral module is created to learn the novel behavior being demonstrated. A somewhat similar model of human motor control, also using multiple forward and inverse models has been proposed by Wolpert & Kawato (2001), with the main difference being that in this work, all models (rather than simply the one with the highest confidence value) contribute to the final motor command (albeit in different amounts).

Finally, Wermter et al. (2003; 2005; 2004) developed a self-organizing architecture which "takes as inputs language, vision and actions ... [and] ... is able to associate these so that it can produce or recognize the appropriate action. The architecture either takes a language instruction and produces the behavior or receives the visual input and action at the particular time-step and produces the language representation" (Wermter et al., 2005, cf. Wermter et al., 2004). This architecture was implemented in a wheeled (non-humanoid) robot based on the PeopleBot platform. This robot can thus be seen to "understand" actions by either observing them or from its stored representation related to observing the action. This is therefore an example of a robot control architecture that makes use of embodied representations of actions.

In related work on understanding of concepts/language in mirror-neuron-like neural robotic controllers (Wermter et al., 2005) researchers use the insight that language can be grounded in semantic representations derived from sensory-motor input to construct multimodal neural network controllers for the PeopleBot platform that are capable of learning. The robot in this scenario is capable of locating a certain object, navigating towards it and picking it up. A modular associator-based architecture is used to perform these tasks. One module is used for vision, another one for the execution of the motor actions. A third module is used to process linguistic input while an overall associator network combines the inputs from each module.

What all these examples illustrate is that insights from mirror neuron studies (in particular their potential role in grounding higher-level cognition in an agent's sensory-motor experiences) can be useful in robotics. In terms of using insights from embodied cognition, these are relatively simple examples since the main role of the body in these cases is to enable the grounding of concepts. For instance, a robot would "know" what a grasp is because it can relate it to its own grasping movements via the mirror neurons.

However, from Rob's perspective, there is still substantial work that needs to be done in this direction. In essence, what the field is currently missing are robots that can display higher-level cognition which goes beyond simple toy problems. For example, most of the examples above dealing with learning by imitation understand imitation as reproducing the trajectory of, for instance, the end-effector. However, imitation learning is more complex than that and involves decisions on whether it is really the trajectory that should be copied or the overall goal of the action (see *e.g.* Breazeal & Scassellati, 2002, for a discussion of such issues). Similarly, while there is work on endowing robots with embodied language skills (*e.g.* Wermter et al., 2005), it rarely goes beyond associating verbal labels to sensory-motor experiences (Although see Cangelosi & Riga (2006) for an attempt to build more abstract concepts using linguistic labels for such experiences). Again, while this is a worthwhile exercise, it is not really an example of a robot with true linguistic abilities.

4.2.4 Morphological computation

Closely related to embodiment is the concept of *morphological computation* (Pfeifer & Iida, 2005; Pfeifer et al., 2006). This explicitly takes into account that the body can do more than just provide a grounding for concepts; it can actually, through its very morphology, perform computations which no longer need to be taken care of by the controller itself.

The classic example of this is the passive dynamic walker (McGeer, 1990), a pair of legs that can walk down a slope in a biologically realistic manner with no active controllers (or indeed any form of computation) involved at all. Pfeifer et al. (2006) offer additional examples: the eye of a house-fly is constructed to intrinsically compensate for motion parallax (Franceschini et al., 1992) and a similar design can facilitate the vision of a moving robot. Another example is the "Yokoi hand" ((Yokoi et al., 2004)), whose flexible and elastic material allows it to naturally adapt to an object it is grasping without any need for an external controller to evaluate the shape of the object beforehand.

A completely different type of gripper that fulfils a similar role as the Yokoi hand (namely the ability to grasp an object without the need to evaluate its shape beforehand) is presented by Brown et al. (2010). This gripper is in the shape of a ball filled with a granular material that contracts and hardens when a vacuum is applied. When in contact with an object, it reshapes to fit the object and can lift it. Here, the morphology of the hand goes as far as removing the need for any joints in the gripper, thus significantly reducing the computational requirements associated with it.

From the perspective of Rob's robot, the key insight from the examples given here is that a suitably designed robot can reduce the computations required within the controller itself by offloading them onto the embodiment. At the same time, he also identifies a shortcoming, namely that there are no examples of humanoids that take this into account to a significant degree. Rather, the demonstrations mostly focus on small robots with a very limited behavioral repertoire designed solely to illustrate the utility of the particular embodiment in that specific case. Although Rob finds the approach itself promising, he feels that it still needs to overcome the restrictions in form of the focus on limited behavioral repertoires.

4.2.5 Dynamic Field Theory

Dynamic Field Theory (DFT) is a mathematical framework based on the concepts of Dynamical Systems and the guidelines from Neurophysiology. A field represents a population of neurons and their activations follow continuous responses to external stimuli. Amari (1977) studied the properties of these networks as a model of the activation observed in cortical tissue. Fields have the same structure of a recurrent neural network since their connections can have, depending on the relative location within the network, a local excitation or a global inhibition, Fig. 4.

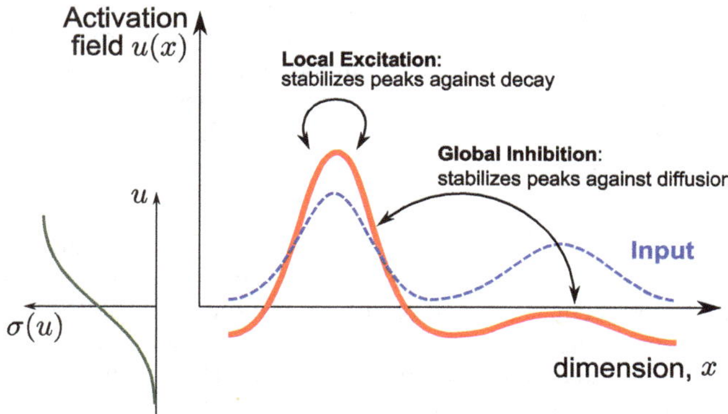

Fig. 4. Typical activations in a dynamics field, from Schöner (2008).

Fields are used to represent perceptual features, motion or cognitive decisions, e.g. position, orientation, color, speed. The dynamics of these fields allow the creation of peaks which are the units of representation in DFT (Schöner, 2008). Different configurations of one or more fields are possible, being the designer responsible for creating a proper connectivity and tuning of parameters. The result of activating this type of network is a continuously adaptive system that responds dynamically to any change coming from external stimuli.

Rob has learned about the different properties and potentials of dynamic fields for using it as part of a robust cognitive architecture. Some of the most attractive features of this approach include the possibility of having a hebbian-type of learning by exploiting the short-term memory features implicit in the dynamics of this algorithm. Long-term memory, decision making mechanism and noise robustness (also implicit in the dynamics of fields), and single-shot learning are all important tools that can and must be included in any cognitive architecture. Several applications modeling experiments on human behavior (Dineva (2005); Johnson et al. (2008); Lowe et al. (2010)) and robotic implementations (Bicho et al. (2000); Erlhagen et al. (2006); Zibner et al. (2011)) have demonstrated DFT's potential.

Nonetheless, from Rob's perspective, the current work with dynamic fields still needs to overcome a number of challenges. Dynamic field controllers are currently designed by hand, through elaborate parameter space explorations, to solve a very specific problem. Although these models nicely illustrate that decision-making can be grounded directly in sensory-motor experiences, their learning ability is limited and any particular model does not generalize well to other tasks, even though modular combinations of different models each solving a particular task seems possible (Johnson et al. (2008); Simmering et al. (2008); Simmering & Spencer (2008)).

4.3 Interim summary 2

To summarize, it seems likely that embodied approaches are the future, both from a theoretical perspective (given insights in the cognitive sciences about the functioning of the human mind) and practically speaking (given the numerous examples of the associated benefits in robotic applications) and thus, this is the approach that Rob would prefer when designing his robot. However, he thinks about the successful applications of symbolic approaches and leaves open the door to possible interactions between both approaches. It could be possible to use dynamic tools to face unexpected circumstances and once that information looks stable let a symbolic algorithm work with it and give some feedback to the body.

There are significant challenges still to overcome for embodied approaches as well. Most of the existing work (while demonstrating that an embodied approach is indeed viable) currently focuses on proof-of-concept solutions to specific problems. These limitations go hand-in-hand with the development of physical platforms to test new models.

5. Discussion

Rob has completed his brief study of the current state of the art in humanoid robotics. Along the way he has found very interesting developments for both hardware and software. In almost all fields he found technology that oftentimes exceeds the abilities found in humans. However, there are several questions that come to his mind: why is it so difficult to create human behavior? Or is it intelligence? Or maybe cognition? From his analysis it seems that the most important parts are there, but still something is missing in the overall picture.

The field of humanoid robotics was formally born when a group of engineers decided to replicate human locomotion. This work is more than 40 years old and we are still unable to replicate the simplest walking gait found in humans. In section 3 Rob found several state of the art platforms that are showing signs of improvement for locomotion. He is confident that in the near future a more robust and stable human-like walking gait will be available for biped platforms. However, he does not feel so positive about other forms of motion and the transitions between them. Rob picks the embodied approach, and more specifically the dynamical systems approach, as the best option for handling instabilities and transitions between stable states such as walking, running, crawling, jumping, etc.

In Rob's opinion, grasping and manipulation do seem to be close to be maturing as well. In hardware terms, several platforms are progressing thanks to the integration of state of the art technology in skin-like materials. The addition of this component to the traditional "naked" robotic hand has improved substantially the results of grasping and manipulation. Rob is convinced that the same strategy should be applied in the lower limbs of his humanoid platform. A large part of controlling balance and motion can be transferred to the interactions between materials and environment as it has been showed by morphological computation in section 4.2.4.

There are however several challenges that worry Rob. He is still not sure about the best way to go regarding his cognitive architecture. He is not worried about the quantity or quality of the information collected through state of the art sensors; his study showed that sensors are not the problem. The problem is how to merge all that information and make some sense out of it. The missing piece seems to be finding the right way of creating and then recovering associations. Finally, a proper way of interacting with humans is needed, e.g. language.

The biggest challenge for the area of humanoid robotics right now is the design and implementation of a large project with the collaboration of an equally large number of researchers. If we continue doing research in pieces, then the development of cognitive abilities in humanoid robots will be very slow and the use of humanoids in the future very limited. The above mentioned mega-project should have as common goal the construction of a humanoid robot with state of the art sensors and actuators. If anything can be learned from the embodiment approach 4.2, it is that a complete and sophisticated body like this is needed mainly because of the biggest challenge for the future of humanoid robotics, i.e. the development of cognitive abilities through dynamic interactions with unconstrained environments.

6. References

Amari, S.-I. (1977). Dynamics of pattern formation in lateral-inhibition type neural fields, *Biological Cybernetics* 27: 77–87.

Ambrose, R. O., Aldridge, H., Askew, R. S., Burridge, R. R., Bluethmann, W., Diftler, M., Lovchik, C., Magruder, D. & Rehnmark, F. (2000). Robonaut: Nasa's space humanoid, *IEEE Intelligent Systems* 15: 57–63.
URL: *http://dx.doi.org/10.1109/5254.867913*

Anderson, M. L. (2003). Embodied cognition: a field guide., *Artificial Intelligence* 149: 91–130.

Anybots (2008). Dexter.
URL: *http://www.anybots.com/*

Bao, Z., McCulloch, I., (Society), S., Staff, S. S., Company, A. C., Staff, A. C. C., Incorporated, C., Staff, C. I., (Firm), S. S. A. & Staff, S. S. A. F. (2009). *Organic Field-Effect Transistors VIII: 3-5 August 2009, San Diego, California, United States*, Proceedings of SPIE–the International Society for Optical Engineering, S P I E-International Society for Optical Engineering.
URL: *http://books.google.com/books?id=IjGvSgAACAAJ*

Bicho, E., Mallet, P. & Schoner, G. (2000). Target representation on an autonomous vehicle with low-level sensors., *The International Journal of Robotics Research* 19(5): 424–447.
URL: *http://dx.doi.org/10.1177/02783640022066950*

Bonini, L., Rozzi, S., Serventi, F. U., Simone, L., Ferrari, P. F. & Fogassi, L. (2010). Ventral premotor and inferior parietal cortices make distinct contributions to action organization and intention understanding, *Cerebral Cortex* 20: 1372–1385.

Breazeal, C. & Scassellati, B. (2002). Robots that imitate humans, *Trends in Cognitive Sciences* 6(11): 481–487.

Brown, E., Rodenberg, N., Amend, J., Mozeika, A., Steltz, E., Zakin, M. R., Lipson, H. & Jaeger, H. M. (2010). Universal robotic gripper based on the jamming of granular material, *Proceedings of the National Academy of Sciences USA* 107(44): 18809–18814.

Bullock, I. & Dollar, A. (2011). Classifying human manipulation behavior, *IEEE International Conference on Rehabilitation Robotics, Proceedings of the.*

Cangelosi, A. & Riga, T. (2006). An embodied model for sensorimotor grounding and grounding transfer: experiments with epigenetic robots, *Cognitive Science* 30: 673–689.

Cattaneo, L., Fabbri-Destro, M., Boria, S., Pieraccini, C., Monti, A., Cossu, G. & Rizzolatti, G. (2007). Impairment of actions chains in autism and its possible role in intention understanding, *Proceedings of the National Academy of Sciences USA* 104(45): 17825–17830.

Chersi, F., Thill, S., Ziemke, T. & Borghi, A. M. (2010). Sentence processing: linking language to motor chains, *Frontiers in Neurorobotics* 4(4): DOI:10.3389/fnbot.2010.00004.

Chrisley, R. & Ziemke, T. (2003). *Encyclopedia of Cognitive Science*, Macmillan Publishers, chapter Embodiment, pp. 1102–1108.

Collette, C., Micaelli, A., Andriot, C. & Lemerle, P. (2007). Dynamic balance control of humanoids for multiple grasps and non coplanar frictional contacts, *Humanoid Robots, 2007 7th IEEE-RAS International Conference on*, pp. 81 –88.

Collins, S., Ruina, A., Tedrake, R. & Wisse, M. (2005). Efficient bipedal robots based on passive-dynamic walkers, *Science* 307(5712): 1082–1085.
URL: *http://www.sciencemag.org/content/307/5712/1082.abstract*

Cui, Q. N., Bachus, L., Knoth, E., O'Neill, W. E. & Paige, G. D. (2008). Eye position and cross-sensory learning both contribute to prism adaptation of auditory space, *in* C. Kennard & R. J. Leigh (eds), *Using Eye Movements as an Experimental Probe of Brain function - A Symposium in Honor of Jean BÃijttner-Ennever*, Vol. 171 of *Progress in Brain Research*, Elsevier, pp. 265 – 270.
URL: *http://www.sciencedirect.com/science/article/pii/S0079612308006377*

Dahiya, R., Metta, G., Valle, M. & Sandini, G. (2010). Tactile sensing - from humans to humanoids, *Robotics, IEEE Transactions on* 26(1): 1 –20.

Davis, S., Tsagarakis, N. & Caldwell, D. (2008). The initial design and manufacturing process of a low cost hand for the robot icub, *IEEE International Conference on Humanoid Robots*.

Demiris, Y. & Hayes, G. (2002). *Imitation in animals and artefacts*, MIT Press, Cambridge, MA, chapter Imitation as a doual-route process featuring predictive and learning components: A biologically-plausible computational model.

Demiris, Y. & Johnson, M. (2003). Distributed, predictive perception of actions: A biologically inspired robotics architecture for imitation and learning., *Connection Science* 15(4): 231–243.

Dineva, E. (2005). *Dynamical Field Theory of Infant Reaching and its Dependence on Behavioral History and Context.*, PhD thesis, Institut für Neuroinformatik & Intrnational Graduate School for Neuroscience, Ruhr-Universiät-Bochum, Germany.

Erlhagen, W., Mukovski, A. & Bicho, E. (2006). A dynamic model for action understanding and goal-directed imitation, *Brain Research* 1083: 174–188.

Faisal, A., Stout, D., Apel, J. & Bradley, B. (2010). The manipulative complexity of lower paleolithic stone toolmaking, *PLoS ONE* 5(11): e13718.
URL: *http://dx.doi.org/10.1371/journal.pone.0013718*

Festo (2011).
URL: *http://www.festo.com/*

Fischer, M. H. (2008). Finger counting habits modulate spatial-numerical associations, *Cortex* 44: 386–392.

Fogassi, L., Ferrari, P. F., Gesierich, B., Rozzi, S., Chersi, F. & Rizzolatti, G. (2005). Parietal lobe: from action organization to intention understanding, *Science* 308: 662–667.

Franceschini, N., Pichon, J. M. & Blanes, C. (1992). From insect vision to robot vision, *Philosophical Transactions of The Royal Society Of London. Series B: Biological Sciences* 337: 283–294.

Gallagher, S. (2005). *How the body shapes the mind*, Oxford University Press, New York.

Gallese, V., Fadiga, L., Fogassi, L. & Rizzolatti, G. (1996). Action recognition in the premotor cortex, *Brain Research* 119: 593–609.

Harnad, S. (1990). The symbol grounding problem, *Physica D: Nonlinear Phenomena* 42(1-3): 335–346.

Haykin, S. & Chen, Z. (2005). The cocktail party problem, *Neural Computation* 17(9): 1875–1902.
URL: *http://www.mitpressjournals.org/doi/abs/10.1162/0899766054322964*

Hickok, G. (2008). Eight problems for the mirror neuron theory of action understanding in monkeys and humans, *Journal of Cognitive Neuroscience* 21(7): 1229–1243.

Hirose, M., Haikawa, Y., Takenaka, T. & Hirai, K. (2001). Development of humanoid robot asimo, *International Conference on Intelligent Robots and Systems*.

Hobbelen, D., de Boer, T. & Wisse, M. (2008). System overview of bipedal robots flame and tulip: Tailor-made for limit cycle walking, *Intelligent Robots and Systems, 2008. IROS 2008. IEEE/RSJ International Conference on*, pp. 2486 –2491.

Ishikawa, M. & Oku, H. (2011). Ishikawa oku laboratory, university of tokyo.
URL: *http://www.k2.t.u-tokyo.ac.jp/*

Iwata, H., Kobashi, S., Aono, T. & Sugano, S. (2005). Design of anthropomorphic 4-dof tactile interaction manipulator with passive joints, *2005 IEEERSJ International Conference on Intelligent Robots and Systems* pp. 1785–1790.
URL: *http://ieeexplore.ieee.org/lpdocs/epic03/wrapper.htm?arnumber=1545261*

Johnson, J. S., Spencer, J. P. & Schï£¡ner, G. (2008). Moving to higher ground: The dynamic field theory and the dynamics of visual cognition, *New Ideas in Psychology* 26(2): 227 – 251. Dynamics and Psychology.
URL: *http://www.sciencedirect.com/science/article/pii/S0732118X07000505*

Kaneko, K., Harada, K., Kanehiro, F., Miyamori, G. & Akachi, K. (2008). Humanoid robot hrp-3, *Intelligent Robots and Systems, 2008. IROS 2008. IEEE/RSJ International Conference on*, pp. 2471 –2478.

Keyrouz, F. (2008). Efficient binaural sound localization for humanoid robots and telepresence applications.

Ko, H. C., Stoykovich, M. P., Song, J., Malyarchuk, V., Choi, W. M., Yu, C.-J., III, J. B. G., Xiao, J., Wang, S., Huang, Y. & Rogers, J. A. (2008). A hemispherical electronic eye camera based on compressible silicon optoelectronics, *Nature* 454: 748–753.

Lipomi, D., Tee, B., Vosgueritchian, M. & Bao, Z. (2011). Stretchable organic solar cells., *Adv Mater* 23(15): 1771–5.

Liu, H., Meusel, P., Seitz, N., Willberg, B., Hirzinger, G., Jin, M., Liu, Y., Wei, R. & Xie, Z. (2007). The modular multisensory dlr-hit-hand, *Mechanism and Machine Theory* 42(5): 612 – 625.
URL: *http://www.sciencedirect.com/science/article/pii/S0094114X0600098X*

Lowe, R., Duran, B. & Ziemke, T. (2010). A dynamic field theoretic model of iowa gambling task performance, *Development and Learning (ICDL), 2010 IEEE 9th International Conference on*, pp. 297 –304.

Mahmoudi, E. (2009). Electronic nose technology and its applications, *Sensors & Transducers* 107: 17–25.
URL: *http://www.sensorsportal.com/HTML/DIGEST/august_09/P_470.pdf*

Mahon, B. Z. & Caramazza, A. (2008). A critical look at the embodied cognition hypothesis and a new proposal for grounding conceptual content, *Journal of Physiology Paris* 102: 59–70.

McGeer, T. (1990). Passive dynamic walking, *The International Journal of Robotics Research* 9(2): 62–82.

Niiyama, R. & Kuniyoshi, Y. (2010). Design principle based on maximum output force profile for a musculoskeletal robot, *Industrial Robot: An International Journal* 37(3).

Oztop, E., Kawato, M. & Arbib, M. A. (2006). Mirror neurons and imitation: A computationally guided review, *Neural Networks* 19: 254–271.

Park, I.-W., Kim, J.-Y., Lee, J. & Oh, J.-H. (2005). Mechanical design of humanoid robot platform khr-3 (kaist humanoid robot 3: Hubo), pp. 321 –326.

Peratech (2011).
URL: *http://www.peratech.com/*

Petman (2011). Boston dynamics.
URL: *http://www.bostondynamics.com/*

Pezzulo, G., Barsalou, L. W., Cangelosi, A., Fischer, M. H., McRae, K. & Spivey, M. J. (2011). The mechanics of embodiment: a dialog on embodiment and computational modeling, *Frontiers in Psychology* 2(5).

Pfeifer, R. & Bongard, J. (2007). *How the body shapes the way we think*, MIT Press, Cambridge.

Pfeifer, R. & Iida, F. (2005). Morphological computation: conecting brain, body and environment, *Japanese Scientific Monthly* 58(2): 48–54.

Pfeifer, R., Iida, F. & Gómez, G. (2006). Morphological computing for adaptiver behavior and cognition, *International Congress Series* 1291: 22–29.

Pons, J., Rocon, E., Ceres, R., Reynaerts, D., Saro, B., Levin, S. & Van Moorleghem, W. (2004). The manus-hand dextrous robotics upper limb prosthesis: Mechanical and manipulation aspects, *Autonomous Robots* 16: 143–163. 10.1023/B:AURO.0000016862.38337.f1.
URL: *http://dx.doi.org/10.1023/B:AURO.0000016862.38337.f1*

Righetti, L. & Ijspeert, A. J. (2006). Design methodologies for central pattern generators: an application to crawling humanoids, *Proceedings of Robotics: Science and Systems*, pp. 191–198.

Rizzolatti, G. & Sinigaglia, C. (2010). The functional role of the parieto-frontal mirror circuit: interpretations and misinterpretations, *Nature Reviews Neuroscience* 11(4): 264–274.

Ruesch, J., Lopes, M., Bernardino, A., Hornstein, J., Santos-Victor, J. & Pfeifer, R. (2008). Multimodal saliency-based bottom-up attention a framework for the humanoid robot icub, *Robotics and Automation, 2008. ICRA 2008. IEEE International Conference on*, pp. 962 –967.

Schöner, G. (2008). *Toward a New Grand Theory of Development? Connectionism and Dynamic Systems Theory Re-Considered.*, Oxford University Press, New York, chapter Development as Change of System Dynamics: Stability, Instability, and Emergence., p. In press.

Shadow Robot Company (2003). Design of a dextrous hand for advanced CLAWAR applications, *Climbing and Walking Robots and the Supporting Technologies for Mobile Machines*, pp. 691–698.
URL: *http://www.shadowrobot.com/*

Shotton, J. & Sharp, T. (2011). Real-time human pose recognition in parts from single depth images, *statosuedu* 2: (to appear).
URL: *http://www.stat.osu.edu/ dmsl/BodyPartRecognition.pdf*

Simmering, V. R., Schutte, A. R. & Spencer, J. P. (2008). Generalizing the dynamic field theory of spatial cognition across real and developmental time scales, *Brain Research* 1202: 68 – 86. Computational Cognitive Neuroscience.
URL: *http://www.sciencedirect.com/science/article/pii/S0006899307013807*

Simmering, V. R. & Spencer, J. P. (2008). Generality with specificity: the dynamic field theory generalizes across tasks and time scales, *Developmental Science* 11(4): 541–555.
URL: *http://dx.doi.org/10.1111/j.1467-7687.2008.00700.x*

Strack, F., Marin, L. & Stepper, S. (1988). Inhibiting and facilitating conditions of the human smile: A nonobtrusive test of the facial feedback hypothesis, *Journal of Personality and Social Psychology* 54: 768–777.

Takei, K., Takahashi, T., Ho, J. C., Ko, H., Gillies, A. G., Leu, P. W., Fearing, R. S. & Javey, A. (2010). Nanowire active-matrix circuitry for low-voltage macroscale artificial skin, *Nature Materials* 9(10): 821âĂŞ826.

Tani, J., Ito, M. & Sugita, Y. (2004). Self-organization of distributedly represented multiple behavior schemata in a mirror system: reviews of robot experiments using RNNPB, *Neural Networks* 17: 1273–1289.

Thill, S. (2011). Considerations for a neuroscience-inspired approach to the design of artificial intelligent systems, *in* J. Schmidhuber, K. R. Thórisson & M. Looks (eds), *Proceedings of the Fourth Conference on Artificial General Intelligence, LNAI 6830*, Springer, Heidelberg, pp. 247–254.

Thill, S., Svensson, H. & Ziemke, T. (In Press). Modeling the development of goal-specificity in mirror neurons, *Cognitive Computation* .

Thill, S. & Ziemke, T. (2010). Learning new motion primitives in the mirror neuron system: A self-organising computational model., *in* S Doncieux et al (ed.), *SAB 2010, LNAI 6226*, Springer, Heidelberg, pp. 413–423.

Traver, J. & Bernardino, A. (2010). A review of log-polar imaging for visual perception in robotics., *Robotics and Autonomous Systems* pp. 378–398.

Tsakarakis, N., Metta, G., Sandini, G., Vernon, D., Beira, R., Becchi, F., Righetti, L., Santos-Victor, J., Ijspeert, A., Carrozza, M. & Caldwell, D. (2007). icub - the design and realization of an open humanoid platform for cognitive and neuroscience research, *Journal of Advanced Robotics, Special Issue on Robotic platforms for Research in Neuroscience* 21(10): 1151–1175.

Ueno, T. & Oda, M. (2009). Robotic hand developed for both space missions on the international space station and commercial applications on the ground, *Proceedings of the 2009 IEEE/RSJ international conference on Intelligent robots and systems*, IROS'09, IEEE Press, Piscataway, NJ, USA, pp. 1791–1796.
URL: *http://portal.acm.org/citation.cfm?id=1733023.1733039*

Umiltà, M. A., Escola, L., Intskirveli, I., Grammont, F., Rochat, M., Caruana, F., Jezzini, A., Gallese, V. & Rizzolatti, G. (2008). When pliers become fingers in the monkey motor system, *Proceedings of the National Academy of Sciences USA* 105(6): 2209–2213.

Vernon, D. (2008). Cognitive vision: The case for embodied perception, *Image Vision Comput.* 26(1): 127–140.

Vukobratovic, M. & Borovac, B. (2004). Zero-moment point - thirty five years of its life, *International Journal of Humanoid Robotics* 1: 157–173.

Wermter, S., Elshaw, M. & Farrand, S. (2003). A modular approach to self-organization of robot control based on language instruction, *Connection Science* 15(2-3): 73–94.

Wermter, S., Weber, C. & Elshaw, M. (2005). Associative neural models for biomimetic multimodal learning in a mirror-neuron based robot, *Progess in Neural Processing* 16: 31–46.

Wermter, S., Weber, C., Elshaw, M., Panchev, C., Erwin, H. & Pulvermüller, F. (2004). Towards multimodal neural robot learning, *Robotics and Autonomous Systems* 47(2-3): 171–175.

Wolpert, D. & Kawato, M. (2001). Multiple paired forward and inverse models for motor control, *Neural Networks* 11: 1317–1329.

Yamashita, Y. & Tani, J. (2008). Emergence of functional hierarchy in a multiple timescale neural network model: a humanoid robot experiment, *PLoS Computational Biology* 4(11): 1–18.

Yokoi, H., Arieta, A. H., Katoh, R., Yu, W., Watanabe, I. & Maruishi, M. (2004). Mutual adaption in a prosthetics application, *in* F. Iida, R. Pfeifer, L. Steels & Y. Kuniyoshi (eds), *Embodied Artificial Intelligence, LNAI 3139,* Springer, Heidelberg, pp. 146–159.

Zibner, S., Faubel, C., Iossifidis, I. & Schoner, G. (2011). Dynamic neural fields as building blocks of a cortex-inspired architecture for robotic scene representation, *Autonomous Mental Development, IEEE Transactions on* 3(1): 74 –91.

Permissions

The contributors of this book come from diverse backgrounds, making this book a truly international effort. This book will bring forth new frontiers with its revolutionizing research information and detailed analysis of the nascent developments around the world.

We would like to thank Riadh Zaier, for lending his expertise to make the book truly unique. He has played a crucial role in the development of this book. Without his invaluable contribution this book wouldn't have been possible. He has made vital efforts to compile up to date information on the varied aspects of this subject to make this book a valuable addition to the collection of many professionals and students.

This book was conceptualized with the vision of imparting up-to-date information and advanced data in this field. To ensure the same, a matchless editorial board was set up. Every individual on the board went through rigorous rounds of assessment to prove their worth. After which they invested a large part of their time researching and compiling the most relevant data for our readers. Conferences and sessions were held from time to time between the editorial board and the contributing authors to present the data in the most comprehensible form. The editorial team has worked tirelessly to provide valuable and valid information to help people across the globe.

Every chapter published in this book has been scrutinized by our experts. Their significance has been extensively debated. The topics covered herein carry significant findings which will fuel the growth of the discipline. They may even be implemented as practical applications or may be referred to as a beginning point for another development. Chapters in this book were first published by InTech; hereby published with permission under the Creative Commons Attribution License or equivalent.

The editorial board has been involved in producing this book since its inception. They have spent rigorous hours researching and exploring the diverse topics which have resulted in the successful publishing of this book. They have passed on their knowledge of decades through this book. To expedite this challenging task, the publisher supported the team at every step. A small team of assistant editors was also appointed to further simplify the editing procedure and attain best results for the readers.

Our editorial team has been hand-picked from every corner of the world. Their multi-ethnicity adds dynamic inputs to the discussions which result in innovative outcomes. These outcomes are then further discussed with the researchers and contributors who give their valuable feedback and opinion regarding the same. The feedback is then collaborated with the researches and they are edited in a comprehensive manner to aid the understanding of the subject.

Apart from the editorial board, the designing team has also invested a significant amount of their time in understanding the subject and creating the most relevant covers. They scrutinized every image to scout for the most suitable representation of the subject and create an appropriate cover for the book.

The publishing team has been involved in this book since its early stages. They were actively engaged in every process, be it collecting the data, connecting with the contributors or procuring relevant information. The team has been an ardent support to the editorial, designing and production team. Their endless efforts to recruit the best for this project, has resulted in the accomplishment of this book. They are a veteran in the field of academics and their pool of knowledge is as vast as their experience in printing. Their expertise and guidance has proved useful at every step. Their uncompromising quality standards have made this book an exceptional effort. Their encouragement from time to time has been an inspiration for everyone.

The publisher and the editorial board hope that this book will prove to be a valuable piece of knowledge for researchers, students, practitioners and scholars across the globe.

List of Contributors

Andrej Gams, Tadej Petric, Aleš Ude and Leon Žlajpah
Jožef Stefan Institute, Ljubljana, Slovenia

Riadh Zaier
Department of Mechanical and Industrial Engineering, Sultan Qaboos University, Sultanate of Oman

Yuki Funabora, Shinji Doki and Shigeru Okuma
Nagoya University, Japan

Yoshikazu Yano
Aichi Institute of Technology, Japan

Tokuo Tsuji, Kensuke Harada, Kenji Kaneko, Fumio Kanehiro and Kenichi Maruyama
Intelligent Systems Research Institute, National Institute of Advanced, Industrial Science and Technology (AIST), Japan

Dongwoon Choi, Dong-Wook Lee, Woonghee Shon and Ho-Gil Lee
Department of Applied Robot Technology, Korea Institute of Industrial Technology, Republic of Korea

Ignazio Infantino
Istituto di Calcolo e Reti ad Alte Prestazioni, Consiglio Nazionale delle Ricerche, Italy

Yosuke Matsusaka
National Institute of Advanced Industrial Science and Technology (AIST), Japan

T. Bock, T. Linner and W. Ikeda
Technische Universität München, Germany

Huan Tan
Vanderbilt University, USA

Md. Hasanuzzaman
Department of Computer Science & Engineering, University of Dhaka, Dhaka, Bangladesh

Haruki Ueno
National Institute of Informatics (NII), Tokyo, Japan

Muhammad Attamimi, Tomoaki Nakamura and Takayuki Nagai
The University of Electro-Communications, Japan

Komei Sugiura and Naoto Iwahashi
National Institute of Information and Communications Technology, Japan

Ravi Sarvadevabhatla and Victor Ng-Thow-Hing
Honda Research Institute USA, 425 National Ave Suite 100, Mountain View CA 94043, USA

Boris Durán and Serge Thill
University of Skövde, Sweden